Lecture Notes in Physics

Volume 989

The Lecture Notes in Physics

The series Lecture Notes in Physics (LNP), founded in 1969, reports new developments in physics research and teaching - quickly and informally, but with a high quality and the explicit aim to summarize and communicate current knowledge in an accessible way. Books published in this series are conceived as bridging material between advanced graduate textbooks and the forefront of research and to serve three purposes:

- to be a compact and modern up-to-date source of reference on a well-defined topic;
- to serve as an accessible introduction to the field to postgraduate students and non-specialist researchers from related areas;
- to be a source of advanced teaching material for specialized seminars, courses and schools.

Both monographs and multi-author volumes will be considered for publication. Edited volumes should however consist of a very limited number of contributions only. Proceedings will not be considered for LNP.

Volumes published in LNP are disseminated both in print and in electronic formats, the electronic archive being available at springerlink.com. The series content is indexed, abstracted and referenced by many abstracting and information services, bibliographic networks, subscription agencies, library networks, and consortia.

Proposals should be sent to a member of the Editorial Board, or directly to the responsible editor at Springer:

Dr Lisa Scalone
Springer Nature
Physics
Tiergartenstrasse 17
69121 Heidelberg, Germany
lisa.scalone@springernature.com

More information about this series at http://www.springer.com/series/5304

Wolfgang Cassing

Transport Theories for Strongly-Interacting Systems

Applications to Heavy-Ion Collisions

 Springer

Wolfgang Cassing
Theoretical Physics
University of Gießen
Gießen, Germany

ISSN 0075-8450   ISSN 1616-6361   (electronic)
Lecture Notes in Physics
ISBN 978-3-030-80294-3   ISBN 978-3-030-80295-0   (eBook)
https://doi.org/10.1007/978-3-030-80295-0

This Springer imprint is published by the registered company Springer Nature Switzerland AG.
The registered company address is: Gewerbestrasse 11, 6330 Cham, Switzerland

Preface

This book provides an overview on transport theories for readers from different fields although the actual applications and especially the relativistic off-shell transport theory are of particular interest for physicists working in the field of relativistic strong-interaction physics such as relativistic or ultra-relativistic heavy-ion collisions or the evolution of the early universe. Instead of giving hand-waving arguments for transport models in various fields, the focus here is on a thorough derivation of the transport equations and a careful analysis of the approximations employed. In order to keep the arguments and extensions in line, a multitude of Appendices is added that partly recall elements of elementary lectures on quantum mechanics or present examples for specific models. As for basic knowledge, the reader should be familiar with quantum mechanics and its principles as well as some basic concepts of the quantum many-body physics and field theory. Furthermore, the reader should not be afraid of sometimes lengthy equations and derivations that are mandatory for a stringent mathematical derivation and allow to point out relevant approximations. The detailed formulations allow for independent numerical studies that provide a space-time 'movie' of nonequilibrium dynamics of weakly and strongly interacting many-body systems. Exercises are incorporated throughout the chapters and are expected to deepen an understanding of the material presented.

The contents of this work have been developed with my PhD students and collaborators over a period of about 35 years and has led to some deeper insight into the physics of strongly interacting systems in and out of equilibrium, especially in comparison to experimental data from worldwide accelerator facilities. I hope the reader will enjoy reading and find helpful sections for his or her own research.

Gießen, Germany
May 2021

Wolfgang Cassing

Acknowledgements

This book results from the collaboration with many friends and collaborators throughout roughly 35 years of common research. These are Wolfgang Bauer, Tamas Biro, Bernhard Blättel, Elena Bratkovskaya, Wolfgang Ehehalt, Carsten Greiner, Jochen Geiss, Jörn Häuser, Sascha Juchem, Volker Koch, Volodya Konchakovski, Andreas Lang, Olena Linnyk, Tomoyuki Maruyama, Pierre Moreau, Ulrich Mosel, Koji Niita, Vitalii Ozvenchuk, Alessia Palmese, Andreas Peter, Alfred Pfitzner, Pradip Sahu, Eduard Seifert, Taesoo Song, Thorsten Steinert, Stefan Teis, Markus Thoma, Viacheslav Toneev, Vadim Voronyuk, Shun-Yin Wang, Klaus Weber and Gyuri Wolf. Furthermore, I am indebted to Elena Bratkovskaya and Olga Soloveva for valuable suggestions throughout the preparation of this work and in particular to Elena Bratkovskaya for her help in preparing the illustrations and figures.

Contents

About the Author

Wolfgang Cassing is a retired professor of the Institute of Theoretical Physics at the University of Gießen, Germany. He obtained his diploma degree in physics from the Institute of Theoretical Physics at the University of Münster in 1977, and a PhD degree in philosophy in 1978 and a PhD degree in physics in 1980 from the University of Münster. During the period 1980-1984, he worked as a postdoctoral researcher at the GSI in Darmstadt, Germany, on the dynamics of low-energy heavy-ion reactions and obtained a permanent position in theoretical physics at the University of Gießen starting from 1985. In spring 1986, he finished his habilitation at the Technical University of Darmstadt and received an offer for the post of professor in theoretical physics at the University of Gießen in 1991. Since spring 1992, he is professor for theoretical heavy-ion physics at the University of Gießen, and has served repeatedly as a director of the Institute of Theoretical Physics, as a dean of studies for the Faculty of Mathematics and Informatics, Physics and Geography as well as senator of the University of Gießen until retirement in spring 2019. His research interests include nonrelativistic and relativistic many-body theory, transport theories for heavy-ion collisions, field theories as well as computational physics.

Acronyms

AMD	Antisymmetric molecular dynamics
BBGKY	Bogolyubov, Born, Green, Kirkwood, Yvon
BM	Botermans and Malfliet
BUU	Boltzmann-Uehling-Uhlenbeck
CMD	Classical molecular dynamics
CTP	Closed time path
FMD	Fermionic molecular dynamics
KB	Kadanoff-Baym
LHC	Large Hadron Collider
MeV/u	Kinetic energy per nucleon in MeV in case of fixed target experiments
QHD	Quantum hadro-dynamics
QMD	Quantum molecular dynamics
SIS	SchwerIonen Synchrotron
SPS	Super-Proton Synchrotron
RBUU	Relativistic Bolzmann-Uehling-Uhlenbeck
RHIC	Relativistic Heavy-ion Collider
TDHF	Time Dependent Hartree-Fock
VUU	Vlasov-Uehling-Uhlenbeck

List of Figures

Introduction

Nonequilibrium many-body theory or quantum-field theory has become a major topic of research for transport processes in nuclear physics, in cosmological particle physics as well as condensed matter physics. The multidisciplinary aspect arises due to a common interest to understand the various relaxation phenomena of quantum dissipative systems. Recent progress in cosmological observations has also intensified the research on quantum fields out-of equilibrium. Important questions in high-energy nuclear or particle physics at the highest energy densities are: (1) how do nonequilibrium systems in extreme environments evolve, (2) how do they eventually thermalize, (3) how phase transitions do occur in real time with possibly nonequilibrium remnants, and (4) how do such systems evolve for unprecedented short and nonadiabatic timescales?

The very early history of the universe provides important scenarios, where nonequilibrium effects might have played an important role, like in the (post-) inflationary epoque [1–5], for the understanding of baryogenesis and also for the general phenomena of cosmological decoherence [6]. Referring to modern nuclear physics the dynamics of heavy-ion collisions at various bombarding energies has always been a major motivation for research on nonequilibrium quantum many-body physics and relativistic quantum-field theories, since the initial state of a collision resembles an extreme nonequilibrium situation, i.e. imping nuclei in their ground states, while the final state might even exhibit a certain degree of thermalization including a multitude of newly produced particles [7–9]. Indeed, at the presently highest energy heavy-ion collider experiments at the Relativistic Heavy-Ion Collider (RHIC) in Brookhaven or the Large Hadron Collider (LHC) at CERN a variety of nonequilibrium phenomena and possible phase transitions are expected or already have been seen. Presently it is evident that one observes a transient deconfined state of matter denoted as quark-gluon plasma (QGP), which shows up in the buildup of collective flow of hadrons due to a very large pressure in the initial reaction phase [10]. These examples demonstrate that one needs an

© The Author(s), under exclusive license to Springer Nature Switzerland AG 2021
W. Cassing, *Transport Theories for Strongly-Interacting Systems*, Lecture Notes in Physics 989, https://doi.org/10.1007/978-3-030-80295-0_1

ab-initio understanding of the dynamics of out-of-equilibrium many-body systems or quantum-field theory.

Especially the powerful method of the "Schwinger-Keldysh" or "closed-time-path" (CTP) (nonequilibrium) real-time Green's functions [11–15] has been shown to provide a suitable basis for the formulation of the complex problems in the various areas of nonequilibrium quantum many-body physics. Within this framework one can derive valid approximations—depending, of course, on the problem under consideration—by preserving overall consistency relations [16]. Originally, the resulting causal Dyson-Schwinger equation of motion for the one-particle Green's functions (or two-point functions), i.e. the Kadanoff–Baym (KB) equations [17], has served as the underlying scheme for deriving various transport phenomena and generalized transport equations.

Furthermore, kinetic transport theory is a convenient tool to study many-body nonequilibrium systems, nonrelativistic or relativistic. Kinetic equations, which do play the central role in more or less all practical simulations, can be derived by means of appropriate Kadanoff–Baym equations [17] within suitable approximations or from the density-matrix hierarchy [18]. Hence, a major impetus in the past has been to derive semiclassical Boltzmann-like transport equations within the standard quasiparticle approximation. Additionally, off-shell extensions by means of a gradient expansion in the space-time inhomogenities - as already introduced by Kadanoff and Baym—have been formulated for a relativistic electron-photon plasma, for transport of electrons in a metal with external electrical field, for transport of nucleons at intermediate heavy-ion reactions, for transport of particles in ϕ^4-theory, for transport of electrons in semiconductors, for transport of partons or fields in high-energy heavy-ion reactions, or for a trapped Bose system described by effective Hartree–Fock–Bogolyubov kinetic equations. We recall that on the formal level of the Kadanoff–Baym equations the various forms assumed for the selfenergy have to fulfill consistency relations in order to preserve symmetries of the fundamental Lagrangian [17, 19, 20]. This allows also for a unified treatment of stable and unstable (resonance) particles which are encountered especially in high-energy heavy-ion reactions. For review articles on the Kadanoff–Baym equations in the various areas of nonequilibrium quantum physics we refer the reader to Refs. [21–29].

In nonequilibrium quantum-field theory typically the nonperturbative description of (second-order) phase transitions has been in the foreground of interest by means of mean-field (Hartree) descriptions, with applications for the evolution of disoriented chiral condensates or the decay of the (oscillating) inflaton in the early reheating era of the universe. "Effective" mean-field dissipation (and decoherence)—solving the so-called backreaction problem—was incorporated by particle production through order parameters explicitly varying in time. However, it had been then realized that such a dissipation mechanism, i.e. transferring collective energy from the time-dependent order parameter to particle degrees of freedom, cannot lead to true dissipation and thermalization. Such a conclusion has already been known for quite some time within the effective description of heavy-ion collisions at low energy. Full time-dependent Hartree or Hartree–Fock descriptions were insufficient to describe the reactions with increasing collision energy; additional Boltzmann-like collision terms had to be incorporated in order to provide a more adequate description of the reaction dynamics [30, 31].

The incorporation of true collisions then has been formulated also for various quantum-field theories. Here, a systematic $1/N$ expansion of the "two-particle irreducible (2PI) effective action" is conventionally invoked serving as a nonperturbative expansion parameter, where N denotes the number of intrinsic degrees of freedom (such as colors). Of course, only for large N this might be a controlled expansion and 1/3 is definitely nonzero.[1] In any case, the understanding and the influence of dissipation with the chance of true thermalization—by incorporating collisions—has become a major focus of investigations. The resulting equations of motion always do resemble the Kadanoff–Baym equations; in their general form (beyond the mean field or Hartree(–Fock) approximation) they do break time invariance and thus lead to irreversibility. This macroscopic irreversibility arises from the truncations of the full theory to obtain the selfenergy operators in a specific limit. As an example we mention the truncation of the (exact) Martin–Schwinger hierarchy in the derivation of the collisional operator or the truncation of the (exact) BBGKY hierarchy [2] in terms of n-point functions and n-point correlation functions [32].

Apart from the transport theories presented in this book there exists a multitude of approaches that also are denoted as transport theories or models [33]. A simple example is Classical Molecular Dynamics (CMD) solving nonrelativistic Hamilton equations for the constituents on the basis of some appropriate two- or three-body Hamiltonian. These models, however, are not suitable for nuclear systems (of fermions) due to the lack of antisymmetry leading to a false equilibrium state. Extensions of the molecular models—based on a variational principle for a system of Slater determinants—have been developed and denoted by Antisymmetrized Molecular Dynamics (AMD) [34] or Fermionic Molecular Dynamics (FMD) [35]. Due to the numerical complexity these approaches are limited to low energy problems and light or medium nuclei. Further extensions include—on top of the molecular equations of motion—an explicit Boltzmann-like collision term that is attributed to the short-range part of the strong interaction. Additionally some mechanisms have to be introduced to avoid binary collisions with final states that are forbidden by the Pauli principle (Pauli blocking). These models are denoted as Quantum Molecular Dynamics (QMD) models [36], which are realized on different levels of sophistication. Their advantage is that they propagate many-body correlations of the system which appear mandatory for cluster formation, however, it is not yet clear if these correlations are identical to those of few-particle quantum systems. Recently the QMD model has been extended also to the relativistic energy regime [37]. In spite of frequent applications these approaches will not be presented in this work. An overview is given in the actual review by Ono [38].

---

[1] Here $N = 3$ refers to the number of colors.

[2] According to the authors **Bogolyubov**, **Born**, **Green**, **Kirkwood** and **Y**von.

The layout of this book is as follows: We will start with the nonrelativistic many-body problem for a system of fermions and introduce a suitable approximation scheme for the BBGKY hierarchy in terms of correlation functions within the Schrödinger picture. The lowest order truncation scheme will lead to equations of motion which are equivalent to the Time-Dependent Hartree–Fock (TDHF). The semiclassical limit of this truncation scheme will lead to the Vlasov equation for the single-particle phase-space distribution. However, this limit is inappropriate for the description of heavy-ion collisions in the energy range above 10–20 MeV/u[3] as will be demonstrated by explicit calculations in the framework of TDHF or the Vlasov equation. Accordingly, we will consider the next order truncation scheme which will be shown to incorporate two-body collisions in the Born approximation while keeping track of Pauli blocking for the final states. This limit leads to the Vlasov–Uehling–Uhlenbeck (VUU) or Boltzmann–Uehling–Uhlenbeck (BUU) equation, however, with mean fields and cross sections based on the bare interaction which is inappropriate for nuclear physics problems.

We will step on with the next order truncation scheme that includes a resummation of the strong interaction in terms of ladder diagrams (in the quasi-stationary limit) and incorporates the matrix elements of the $\mathcal{G}$-matrix for the mean fields as well as for two-body scatterings. The semiclassical limit of this truncation scheme leads again to a VUU or BUU transport equation but with nonperturbative mean fields (or selfenergies) and scattering cross sections. Actual results for central nucleus-nucleus collisions at 40 MeV/u will be reported and discussed in phase-space representation. An extension to coupled-channel transport equations is given briefly that incorporate either explicit spin and isospin degrees of freedom, excited states of nucleons as well as pions or η-mesons which dominantly stem from the hadronic decays of baryon resonances at these energies.

These nonrelativistic transport theories are Galilei but not Lorentz-invariant such that an extension to higher bombarding energies has to employ a covariant formulation. Whereas the kinematics are easily formulated in a covariant fashion the mean fields (or selfenergies) have to obey explicit properties under Lorentz transformations, i.e. to be of scalar, vector, or tensor type. A suitable extension of the nonrelativistic VUU equations will be formulated in Chap. 3 on the basis of Quantum-Hadro-Dynamics (QHD) in mean-field approximation (cf. Appendix G), which provides a flexible covariant approach for the description of nucleonic degrees of freedom. The semiclassical limit of the dynamical equations in phase-space representation will lead to Relativistic BUU (RBUU) equations that allow to extend the range of applicability to higher energies where also multi-particle production from binary scatterings comes into play. Illustrative examples are presented for Au+Au collisions at 1 A GeV with a particular emphasis on the role of the nucleon selfenergies for the formation of collective flow. Now the question comes up how to obey "detailed-balance," i.e. how to describe the backward

---

[3] This notation is conventionally used for fixed-target experiments and denotes the kinetic energy per nucleon of the projectile in the laboratory frame.

reactions $n \rightarrow 2$ for $n > 2$. A solution to this problem will be presented in a transport theoretical framework for transition matrix elements that are not dependent on the angular distribution of the final states and essentially depend on the invariant energy $\sqrt{s}$ of the collision partners.

At bombarding energies of 1–2 GeV/u (or A GeV) additional strangeness degrees of freedom appear $(\Lambda, \Sigma, K, \bar{K})$ as well as short lived vector mesons such as the ρ-meson which either stem from pion-pion scattering or the decay of high-mass baryonic resonances. Furthermore, the local baryon density achieves a couple of times the nuclear saturation density $(\rho_0 \approx 0.166 \text{ fm}^{-3})$ such that the interaction rate (or width) $\Gamma_{coll}(j)$ of particle type j becomes large. Coupled-channel $\mathcal{G}$-matrix calculations—as well as related nonperturbative approaches—show that the spectral functions of the antikaon $\bar{K}$ as well as the ρ-meson become sizeably broadened in the dense medium such that an on-shell quasiparticle approximation loses validity. This also holds for nucleons which achieve a sizeable spectral width due to frequent collisions. Accordingly, a covariant transport theory has to be formulated that incorporates dynamical spectral functions and describes the "particle" properties as well as the dynamical evolution in a medium also out-of equilibrium . This task is addressed in Chap. 4 on the basis of the Kadanoff–Baym theory [17], which is studied in detail for a scalar field theory, i.e. the ϕ^4-theory in weak and strong coupling. Apart from exploring the spectral properties of the degrees of freedom we will derive the quantum Boltzmann limit from the Kadanoff–Baym equations and compare the solutions as a function of the coupling strength. Furthermore, we will derive covariant off-shell transport equations and propose an extended quasiparticle Ansatz for a practical solution of these equations. A related formulation of off-shell transport equations for fermions is briefly reported. Numerical simulations are provided for the spectral evolution of vector mesons in heavy-ion collisions at 2 A GeV as well as for retarded electromagnetic fields in ultra-relativistic collisions of $Au + Au$ and $Cu + Au$ at 21,300 A GeV, i.e. at invariant energies per nucleon $\sqrt{s_{NN}} \approx 200$ GeV.

A couple of Appendices follow the text in order to recall basic elements of scattering theory, many-body theory and field theory within the notation used throughout this book. Particular examples and model cases are included in order to illustrate various physical aspects and to provide actual numbers for orientation.

This book includes a variety of exercises throughout all chapters that further introduce technical aspects and helpful model studies. Solutions for the exercises can be found at the end of each chapter.

References

1. D.H. Lyth, A. Riotto, Phys. Rep. **314**, 1 (1999)
2. A. Riotto, M. Trodden, Ann. Rev. Nucl. Part. Sci. **49**, 35 (1999)
3. M. Trodden, Rev. Mod. Phys. **71**, 1463 (1999)
4. D. Boyanovsky, H.J. de Vega, R. Holman, J.F.J. Salgado, Phys. Rev. **D54**, 7570 (1996)
5. L. Kofman, A. Linde, A.A. Starobinsky, Phys. Rev. **D56**, 3258 (1997)
6. M. Gell-Mann, J.B. Hartle, Phys. Rev. **D47**, 3345 (1993)

7. B. Müller, *The Physics of the Quark-Gluon Plasma*. Lecture Notes in Physics, vol. 225 (Springer, Berlin, 1985)
8. H. Stöcker, Nucl. Phys. A **750**, 121 (2005)
9. P. Braun-Munzinger, V. Koch, T. Schäfer, J. Stachel, Phys. Rept. **621**, 76 (2016)
10. U. Heinz, R. Snellings, Ann. Rev. Nucl. Part. Sci. **63**, 123 (2013)
11. J. Schwinger, Phys. Rev. **83**, 664 (1951)
12. J. Schwinger, J. Math. Phys. **2**, 407 (1961)
13. P.M. Bakshi, K.T. Mahanthappa, J. Math. Phys. **4**, 12 (1963)
14. L.V. Keldysh, Zh. Eks. Teor. Fiz. **47**, 1515 (1964); Sov. Phys. JETP **20**, 1018 (1965)
15. R.A. Craig, J. Math. Phys. **9**, 605 (1968)
16. M. Bonitz, *Quantum Kinetic Theory* (B.G. Teubner, Stuttgart, 1998)
17. L.P. Kadanoff, G. Baym, *Quantum Statistical Mechanics* (Benjamin, New York, 1962)
18. W. Cassing, K. Niita, S.J. Wang, Z. Physik A **331**, 439 (1988)
19. Y.B. Ivanov, J. Knoll, D.N. Voskresensky, Nucl. Phys. A **657**, 413 (1999)
20. J. Knoll, Y.B. Ivanov, D.N. Voskresensky, Ann. Phys. **293**, 126 (2001)
21. W. Botermans, R. Malfliet, Phys. Rep. **198**, 115 (1990)
22. S. Mrowczynski, P. Danielewicz, Nucl. Phys. B **342**, 345 (1990)
23. D.F. DuBois, in *Lectures in Theoretical Physics*, ed. by W.E. Brittin (Gordon and Breach, New York, 1967), pp. 469–619
24. P. Danielewicz, Ann. Phys. **152**, 305 (1984)
25. P. Danielewicz, Ann. Phys. **197**, 154 (1990)
26. K. Chou, Z. Su, B. Hao, L. Yu, Phys. Rep. **118**, 1 (1985)
27. J. Rammer, H. Smith, Rev. Mod. Phys. **58**, 323 (1986)
28. E. Calzetta, B.L. Hu, Phys. Rev. D **37**, 2878 (1988)
29. H. Haug, A.P. Jauho, *Quantum Kinetics in Transport and Optics of Semiconductors* (Springer, New York, 1999)
30. G. F. Bertsch, S. Das Gupta, Phys. Rept. **160**, 189 (1988)
31. W. Cassing, U. Mosel, Prog. Part. Nucl. Phys. **25**, 235 (1990)
32. S. J. Wang, W. Cassing, Ann. Phys. **159**, 328 (1985)
33. K. Morawetz, *Interacting Systems Far from Equilibrium* (Oxford University Press, Oxford, 2018)
34. A. Ono, H. Horiuchi, T. Maruyama, Prog. Theo. Phys. **87**, 1185 (1992)
35. H. Feldmeier, Nucl. Phys. A **515**, 147 (1990)
36. J. Aichelin, Phys. Rep. **202**, 233 (1991)
37. J. Aichelin, E.L. Bratkovskaya, A. Le Fevre, V. Kireyeu, V. Kolesnikov, Phys. Rev. C **101**, 044905 (2020)
38. A. Ono, Prog. Part. Nucl. Phys. **105**, 139 (2019)

Nonrelativistic On-Shell Kinetic Theories

2

This chapter is devoted to the derivation of nonrelativistic transport equations on the basis of many-body theory for different levels of truncation. Contrary to classical transport equations this derivation starts from the quantum mechanics of wavefunctions incorporating the antisymmetry of the N-body wavefunction with respect to particle exchange for systems of fermions. We will use the conventional Schrödinger picture of quantum mechanics[1] and present a coupled set of equations of motion for density matrices of rank $1 \leq n \leq N$, where N denotes the conserved number of fermions in the system. By introducing a cluster expansion for the n-body density matrices different truncation schemes can be formulated that are compatible with conservation laws and preserve the antisymmetry of the system. A Wigner transformation to phase-space variables then provides the basis for the formulation of kinetic theories and their approximations employed. The lower order truncation schemes are presented in detail and numerical examples are given to provide some idea about the relative range of validity or applicability. To this end we will derive equations of motion for "testparticles" that simulate the time evolution of the system in phase space. A brief summary and critical discussion of the results achieved will close this chapter.

2.1 Time Evolution of N-body Interacting Fermi Systems

In case of nonrelativistic systems, i.e. when the mass of the particles is large compared to their kinetic energy, a many-body system consisting of N-interacting fermions is most conveniently described in the Schrödinger picture of quantum mechanics, where the time evolution of the system is governed by a N-body

---

[1] A reminder of the different pictures of quantum mechanics is given in Appendix A.

© The Author(s), under exclusive license to Springer Nature Switzerland AG 2021
W. Cassing, *Transport Theories for Strongly-Interacting Systems*, Lecture Notes
in Physics 989, https://doi.org/10.1007/978-3-030-80295-0_2

Hamiltonian $\hat{H}_N$. We thus may start with the N-body Schrödinger equation[2]

$$i\frac{\partial}{\partial t}\,\Psi_N(1,\ldots,N;t) = \hat{H}_N(1,\ldots,N)\Psi_N(1,\ldots,N;t) \tag{2.1}$$

and its Hermitian conjugate

$$-i\frac{\partial}{\partial t'}\,\Psi_N^*(1',\ldots,N';t') = \hat{H}_N(1',\ldots,N')\Psi_N^*(1',\ldots,N';t') \tag{2.2}$$

using $\hat{H}_N = \hat{H}_N^\dagger$ as mandatory for a Hamiltonian. In Eqs. (2.1) and (2.2) the lower index N stands for the number of fermions in the system which should not be mixed up with the variable N of the wavefunction since we use the notation i for the coordinates of particle i, (e.g.: $i \equiv \mathbf{r}_i, \sigma_i, \tau_i \equiv$ space $\mathbf{r}_i$, spin, isospin, etc.). The variable N then stands for $N \equiv (\mathbf{r}_N, \sigma_N, \tau_N)$. Alternatively, one might use the momentum representation for particle i or any other unitary transformation of coordinates.

As an example we consider the case of 3 particles where the 3-body wavefunction Ψ_3 in coordinate-space representation reads explicitly

$$\Psi_3(\mathbf{r}_1, \sigma_1, \mathbf{r}_2, \sigma_2, \mathbf{r}_3, \sigma_3; t), \tag{2.3}$$

and its conjugate

$$\Psi_3^*(\mathbf{r}_1', \sigma_1', \mathbf{r}_2', \sigma_2', \mathbf{r}_3', \sigma_3'; t'), \tag{2.4}$$

with σ_i denoting the spin projections. In case of additional discrete internal quantum numbers (like isospin τ_i or flavor f_i) the list of continuous or discrete variables has to be increased to $(\mathbf{r}_i, \sigma_i, \tau_i, \ldots \equiv i)$. Note that the wavefunction Ψ_N has to be antisymmetric with respect to particle exchange, i.e. (for $N = 3$)

$$\Psi_3(1, 2, 3; t) = -\Psi_3(2, 1, 3; t) = \Psi_3(3, 1, 2; t) = -\Psi_3(3, 2, 1; t), \text{ etc.} \tag{2.5}$$

Multiplying (2.1) by Ψ_N^*, (2.2) by Ψ_N and taking the difference we obtain

$$i\left(\frac{\partial}{\partial t} + \frac{\partial}{\partial t'}\right)\,\Psi_N(1,\ldots,N;t)\,\Psi_N^*(1',\ldots,N';t') \tag{2.6}$$

$$= \left(\hat{H}_N(1,\ldots,N) - \hat{H}_N(1',\ldots,N')\right)\Psi_N(1,\ldots,N;t)\Psi_N^*(1',\ldots,N';t').$$

---

[2] For convenience we will use natural units $\hbar = c = 1$ throughout this work if not specified explicitly.

Defining two-time density-matrix elements by

$$\rho_N(1, \ldots, N, 1', \ldots, N'; t, t') = \Psi_N(1, \ldots, N; t)\Psi_N^*(1', \ldots, N'; t'), \quad (2.7)$$

which are the matrix elements of the density-matrix operator $\hat{\rho}_N(t, t')$,

$$\rho_N(1, \ldots, N, 1', \ldots, N'; t, t') = \langle 1', \ldots, N' | \hat{\rho}_N(t, t') | 1, \ldots, N \rangle, \quad (2.8)$$

we obtain

$$i \left(\frac{\partial}{\partial t} + \frac{\partial}{\partial t'} \right) \hat{\rho}_N(t, t') = \left(\hat{H}_N(1, \ldots, N) \hat{\rho}_N(t, t') - \hat{\rho}_N(t, t') \hat{H}_N(1', \ldots, N') \right)$$
$$(2.9)$$

or in shorthand form as an operator equation

$$i \left(\frac{\partial}{\partial t} + \frac{\partial}{\partial t'} \right) \hat{\rho}_N = [\hat{H}_N, \hat{\rho}_N], \quad (2.10)$$

where $[., .]$ denotes the usual commutator. When performing a Fourier transformation with respect to t and t',

$$\hat{\rho}_N(\omega, \omega') = \int dt \int dt' \, \exp(-i\omega t + i\omega' t') \, \hat{\rho}_N(t, t'), \quad (2.11)$$

Eq. (2.10) turns to

$$\left[(\omega - \hat{H}_N) - (\omega' - \hat{H}_N') \right] \hat{\rho}_N(\omega, \omega') = 0, \quad (2.12)$$

where $\hat{H}_N'$ acts on the coordinates i' and $\hat{H}_N$ on the coordinates i. Since $\hat{H}_N = \hat{H}_N^\dagger$ the Hamiltonian has eigenvalues on the real axis in ω, which gives singularities in ω for the inverse operator $(\omega - \hat{H}_N)^{-1}$. Adding some infinitesimal $2i\epsilon\hat{\rho}_N$ we get

$$\left[(\omega - \hat{H}_N + i\epsilon) - (\omega' - \hat{H}_N' - i\epsilon) \right] \hat{\rho}_N(\omega, \omega') = 0. \quad (2.13)$$

With the retarded N-body Green's operator defined by

$$\hat{G}_N(\omega) := \left[\omega - \hat{H}_N + i\epsilon \right]^{-1} \quad (2.14)$$

we can rewrite Eq. (2.13) as

$$\left[\hat{G}_N(\omega)^{-1} - \hat{G}_N^\dagger(\omega')^{-1} \right] \hat{\rho}_N(\omega, \omega') = 0, \qquad (2.15)$$

where $\hat{G}_N(\omega)$ acts on coordinates i and $\hat{G}_N(\omega')$ on coordinates i'.

It is worth to comment on the relation of $\hat{\rho}_N$ to the statistical operator ρ_{St} in quantum statistics. To establish this connection we expand the N-body wavefunction in terms of complete antisymmetric N-body basis states (e.g. Slater determinants $\Phi_k(1, \ldots, N)$),

$$\Psi_N(1, \ldots, N; t) = \sum_k c_k(t) \Phi_k(1, \ldots, N), \qquad (2.16)$$

where the $c_k(t)$ are complex time-dependent expansion coefficients. The N-body density ρ_N then has the following matrix elements:

$$\rho_N(1, \ldots, N, 1' \ldots N'; t, t') = \sum_k \sum_{k'} c_k(t) c_{k'}^*(t') \, \Phi_k(1, \ldots, N) \Phi_{k'}^*(1', \ldots, N').$$

$$(2.17)$$

This will not change Eq. (2.10) due to the unitary transformation (2.16). When keeping only diagonal terms in the expansion of ρ_N, i.e. the coefficients $c_k(t) c_k^*(t') =: \mathcal{P}_k(t) \delta(t - t')$ one loses information and thus introduces a finite entropy in the system, which will increase with time according to the second law of thermodynamics. In this case the real quantities $\mathcal{P}_k(t)$ have the interpretation of a time-dependent probability for the occupation of the state Φ_k in the system. Furthermore, in thermal equilibrium the occupation numbers become time-independent and ρ_N leads to the statistical operator ρ_{St} in quantum statistics, although with a different normalization, i.e. divided by $N!$. On the other hand a two-time formulation will be needed for the formulation of off-shell transport theories on the basis of Green's functions in case of strongly-interacting systems (cf. Chap. 4).

2.2   The Density-Matrix Formalism

The quantum mechanical N-body problem (2.9) in practice cannot be solved in case of interacting systems involving many particles. Accordingly, one has to give up information and restrict to the dynamics of lower complexity, i.e. to **reduced density matrices** of lower rank and ultimately derive nonlinear equations of motion for the one-body density matrix $\rho_1(1, 1'; t)$ while shifting higher order correlations to resummed matrix elements of the interaction. This strategy leads to the density-matrix formalism.

Within this formalism one conventionally restricts to a single time t, i.e. one considers only the time-diagonal information in ρ_N,

$$\rho_N(1,\ldots,N,1',\ldots,N';t) = \rho_N(1,\ldots,N,1',\ldots,N';t,t')\,\delta(t-t'),\quad (2.18)$$

which gives the **von-Neumann equation** for the density matrix $\rho_N(t)$,

$$i\frac{\partial}{\partial t}\,\rho_N(1,\ldots,N;1'..N';t) = [\hat{H}_N,\rho_N(1,\ldots,N;1'..N';t)]\,.\qquad (2.19)$$

In the following we will assume that the Hamiltonian is approximately given by a mean-field part and mutual two-body interactions,

$$\hat{H}_N = \sum_{i=1}^{N} h^0(i) + \sum_{i<j}^{N-1} v(ij),\qquad (2.20)$$

where

$$h^0(i) = t(i) + U^0(i)\qquad (2.21)$$

gives the one-body part of $\hat{H}_N$, consisting of the kinetic energy of particle i and a (possible) external mean-field potential $U^0(i)$, which e.g. might be an external electromagnetic field, while $v(ij)$ describes a two-body interaction. Equation (2.19) is practically not solvable for many-body systems such that suitable approximation schemes are required.

To this aim we introduce **reduced density matrices** $\rho_n(1\ldots n, 1'\ldots n';t)$, which are defined by taking the trace over particles $n+1,\ldots,N$ of the N-body density matrix ρ_N:

$$\rho_n = \frac{1}{(N-n)!}Tr_{(n+1,\ldots,N)}\rho_N = \frac{1}{(n+1)}Tr_{(n+1)}\rho_{n+1}.\qquad (2.22)$$

In the reduced density matrices ρ_n all information about particles from $n+1$ to N thus is integrated out. In (2.22) the relative normalization between ρ_n and ρ_{n+1} is fixed and it is useful to choose the normalization

$$Tr_{(1,\ldots,N)}\,\rho_N = N!,\qquad (2.23)$$

which leads to the following normalization for the one-body density matrix,

$$Tr_{(1=1')}\rho(11';t) = \sum_i \langle a_i^\dagger a_i\rangle = N,\qquad (2.24)$$

i.e. the particle number for the N-body Fermi system. In Eq. (2.24) $a_i^\dagger$ and a_i denote Fermion creation and annihilation operators that follow the Fermi anti-commutation laws. The normalization of the two-body density matrix then reads

$$Tr_{(1,2)}\rho_2 = \sum_{i,j}\langle a_i^\dagger a_j^\dagger a_j a_i\rangle = -\sum_{i,j}\langle a_i^\dagger a_j^\dagger a_i a_j\rangle = \sum_{i,j}\{\langle a_i^\dagger a_i a_j^\dagger a_j\rangle - \langle a_i^\dagger a_j\rangle\delta_{ij}\}$$

(2.25)

$$= (N-1)\sum_j\langle a_j^\dagger a_j\rangle = N(N-1).$$

In analogy we obtain for the traces of the density matrices ρ_n (for $n \leq N$),

$$Tr_{(1,\ldots,n)}\rho_n = \frac{N!}{(N-n)!},$$

(2.26)

since for $n = N$ the density matrix ρ_N is normalized to $N!$ according to (2.26).

Exercise 2.1: Prove Eq. (2.26) starting from (2.23).

Now taking the trace $Tr_{(n+1,\ldots,N)}$) of the von-Neumann equation (2.19), one obtains a system of coupled differential equations of first-order in time t, which is denoted as BBGKY hierarchy (according to the authors **Bogolyubov**, **Born**, **Green**, **Kirkwood**, and **Yvon**),

$$i\frac{\partial}{\partial t}\rho_n = \sum_{i=1}^{n}\left[h^0(i), \rho_n\right] + \sum_{1=i<j}^{n-1}[v(ij), \rho_n] + \sum_{i=1}^{n}Tr_{n+1}[v(i, n+1), \rho_{n+1}]$$

(2.27)

for $1 \leq n \leq N$. This system is of finite order in case of nonrelativistic systems since $\rho_{N+1} = 0$. In (2.27) the time evolution of ρ_n couples to the time evolution of ρ_{n+1} by the interaction $v(ij)$ for $1 \leq i \leq n$ and $j = n+1$. Accordingly, the system of equations is not closed in any order $n < N$ and becomes equivalent to the von-Neumann equation (2.19) for $n = N$.

At first sight one might not gain anything but let us have a closer look at the equations of lowest order. The equations for $n = 1$ and $n = 2$ read explicitly:

$$i\frac{\partial}{\partial t}\rho_1 = [h^0(1), \rho_1] + Tr_2[v(12), \rho_2],$$

(2.28)

$$i\frac{\partial}{\partial t}\rho_2 = \sum_{i=1}^{2}\left[h^0(i), \rho_2\right] + [v(12), \rho_2] + Tr_3[v(13) + v(23), \rho_3],$$

(2.29)

which are not closed—as mentioned above—since the time evolution of ρ_2 is still determined by the 3-body density matrix ρ_3. In order to obtain a closed set of equations one has to introduce a suitable approximation for ρ_3, which must be fully antisymmetric in the particle indices as required for fermions in quantum mechanics.

2.2.1 Separation of Correlation Functions

The next step is to introduce a **cluster expansion** in the antisymmetrized form (omitting the explicit t-dependence) [1]:

$$\rho_1(11') = \rho(11'), \tag{2.30}$$

$$
\begin{aligned}
\rho_2(12, 1'2') &= \rho(11')\rho(22') - \rho(12')\rho(21') + c_2(12, 1'2') \\
&= \rho_{20}(12, 1'2') + c_2(12, 1'2') \\
&= \mathcal{A}_{12}\rho(11')\rho(22') + c_2(12, 1'2'),
\end{aligned} \tag{2.31}
$$

with the two-body antisymmetrization operator $\mathcal{A}_{ij} = 1 - P_{ij}$. Here the permutation operator P_{ij} describes the exchange of particle indices as e.g. for the product $\rho(i, 1')\rho(j, 2')$:

$$P_{ij}\rho(i, 1')\rho(j, 2') = \rho(j, 1')\rho(i, 2'). \tag{2.32}$$

The antisymmetric expansion for the three-body density matrix reads

$$
\begin{aligned}
\rho_3(123, 1'2'3') &= \rho(11')\rho(22')\rho(33') - \rho(12')\rho(21')\rho(33') \\
&\quad - \rho(13')\rho(22')\rho(31') - \rho(11')\rho(32')\rho(23') \\
&\quad + \rho(13')\rho(21')\rho(32') + \rho(12')\rho(31')\rho(23') \\
&\quad + \rho(11')c_2(23, 2'3') - \rho(12')c_2(23, 1'3') \\
&\quad - \rho(13')c_2(23, 2'1') + \rho(22')c_2(13, 1'3') \\
&\quad - \rho(21')c_2(13, 2'3') - \rho(23')c_2(13, 1'2') \\
&\quad + \rho(33')c_2(12, 1'2') - \rho(31')c_2(12, 3'2') \\
&\quad - \rho(32')c_2(12, 1'3') + c_3(123, 1'2'3'),
\end{aligned} \tag{2.33}
$$

where c_3 incorporates explicit 3-body correlations that are contained in ρ_3 but are not covered by the product terms in (2.33). For fermions the exchange symmetries of the correlation matrix c_2 read:

$$c_2(12, 1'2') = -c_2(12, 2'1') = -c_2(21, 1'2') = c_2^*(1'2', 12), \qquad \text{etc.} \tag{2.34}$$

The trace relations (2.26) allow to calculate explicit trace relations for the correlations c_n which are included in Appendix B and evaluated for Fermi systems in thermal and chemical equilibrium for orientation.

By neglecting c_2 in (2.31) we get the limit of independent particles which is also denoted as **Time-Dependent Hartree–Fock** (TDHF):

$$i\frac{\partial}{\partial t}\,\rho(11';t) = [h^0(1) - h^0(1')]\rho(11';t) + \mathrm{Tr}_{(2=2')}[v(12)\mathcal{A}_{12} - v(1'2')\mathcal{A}_{1'2'}]$$

$$\times\,\rho(11';t)\rho(22';t). \tag{2.35}$$

This implies that all effects from collisions or correlations are incorporated in c_2 and higher orders in c_3, etc.

Exercise 2.2: Show that solutions of the TDHF equation

$$i\frac{\partial}{\partial t}\psi_\alpha(\mathbf{r};t) = \left(-\frac{1}{2m}\nabla_r\cdot\nabla_r + \sum_\beta\int d^3r_2\,\psi_\beta^*(\mathbf{r}_2;t)v(\mathbf{r}-\mathbf{r}_2)\psi_\beta(\mathbf{r}_2;t)n_\beta\right)$$

$$\times\psi_\alpha(\mathbf{r};t) \tag{2.36}$$

$$-\sum_\beta\int d^3r_2\,\psi_\beta^*(\mathbf{r}_2;t)v(\mathbf{r}-\mathbf{r}_2)\psi_\beta(\mathbf{r};t)\psi_\alpha(\mathbf{r}_2;t)n_\beta$$

$$=: h^{HF}\psi_\alpha(\mathbf{r};t),$$

where n_β denote time-independent occupation numbers with $N = \sum_\beta n_\beta$, provide a solution for Eq. (2.35) with $\rho(1,1';t)$ given by

$$\rho(1,1';t) = \rho(\mathbf{r},\mathbf{r}';t) = \sum_\alpha n_\alpha\psi_\alpha^*(\mathbf{r}';t)\psi_\alpha(\mathbf{r};t). \tag{2.37}$$

When discarding explicit three-body correlations c_3 in (2.33) the remaining set of equations is closed and we obtain for the one-body density matrix

$$i\frac{\partial}{\partial t}\,\rho(11';t) = [h^0(1) - h^0(1')]\rho(11';t)$$

$$+ \mathrm{Tr}_{(2=2')}[v(12)\mathcal{A}_{12} - v(1'2')\mathcal{A}_{1'2'}]\rho(11';t)\rho(22';t)$$

$$+ \mathrm{Tr}_{(2=2')}[v(12) - v(1'2')]c_2(12,1'2';t) \tag{2.38}$$

and for the two-body correlation matrix (after some tedious analytic work)

$$i\frac{\partial}{\partial t} c_2(12, 1'2'; t) = [h^0(1) + h^0(2) - h^0(1') - h^0(2')]c_2(12, 1'2'; t) \qquad (2.39)$$

$$+ \, \text{Tr}_{(3=3')}[v(13)\mathcal{A}_{13} + v(23)\mathcal{A}_{23}$$

$$- v(1'3')\mathcal{A}_{1'3'} - v(2'3')\mathcal{A}_{2'3'}]\rho(33'; t)c_2(12, 1'2'; t)$$

$$+ [v(12) - v(1'2')]\rho_{20}(12, 1'2')$$

$$- \, \text{Tr}_{(3=3')}\{v(13)\rho(23'; t)\rho_{20}(13, 1'2'; t) - v(1'3')\rho(32'; t)\rho_{20}(12, 1'3'; t)$$

$$+ v(23)\rho(13'; t)\rho_{20}(32, 1'2'; t) - v(2'3')\rho(31'; t)\rho_{20}(12, 3'2'; t)\}$$

$$+ [v(12) - v(1'2')]c_2(12, 1'2'; t) \qquad (2.40)$$

$$- \, \text{Tr}_{(3=3')}\{v(13)\rho(23'; t)c_2(13, 1'2'; t) - v(1'3')\rho(32'; t)c_2(12, 1'3'; t)$$

$$+ v(23)\rho(13'; t)c_2(32, 1'2'; t) - v(2'3')\rho(31'; t)c_2(12, 3'2'; t)\}$$

$$+ \, \text{Tr}_{(3=3')}\{[v(13)\mathcal{A}_{13}\mathcal{A}_{1'2'} - v(1'3')\mathcal{A}_{1'3'}\mathcal{A}_{12}] \, \rho(11'; t)c_2(32, 3'2'; t) \quad (2.41)$$

$$+ [v(23)\mathcal{A}_{23}\mathcal{A}_{1'2'} - v(2'3')\mathcal{A}_{2'3'}\mathcal{A}_{12}] \, \rho(22'; t)c_2(13, 1'3'; t)\},$$

which at first sight looks more complex than the BBGKY hierarchy.

To reduce the complexity (and to shorten the lengthy equations) we introduce a one-body Hamiltonian for particle i by

$$h(i) = t(i) + U^s(i) = t(i) + \text{Tr}_{(n=n')}v(in)\mathcal{A}_{in}\rho(nn'; t), \qquad (2.42)$$

$$h(i') = t(i') + U^s(i') = t(i') + \text{Tr}_{(n=n')}v(i'n')\mathcal{A}_{i'n'}\rho(nn'; t)$$

that includes the interaction in the self-generated time-dependent mean field. With the abbreviations

$$Q_{ij}^=(t) = 1 - \text{Tr}_{(n=n')}(P_{in} + P_{jn})\rho(nn'; t),$$

$$Q_{i'j'}^=(t) = 1 - \text{Tr}_{(n=n')}(P_{i'n'} + P_{j'n'})\rho(nn'; t) \qquad (2.43)$$

we have introduced operators that account for medium corrections due to antisymmetrization and will turn out to describe **Pauli blocking**. Accordingly an effective interaction in the medium turns out to be given by

$$V^=(ij) = Q_{ij}^=v(ij), \qquad\qquad V^=(i'j') = Q_{i'j'}^=v(i'j'), \qquad (2.44)$$

with all exchange operators here acting to the right.

The equations for ρ and c_2 achieve the more compact form:

$$i\frac{\partial}{\partial t}\rho(11';t) = [h(1) - h(1')]\rho(11';t) + \mathrm{Tr}_{(2=2')}[v(12) - v(1'2')]c_2(12, 1'2';t),$$
(2.45)

and

$$i\frac{\partial}{\partial t}c_2(12, 1'2';t) = \left[\sum_{i=1}^{2}h(i) - \sum_{i'=1'}^{2'}h(i')\right]c_2(12, 1'2';t)$$

$$+ [V^=(12) - V^=(1'2')]\rho_{20}(12, 1'2';t)$$
(2.46)

$$+ [V^=(12) - V^=(1'2')]c_2(12, 1'2';t)$$
(2.47)

$$+ \mathrm{Tr}_{(3=3')}\{[v(13)\mathcal{A}_{13}\mathcal{A}_{1'2'} - v(1'3')\mathcal{A}_{1'3'}\mathcal{A}_{12}]$$

$$\times \rho(11';t)c_2(32, 3'2';t)$$
(2.48)

$$+ [v(23)\mathcal{A}_{23}\mathcal{A}_{1'2'} - v(2'3')\mathcal{A}_{2'3'}\mathcal{A}_{12}]\,\rho(22';t)c_2(13, 1'3';t)\}.$$

Equation (2.45) describes the propagation of a particle in the self-generated mean field $U^s(i)$ with additional two-body correlations that are further specified in (2.46). In Eq. (2.46) the first line describes the propagation of two particles in the mean field U^s, the second line incorporates off-shell collisions in the Born approximation while the third line (2.47) incorporates a resummation of the in-medium interaction in the sense of a $\mathcal{G}$-matrix ladder resummation [2–4].[3] After Fourier transformation from time t to frequency/energy ω one gets

$$v(ij) \to \mathcal{G}(ij) = v + vg_{20}^+(\omega)Q^=\mathcal{G} = v\sum_{n=0}^{\infty}(g_{20}^+(\omega)Q^=v)^n = v\frac{1}{1 - g_{20}^+(\omega)Q^=v}$$
(2.49)

in analogy to the T-matrix equation[4] but with an additional intermediate Pauli-blocking operator $Q^=$ (2.43) according to (2.44). In this case the bare retarded propagator includes the mean fields U^s and reads (including the center-of-mass motion)

$$G_0^+(\omega) \to g_{20}^+(\omega) = \frac{1}{\omega - h(1) - h(2) + i\epsilon}$$
(2.50)

---

[3] We will come back to this problem in Sect. 2.5.
[4] cf. Appendix C.

instead of G_0^+. (cf. (C.6) in Appendix C, where the center-of-mass motion has been separated.) These relations are readily derived in the limit $i\partial/\partial t \to \tilde{\omega} \equiv \omega - \omega'$ (see below).

The last two lines in (2.41) describe additional particle—hole interactions that are important for ground state correlations (or vacuum correlations) and the damping of low energy collective modes but might be neglected for configurations at high temperatures where also the Pauli-blocking operator plays a minor role [5].

2.2.2   Expansion Within a Single-Particle Basis

The coupled equations for ρ and c_2 (2.45) and (2.46) are not well suited in their present form to allow for an analytical or even numerical solution. To this aim we expand ρ and c_2 within a complete orthonormal single-particle basis $\varphi_\lambda(\mathbf{r}) \equiv \langle \mathbf{r}|\lambda \rangle$ as

$$\rho(11'; t) = \sum_{\lambda\lambda'} \rho_{\lambda\lambda'}(t)\, \varphi_\lambda(\mathbf{r})\varphi_{\lambda'}^*(\mathbf{r}'), \tag{2.51}$$

$$c_2(12, 1'2'; t) = \sum_{\lambda\gamma\lambda'\gamma'} C_{\lambda\gamma\lambda'\gamma'}(t)\, \varphi_\lambda(\mathbf{r}_1)\varphi_\gamma(\mathbf{r}_2)\varphi_{\lambda'}^*(\mathbf{r}_1')\varphi_{\gamma'}^*(\mathbf{r}_2'). \tag{2.52}$$

The single-particle states $\varphi_\lambda(\mathbf{r})$ here are arbitrary and for actual applications have to be selected properly. As an example we consider the excitation of a nucleus with mass number A by some external probe. In this case the basis $\varphi_\lambda(\mathbf{r})$ may be taken as the single-particle basis that diagonalizes the stationary Hartree–Fock Hamiltonian, which leads to a set of localized orthonormal states. In case of problems localized in a finite volume with periodic boundary conditions one might use discrete plane waves as in the example in Appendix D. For continuum problems a basis of plane waves is conventionally employed if the localization of the system is no longer fulfilled or the system is at least of sufficient size. Furthermore, a time-dependent Hartree–Fock basis has been employed by Tohyama and collaborators in case of low energy heavy-ion collisions [6–9].

We insert these expansions in (2.45) and (2.46) and multiply from the left with $\varphi_\alpha^*(\mathbf{r}_1)\varphi_{\alpha'}(\mathbf{r}_{1'})$ or $\varphi_\alpha^*(\mathbf{r}_1)\varphi_\beta^*(\mathbf{r}_2)\varphi_{\alpha'}(\mathbf{r}_{1'})\varphi_{\beta'}(\mathbf{r}_{2'})$ and integrate over $d^3r_1 d^3r_{1'}$ or $d^3r_1 d^3r_2 d^3r_{1'} d^3r_{2'}$, respectively. The first two equations of the BBGKY hierarchy then read for the expansion coefficients $\rho_{\alpha\alpha'}(t)$ (omitting the explicit t-dependence in $h_{\alpha\beta}(t)$, $Q^=(t)$, $\rho(t)$ and $C(t)$):

$$i\frac{\partial}{\partial t}\rho_{\alpha\alpha'} = \sum_\lambda [h_{\alpha\lambda}\rho_{\lambda\alpha'} - \rho_{\alpha\lambda}h_{\lambda\alpha'}] \tag{2.53}$$

$$+ \sum_\beta \sum_{\lambda\gamma} \{\langle \alpha\beta|v|\lambda\gamma\rangle C_{\lambda\gamma\alpha'\beta} - C_{\alpha\beta\lambda\gamma}\langle \lambda\gamma|v|\alpha'\beta\rangle\}\,,$$

and $C_{\alpha\beta\alpha'\beta'}(t)$

$$i\frac{\partial}{\partial t}\,C_{\alpha\beta\alpha'\beta'} = \sum_\lambda\{h_{\alpha\lambda}C_{\lambda\beta\alpha'\beta'} + h_{\beta\lambda}C_{\alpha\lambda\alpha'\beta'} - C_{\alpha\beta\lambda\beta'}h_{\lambda\alpha'} - C_{\alpha\beta\alpha'\lambda}h_{\lambda\beta'}\}$$

(2.54)

$$+ \sum_{\lambda\lambda'\gamma\gamma'}\{Q^=_{\alpha\beta\lambda'\gamma'}\langle\lambda'\gamma'|v|\lambda\gamma\rangle(\rho_{20})_{\lambda\gamma\alpha'\beta'} - (\rho_{20})_{\alpha\beta\lambda'\gamma'}\langle\lambda'\gamma'|v|\lambda\gamma\rangle Q^=_{\lambda\gamma\alpha'\beta'}\}$$

(2.55)

$$+ \sum_{\lambda\lambda'\gamma\gamma'}\{Q^=_{\alpha\beta\lambda'\gamma'}\langle\lambda'\gamma'|v|\lambda\gamma\rangle C_{\lambda\gamma\alpha'\beta'} - C_{\alpha\beta\lambda'\gamma'}\langle\lambda'\gamma'|v|\lambda\gamma\rangle Q^=_{\lambda\gamma\alpha'\beta'}\}$$ (2.56)

$$+ \mathcal{A}_{\alpha\beta}\mathcal{A}_{\alpha'\beta'}\sum_{\lambda\lambda'\gamma\gamma'}Q^\perp_{\alpha\gamma'\lambda\lambda'}\langle\lambda\lambda'|v|\alpha'\gamma\rangle C_{\gamma\beta\gamma'\beta'}$$ (2.57)

with

$$Q^=_{\alpha\beta\lambda'\gamma'} = \delta_{\alpha\lambda'}\delta_{\beta\gamma'} - \delta_{\alpha\lambda'}\rho_{\beta\gamma'} - \rho_{\alpha\lambda'}\delta_{\beta\gamma'},$$ (2.58)

$$Q^\perp_{\alpha\beta\alpha'\beta'} = \delta_{\alpha\beta'}\rho_{\beta\alpha'} - \delta_{\beta\alpha'}\rho_{\alpha\beta'}$$

and the one-body Hamiltonian

$$h_{\alpha\lambda} = \langle\alpha|t|\lambda\rangle + \langle\alpha|U^0|\lambda\rangle + \sum_{\gamma\gamma'}\langle\alpha\gamma'|v|\lambda\gamma\rangle_{\mathcal{A}}\rho_{\gamma\gamma'}.$$ (2.59)

In (2.59) we have, furthermore, used

$$(\rho_{20})_{\alpha\beta\alpha'\beta'} = \rho_{\alpha\alpha'}\rho_{\beta\beta'} - \rho_{\alpha\beta'}\rho_{\beta\alpha'} = \mathcal{A}_{\alpha\beta}\rho_{\alpha\alpha'}\rho_{\beta\beta'}$$ (2.60)

and

$$\langle\alpha\beta|v|\alpha'\beta'\rangle_{\mathcal{A}} = \langle\alpha\beta|v|\alpha'\beta'\rangle - \langle\alpha\beta|v|\beta'\alpha'\rangle$$ (2.61)

for the antisymmetric matrix elements of the interaction v.

Exercise 2.3: Prove Eq. (2.53) for the expansions (2.51) and (2.52).

Equations (2.53) and (2.54)–(2.57) are the starting points for the derivation of nonrelativistic kinetic theories; these equations are fully antisymmetric in the matrix elements, closed in ρ and c_2 and provide a quantum mechanical description of fermion systems even far from equilibrium. As we will show in the next subsection these equations fulfill the conservation laws for fermion number, total momentum, total angular momentum and energy.

It is, furthermore, instructive to consider these equations in a basis where $\rho_{\alpha\alpha'}(t) =: \delta_{\alpha\alpha'} n_\alpha(t)$,[5] where they reduce to (dropping the explicit t-dependence in $n_\alpha(t)$ and $C_{\alpha\beta\alpha'\beta'}(t)$)

$$i\frac{\partial}{\partial t} n_\alpha = \sum_\beta \sum_{\lambda\gamma} \{\langle\alpha\beta|v|\lambda\gamma\rangle C_{\lambda\gamma\alpha\beta} - C_{\alpha\beta\lambda\gamma}\langle\lambda\gamma|v|\alpha\beta\rangle\}\,, \tag{2.62}$$

and $C_{\alpha\beta\alpha'\beta'}(t)$

$$i\frac{\partial}{\partial t} C_{\alpha\beta\alpha'\beta'} = \{h_\alpha C_{\alpha\beta\alpha'\beta'} + h_\beta C_{\alpha\beta\alpha'\beta'} - C_{\alpha\beta\alpha'\beta'}h_{\alpha'} - C_{\alpha\beta\alpha'\beta'}h_{\beta'}\} \tag{2.63}$$

$$+ [(1 - n_\alpha - n_\beta)\langle\alpha\beta|v|\alpha'\beta'\rangle_A \, n_{\alpha'}n_{\beta'} - n_\alpha n_\beta\langle\alpha\beta|v|\alpha'\beta'\rangle_A(1 - n_{\alpha'} - n_{\beta'})] \tag{2.64}$$

$$+ \sum_{\lambda\gamma} \{(1 - n_\alpha - n_\beta)\langle\alpha\beta|v|\lambda\gamma\rangle C_{\lambda\gamma\alpha'\beta'} - C_{\alpha\beta\lambda\gamma}\langle\lambda\gamma|v|\alpha'\beta'\rangle(1 - n_{\alpha'} - n_{\beta'})\} \tag{2.65}$$

$$+ \sum_{\gamma\gamma'} [(n_{\alpha'}\mathcal{A}_{\alpha\beta} - n_\alpha\mathcal{A}_{\alpha'\beta'})\langle\alpha\gamma'|v|\alpha'\gamma\rangle_A C_{\gamma\beta\gamma'\beta'}$$

$$+ (n_{\beta'}\mathcal{A}_{\alpha\beta} - n_\beta\mathcal{A}_{\alpha'\beta'})\langle\beta\gamma'|v|\beta'\gamma\rangle_A C_{\alpha\gamma\alpha'\gamma'}] \tag{2.66}$$

since

$$Q^=_{\alpha\beta\lambda'\gamma'} = \delta_{\alpha\lambda'}\delta_{\beta\gamma'}(1 - n_\alpha - n_\beta), \tag{2.67}$$

$$Q^\perp_{\alpha\beta\alpha'\beta'} = \delta_{\alpha\beta'}\delta_{\beta\alpha'}(n_{\alpha'} - n_\alpha)$$

and the one-body Hamiltonian reduces to

$$h_\alpha = \langle\alpha|t|\alpha\rangle + \langle\alpha|U^0|\alpha\rangle + \sum_\gamma \langle\alpha\gamma|v|\alpha\gamma\rangle_A n_\gamma. \tag{2.68}$$

Note that the terms in (2.66) $\sim (n_{\alpha'} - n_\alpha)$ and $\sim (n_{\beta'} - n_\beta)$ in case of systems close to the ground state barely contribute if both states α', α or β', β are occupied or unoccupied. Accordingly, those terms are dominantly of particle-hole type and contribute to ground state correlations [5].

---

[5] The quantities $n_\alpha(t)$ denote the occupation numbers of the states φ_α.

2.2.3 Conservation Laws

The rearrangement of terms in Eq. (2.54) at first sight might look arbitrary but any meaningful truncation scheme has to obey some fundamental conservation laws. This will be shown in the following.

(i) Particle Number
The particle number is given by the trace of ρ or in the discrete basis by

$$N(t) = \sum_\alpha \rho_{\alpha\alpha}(t) = \sum_\alpha n_\alpha(t). \tag{2.69}$$

Taking derivatives with respect to time t and inserting the equations of motion for $\rho_{\alpha\alpha}(t)$ gives

$$i\frac{d}{dt} N(t) = i \sum_\alpha \dot\rho_{\alpha\alpha}(t) = \left\{ \sum_{\alpha\lambda}[h_{\alpha\lambda}\rho_{\lambda\alpha} - \rho_{\alpha\lambda}h_{\lambda\alpha}] \right. \tag{2.70}$$

$$\left. + \sum_{\alpha\beta\gamma\lambda}[\langle\alpha\beta|v|\gamma\lambda\rangle C_{\gamma\lambda\alpha\beta} - C_{\alpha\beta\gamma\lambda}\langle\gamma\lambda|v|\alpha\beta\rangle] \right\} = 0,$$

as one easily verifies by redefining the summation indices $\alpha\beta \leftrightarrow \gamma\lambda$. Accordingly, the fermion number is a conserved quantity.

(ii) Total Momentum (Angular Momentum)
The expectation value of the total momentum of the system is given by

$$\langle\mathbf{P}\rangle = Tr(\mathbf{p}\,\rho) = \sum_\alpha \langle\alpha|\mathbf{p}\,\rho|\alpha\rangle = \sum_{\alpha\lambda}\langle\alpha|\mathbf{p}|\lambda\rangle\rho_{\lambda\alpha}, \tag{2.71}$$

since the momentum $\mathbf{p}$ is a single-particle operator. In order to proof the conservation of momentum we first consider the time derivative of (2.71) and insert again the equation of motion for $\rho_{\lambda\alpha}(t)$;

$$i\frac{d}{dt}\langle\mathbf{P}\rangle = \sum_{\alpha\lambda}\langle\alpha|\mathbf{p}|\lambda\rangle i\,\dot\rho_{\lambda\alpha} \tag{2.72}$$

$$= \sum_{\alpha\lambda\lambda'}\langle\alpha|\mathbf{p}|\lambda\rangle[h_{\lambda\lambda'}\rho_{\lambda'\alpha} - \rho_{\lambda\lambda'}h_{\lambda'\alpha}]$$

$$+ \sum_{\alpha\beta\gamma\lambda\lambda'}\langle\alpha|\mathbf{p}|\lambda\rangle[\langle\lambda\beta|v|\lambda'\gamma\rangle C_{\lambda'\gamma\alpha\beta} - C_{\lambda\beta\lambda'\gamma}\langle\lambda'\gamma|v|\alpha\beta\rangle] = 0$$

when redefining the summation indices.

In analogy to the total momentum one proofs the conservation of the total angular momentum since this is again the sum of single-particle operators in case of angular momentum conserving interactions or vanishing commutators $[v, \mathbf{l}_i] = 0$.

(iii) Energy

For all closed systems (without external forces, i.e. $U^0 \equiv 0$) the total energy must be a conserved quantity. It is the sum of the kinetic energy matrix elements

$$E_{kin} = \sum_{\alpha\lambda} \langle \alpha|t|\lambda \rangle \rho_{\lambda\alpha}, \tag{2.73}$$

the contribution from the mean fields

$$E_{MF} = \frac{1}{2} \sum_{\alpha\alpha'\lambda\lambda'} \rho_{\alpha\alpha'} \langle \alpha'\lambda'|v|\alpha\lambda \rangle_A \rho_{\lambda\lambda'}, \tag{2.74}$$

and the correlation energy

$$E_{cor} = \frac{1}{2} \sum_{\alpha\alpha'\lambda\lambda'} \langle \alpha\lambda|v|\alpha'\lambda' \rangle C_{\alpha'\lambda'\alpha\lambda}. \tag{2.75}$$

Since the total energy is a 2-body operator one additionally needs the time evolution of the matrix elements $C_{\alpha\beta\alpha'\beta'}$, i.e.

$$\frac{d}{dt} E = \frac{d}{dt}\{E_{kin} + E_{MF} + E_{cor}\} \tag{2.76}$$

$$= \sum_{\alpha\lambda} \langle \alpha|t|\lambda \rangle \dot{\rho}_{\lambda\alpha} + \frac{1}{2} \sum_{\alpha\alpha'\lambda\lambda'} \langle \alpha'\lambda'|v|\alpha\lambda \rangle_A [\dot{\rho}_{\lambda\lambda'}\rho_{\alpha\alpha'} + \rho_{\lambda\lambda'}\dot{\rho}_{\alpha\alpha'}]$$

$$+ \frac{1}{2} \sum_{\alpha\alpha'\lambda\lambda'} \langle \alpha'\lambda'|v|\alpha\lambda \rangle \dot{C}_{\alpha\lambda\alpha'\lambda'} = \cdots = 0,$$

as obtained from inserting $\dot{\rho}$ from (2.53) and $\dot{C}$ from (2.54)–(2.57) and redefining the summation indices. Accordingly, the total energy is a conserved quantity of the system within the framework of the coupled equations of motion (2.53) and (2.54)–(2.57). Here the different limits for the time evolution of the two-body correlations C in (2.54)–(2.57) fulfill the conservation laws separately which allows to define conserved approximation schemes of different sophistication.

2.3   The Vlasov Equation

In order to clarify the physical meaning of Eqs. (2.53) and (2.54) we first consider Eq. (2.53) in the limit $C_{\alpha\beta\alpha'\beta'} \equiv 0$, i.e.

$$\frac{\partial}{\partial t} \rho_{\alpha\alpha'}(t) + i \left[\sum_{\lambda} h_{\alpha\lambda}(t)\rho_{\lambda\alpha'}(t) - \rho_{\alpha\lambda}(t)h_{\lambda\alpha'}(t) \right] = 0, \tag{2.77}$$

or in space-time representation for $\rho(\mathbf{x}, \mathbf{x}'; t) = \langle \mathbf{x}'|\rho(t)|\mathbf{x}\rangle$:

$$\frac{\partial}{\partial t} \rho(\mathbf{x}, \mathbf{x}'; t) + i \left[-\frac{1}{2m}\nabla_x^2 + U(\mathbf{x}; t) + \frac{1}{2m}\nabla_{x'}^2 - U(\mathbf{x}'; t) \right] \rho(\mathbf{x}, \mathbf{x}'; t) = 0. \tag{2.78}$$

To simplify the notation we have discarded the discrete indices for spin, isospin, etc. in (2.78) since these are not relevant for the following consideration. Furthermore, we focus on local potentials $U(\mathbf{x}; t)$, which include a possible external field $U^0(\mathbf{x}; t)$ (cf. (2.21)) as well as the one-body potential from the self-interactions of the fermions $U^s(\mathbf{x}, t)$,

$$U(\mathbf{x}; t) = U^0(\mathbf{x}; t) + \sum_{spin,\, isospin} \int d^3x_2\, v(\mathbf{x} - \mathbf{x}_2)\, \rho(\mathbf{x}_2, \mathbf{x}_2; t), \tag{2.79}$$

where the exchange part of the interaction (Fock-term) has been omitted for simplicity since this gives a nonlocal interaction term which complicates the notation.[6] For the following examples we will neglect also an external potential $U^0(\mathbf{x}; t)$ such that the mean-field potential (2.79) is given by the local Hartree potential, which is nothing but the interaction integral of the two-body interaction $v(\mathbf{x} - \mathbf{x}_2)$ with the density $\rho(\mathbf{x}_2)$ at position $\mathbf{x}_2$.

As one easily verifies by insertion (Exercise 2.2) a solution of Eq. (2.77) is given by

$$\rho(\mathbf{x}, \mathbf{x}'; t) = \sum_{\alpha} n_{\alpha} \psi_{\alpha}^{HF}(\mathbf{x}; t)\, \psi_{\alpha}^{*HF}(\mathbf{x}'; t), \tag{2.80}$$

e.g. with $n_{\alpha} = 1$ for the N lowest single-particle eigenstates $\psi_{\alpha}^{HF}(t = 0)$ and $n_{\alpha} = 0$ else, such that

$$\sum_{\alpha} n_{\alpha} = N. \tag{2.81}$$

---

[6] When using Skyrme-like local two-body forces ($\sim \delta^3(\mathbf{x} - \mathbf{x}_2)$) the antisymmetrization implies that contributions to the mean field from the same spin and isospin projection are excluded.

The single-particle states $\psi_\alpha^{HF}(t)$ then have to follow the time-dependent Hartree–Fock equation

$$\left(i\frac{\partial}{\partial t} + \frac{1}{2m}\nabla_x^2 - U(\mathbf{x}; t)\right)\psi_\alpha^{HF}(\mathbf{x}; t) = 0, \tag{2.82}$$

which can be solved numerically for a given two-body interaction v. Stationary solutions of Eq. (2.82) are obtained by iterating the Hartree–Fock equations

$$\left(\epsilon_\alpha + \frac{1}{2m}\nabla_x^2 - U(\mathbf{x}; t)\right)\psi_\alpha^{HF}(\mathbf{x}) = 0 \tag{2.83}$$

with respect to the N lowest lying single-particle energies ϵ_α and states $\psi_\alpha^{HF}(\mathbf{x})$.

Some examples and illustrations are in order. To this end we first consider a spin and isospin symmetric system in nuclear physics, i.e. the nucleus ^{40}Ca consisting of 20 protons and 20 neutrons with both spin projections occupied. We employ an effective two-body interaction with a finite-range Yukawa term, a local Skyrme-type interaction—simulating an effective 3-body force—as well as the Coulomb interaction:

$$v(\mathbf{x} - \mathbf{x}_2) = -A_0\frac{\exp(-\mu|\mathbf{x} - \mathbf{x}_2|)}{\mu|\mathbf{x} - \mathbf{x}_2|} + B_0\delta^3(\mathbf{x} - \mathbf{x}_2)\rho(\mathbf{x} - \mathbf{x}_2) + \frac{e^2}{4\pi}\frac{\delta_{pp}}{|\mathbf{x} - \mathbf{x}_2|}, \tag{2.84}$$

where δ_{pp} implies that the Coulomb interaction only is considered between two protons. The range parameter μ is taken to be $2.5\,\text{fm}^{-1}$ while the parameters A_0 and B_0 are fixed to give the binding energy of nuclear matter ($E_B/A \approx -16\,\text{MeV}$) at saturation density $\rho_0 \approx 0.166\,\text{fm}^{-3}$. The solutions of Eq. (2.83) then give a stable nucleus in its ground state—well in line with experimental data for the charge radius and binding energy of ^{40}Ca—which, however, is a trivial stationary solution of Eq. (2.77). In order to explore nonequilibrium phenomena one thus has to generate initial conditions for the TDHF equations (2.82) that involve nonequilibrium configurations in phase space. The systems of choice are heavy-ion collisions , where two nuclei—initially in their mutual ground states—are boosted towards each other (in z-direction) with some impact parameter b (in x-direction) which should be smaller than the sum of the nuclear radii. Very central collisions are described for impact parameter $b = 0$ where the nuclei hit each other head-on. The optimal choice of the reference system is the nucleus-nucleus center-of-mass system such that the total momentum of the system is $\equiv 0$. The magnitude of the boost is determined by the kinetic energy of the projectile nucleons in the laboratory frame—which conventionally is given in MeV/u (or more recently in A MeV)—where the target is at rest. The numerical integration of the TDHF equations (2.82) then provides the time-dependent single-particle states $\psi_\alpha^{HF}(\mathbf{x}; t)$ and consequently $\rho(\mathbf{x}, \mathbf{x}'; t)$ according to (2.80) as well as the mean field $U(\mathbf{x}; t)$ (2.79).

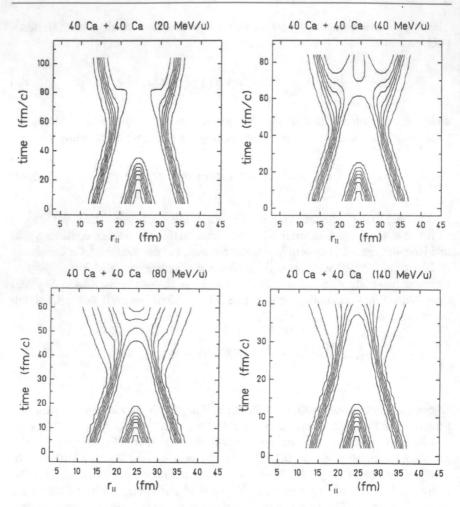

Fig. 2.1 The mean-field potential $U(x = 0, y = 0, z; t)$ $(z = r_\parallel)$ for central collisions of $^{40}Ca + ^{40}Ca$ at beam energies of 20, 40, 80, and 140 MeV/u from TDHF calculations. The contour lines correspond to $0, -5, -10, -15 \ldots$ MeV from outside to inside

As a particular example Fig. 2.1 shows the selfconsistent mean field $U(x = 0, y = 0, z = r_\parallel; t)$ for central collisions of $^{40}Ca + ^{40}Ca$ at beam energies of 20, 40, 80, and 140 MeV/u from a Time-Dependent Hartree–Fock (TDHF) calculation.[7] The contour lines in Fig. 2.1 correspond to $0, -5, -10, -15 \ldots$ MeV from outside to inside. The relative distance of the two nuclei at $t = 0$ was assumed to be 14 fm such that they initially are clearly separated in space. The contour lines indicate a

---

[7] This figure is taken from Ref. [10].

rather simple time dependence of the selfconsistent mean field at these bombarding energies:

1. The boundaries approximately move with a constant velocity in the initial phase which extends to 60, 45, 32, and 25 fm/ at 20, 40, 80, and 140 MeV/u, respectively.
2. The barrier between the mean fields vanishes completely within about 14, 10, 8, 6 fm/c at 20, 40, 80, and 140 MeV/u, respectively. These time scales are of the same order of magnitude as the average collision times of energetic nucleons in the reaction zone.
3. The nucleons move in a common mean field after contact, which is rather flat and changes locally only by a few MeV.

Comment Since the occupation numbers in the expansion (2.80) are constant in time the time evolution of the system in TDHF is isentropic, i.e. does not produce entropy as required for an approach to equilibrium which is characterized by the maximum in entropy.

In order to obtain a closer physical picture of the dynamics incorporated in Eq. (2.78) it is advantageous to transform to the phase-space representation by means of the Wigner transformation

$$\rho(\mathbf{r}, \mathbf{p}; t) = \int d^3 s \ \exp(-i\mathbf{p} \cdot \mathbf{s}) \ \rho(\mathbf{r} + \mathbf{s}/2, \mathbf{r} - \mathbf{s}/2; t) \tag{2.85}$$

with

$$\mathbf{x} = \mathbf{r} + \mathbf{s}/2, \quad \mathbf{x}' = \mathbf{r} - \mathbf{s}/2 \quad \text{or } \mathbf{r} = (\mathbf{x} + \mathbf{x}')/2, \quad \mathbf{s} = \mathbf{x} - \mathbf{x}'. \tag{2.86}$$

The quantum mechanical phase-space density $\rho(\mathbf{r}, \mathbf{p}; t)$ has—in the classical limit—the interpretation of the probability to find a particle at position $\mathbf{r}$ with momentum $\mathbf{p}$ at time t.[8] Independent from the classical limit an integration of (2.85) over momentum[9]

$$\rho(\mathbf{r}; t) = \int \frac{d^3 p}{(2\pi)^3} \ \rho(\mathbf{r}, \mathbf{p}; t) \tag{2.87}$$

---

[8] The Wigner transform (2.85) for quantum mechanical systems in general is **not** a positive definite real function, but a Hermitian operator in phase space. Only in case of systems with large particle number N and averaging over phase-space volumes of order h^3 one should consider the classical limit.

[9] Note that one has to divide additionally by a factor $(\hbar c)^3$ to get the density in length^{-3} if the momentum $\mathbf{p}$ has dimension energy/c.

gives the spatial density $\rho(\mathbf{r}; t)$, while an integration over space gives the momentum density $\rho(\mathbf{p}; t)$,

$$\rho(\mathbf{p}; t) = \int d^3r \, \rho(\mathbf{r}, \mathbf{p}; t); \tag{2.88}$$

the factor $1/(2\pi)^3$ (corresponding to $1/(2\pi\hbar)^3 = h^{-3}$) in (2.87) is responsible for the quantization in phase space for each degree of freedom (spin, isospin, etc.).

Exercise 2.4: The ground state wavefunction of a three-dimensional oscillator is given by a Gaussian

$$\psi(\mathbf{r}) = N_0 \exp(-\mathbf{r} \cdot \mathbf{r}/(2b^2)). \tag{2.89}$$

Calculate the normalization factor N_0 and the Wigner transform of $\psi^*(\mathbf{r}')\psi(\mathbf{r})$. What is the value at the origin? Does the interpretation of a classical phase-space distribution hold?

In order to illustrate the global time evolution of a heavy-ion collision in phase-space representation we show in Fig. 2.2 the Wigner function (with $r_\parallel = z, k_\parallel = k_z = p_z$),

$$\bar{f}(r_\parallel, k_\parallel; t) = \int dx \int dy \int \frac{dk_x}{2\pi} \int \frac{dk_y}{2\pi} \, \rho(\mathbf{r}, \mathbf{k}; t), \tag{2.90}$$

integrated over transverse degrees of freedom in phase space, from TDHF calculations for central collisions of $^{40}Ca +^{40} Ca$ at beam energies of 20, 40, and 80 MeV/u.[10] Due to the integration over transverse degrees of freedom this distribution becomes practically positive definite such that one may interpret the intensity of the dots as the probability to find a nucleon with momentum $k_\parallel$ in beam direction at the position $r_\parallel$ (without knowing their position and momentum in the orthogonal directions). In the plots the actual time is given in units of fm/c. At the early times of 17, 13, and 12 fm/c for the different energies the nuclear distributions are just shifted in momentum (up and down) in line with the actual beam energy. For later times the nuclei overlap in coordinate space but keep approximately separated in phase space and move apart for even longer times (not shown). Apparently not very much happens with the impinging nuclei in the limit of TDHF at these bombarding energies.

---

[10] This figure is taken from: W. Cassing, Z. Phys. A 327 (1987) 87.

$$\overline{f}(r_{\shortparallel}, k_{\shortparallel}; t)$$

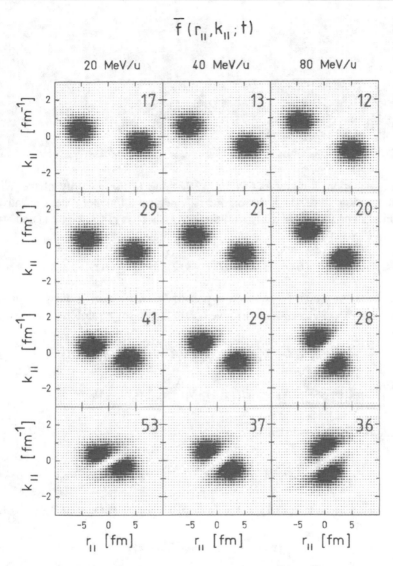

Fig. 2.2 The phase-space density (2.90) for central collisions of $^{40}Ca + {}^{40}Ca$ at beam energies of 20, 40, and 80 MeV/u from TDHF calculations. The times are given in fm/c; the intensity of the dots increases linearly from 0 to 1

After Wigner transformation of Eq. (2.78) (using $\nabla^2_{\mathbf{r}+\mathbf{s}/2} - \nabla^2_{\mathbf{r}-\mathbf{s}/2} = 2\nabla_{\mathbf{s}} \cdot \nabla_{\mathbf{r}}$) and by partial integration we get

$$\frac{\partial}{\partial t} \rho(\mathbf{r}, \mathbf{p}; t) + \frac{\mathbf{p}}{m} \cdot \nabla_{\mathbf{r}}\rho(\mathbf{r}, \mathbf{p}; t) \tag{2.91}$$

$$+ i \int d^3s \, \exp(-i\mathbf{p} \cdot \mathbf{s})[U(\mathbf{r}+\mathbf{s}/2; t) - U(\mathbf{r}-\mathbf{s}/2; t)] \, \rho(\mathbf{r}+\mathbf{s}/2, \mathbf{r}-\mathbf{s}/2; t) = 0.$$

Exercise 2.5: Show that

$$\vec{\nabla}_x^2 - \vec{\nabla}_{x'}^2 = \vec{\nabla}_{\mathbf{r}+\mathbf{s}/2}^2 - \vec{\nabla}_{\mathbf{r}-\mathbf{s}/2}^2 = 2\vec{\nabla}_\mathbf{s} \cdot \vec{\nabla}_\mathbf{r},$$

where $\mathbf{x} = \mathbf{r} + \mathbf{s}/2, \quad \mathbf{x}' = \mathbf{r} - \mathbf{s}/2.$
What is the Wigner transform of $\vec{\nabla}_\mathbf{s} \cdot \vec{\nabla}_\mathbf{r} \, \rho(\mathbf{r} + \mathbf{s}/2, \mathbf{r} - \mathbf{s}/2)$ when assuming that $\rho(\mathbf{r}, \mathbf{s})$ vanishes at $s_i \to \pm\infty$ for $i = x, y, z$?

This equation is equivalent to Eq. (2.77) due to the unitarity of the Wigner transformation. In (2.91), however, one may neglect all higher order derivatives $\geq$ 3rd order in case of sufficiently smooth potentials $U(\mathbf{r}; t)$ (cf. Fig. 2.1) and arrives at

$$[U(\mathbf{r} + \mathbf{s}/2) - U(\mathbf{r} - \mathbf{s}/2)] \approx \mathbf{s} \cdot \nabla_\mathbf{r} U(\mathbf{r}), \tag{2.92}$$

which is even exact for harmonic potentials. After partial integration (with $\mathbf{s} \exp(-i\mathbf{p} \cdot \mathbf{s}) = i\nabla_p \exp(-i\mathbf{p} \cdot \mathbf{s})$) we obtain the **Vlasov equation** :

$$\frac{\partial}{\partial t} \rho(\mathbf{r}, \mathbf{p}; t) + \frac{\mathbf{p}}{m} \cdot \nabla_\mathbf{r} \rho(\mathbf{r}, \mathbf{p}; t) - \nabla_\mathbf{r} U(\mathbf{r}; t) \cdot \nabla_\mathbf{p} \rho(\mathbf{r}, \mathbf{p}; t) \tag{2.93}$$

$$= \left(\frac{\partial}{\partial t} + \frac{\mathbf{p}}{m} \cdot \nabla_\mathbf{r} - \nabla_\mathbf{r} U(\mathbf{r}; t) \cdot \nabla_\mathbf{p} \right) \rho(\mathbf{r}, \mathbf{p}; t) = 0.$$

Equation (2.93) is equivalent to

$$\frac{d}{dt} \rho = 0 = \left\{ \frac{\partial}{\partial t} + \dot{\mathbf{r}} \cdot \nabla_\mathbf{r} + \dot{\mathbf{p}} \cdot \nabla_\mathbf{p} \right\} \rho(\mathbf{r}, \mathbf{p}; t), \tag{2.94}$$

and by comparison with (2.93) we get classical equations of motion for $\dot{\mathbf{r}}$ and $\dot{\mathbf{p}}$, i.e.

$$\dot{\mathbf{r}} = \frac{\mathbf{p}}{m} ; \quad \dot{\mathbf{p}} = -\nabla_\mathbf{r} U(\mathbf{r}; t). \tag{2.95}$$

Accordingly (in the limit $N_t \to \infty$) the "testparticle" distribution

$$\rho_t(\mathbf{r}, \mathbf{p}; t) = \frac{1}{N_t} \sum_{i=1}^{N \cdot N_t} \delta(\mathbf{r} - \mathbf{r}_i(t)) \, \delta(\mathbf{p} - \mathbf{p}_i(t)) \tag{2.96}$$

is a solution of the Vlasov equation (2.93), if the testparticle coordinates and momenta $\mathbf{r}_i(t)$, $\mathbf{p}_i(t)$ are solutions of the classical equations of motion (2.95). This gives a suitable solution of the Vlasov equation in case of high statistics, i.e. in the limit $N_t \to \infty$.

The Ansatz (2.96) with (2.95) is denoted in physics as on-shell **testparticle method** and allows for convenient dynamical simulations of many-body systems in a selfconsistent mean field $U(\mathbf{r}; t)$, which results from the mutual two-body interaction $v(\mathbf{r} - \mathbf{r}_2)$ in (2.79). It is easy to show that the Vlasov equation conserves particle number, total momentum, and angular momentum as well as energy.

For illustration Fig. 2.3 (left column) shows the density distribution of nucleons $\rho(x, y = 0, z; t)$ for a central collision of $^{40}Ca +^{40} Ca$ at a beam energy of 40 MeV/u[11] in the semiclassical limit according to the Vlasov equation (2.93) within the testparticle method (2.96) up to times of 200 fm/c. The nuclei touch at $t \approx 30$ fm/c, completely overlap at $t \approx 50$ fm/c and separate at $t \approx 80$ fm/c such that asymptotically again two moving nuclei are seen although with a slightly lower average momentum. The middle column of Fig. 2.3 displays the momentum distribution $\rho(p_x, p_y = 0, p_z; t)$ for the same collision at the same times. The initial momentum distribution roughly corresponds to overlapping Fermi spheres with about twice the value in the overlap region. In case of full overlap in coordinate space this distribution shows two separate peaks which merge again when the nuclei reseparate in coordinate space. Here the mean-field dynamics leads to some isotropization in momentum space, however, does not equilibrate in momentum space. The right column of Fig. 2.3 finally displays the phase-space distribution $\bar{f}(z, p_z; t)$—integrated over perpendicular degrees of freedom—for the same system and closely resembles the results from the TDHF calculation in Fig. 2.2. Thus the gradient expansion of the mean field up to second order apparently works quite well in case of nucleus-nucleus collisions. This even works better for larger nuclei such as ^{197}Au or ^{208}Pb. Note that the phase-space distributions are clearly separated in phase space even at maximum overlap in coordinate space!

The two-peak structure of the momentum distribution at maximum overlap deserves some more detailed study. To this aim we show in Fig. 2.4 again the momentum distribution $f(p_x, p_y = 0, p_z; t)$ for a central collision of $^{40}Ca +^{40} Ca$ at a beam energy of 40 MeV/u for time intervals of 10 fm/c. Up to the touching configuration at $\sim$30 fm/c the shape is that of two overlapping Fermi spheres as noted before. With the beginning overlap of the nuclei in coordinate space the momentum distribution starts to develop two separate peaks which are most pronounced during the time of maximum overlap at 50–60 fm/c. This is due to the fact that the barrier of the common mean field at touching disappears (cf. Fig. 2.1) and the quasiparticles in this region get accelerated. Since the distributions stay separated in phase space—as pointed out before—they also must be separated in momentum space at full overlap, i.e. after integration of the phase-space distribution over coordinate space. When separating again the nuclei have to build up new surfaces which goes along with a deceleration of the quasiparticles in the region of the new surfaces. Accordingly, the momentum distribution shrinks again to a nonseparated distribution without reaching full isotropy in momentum.

---

[11] The convention for the specification of the bombarding energy per nucleon in the laboratory has changed in time from MeV/u to A MeV.

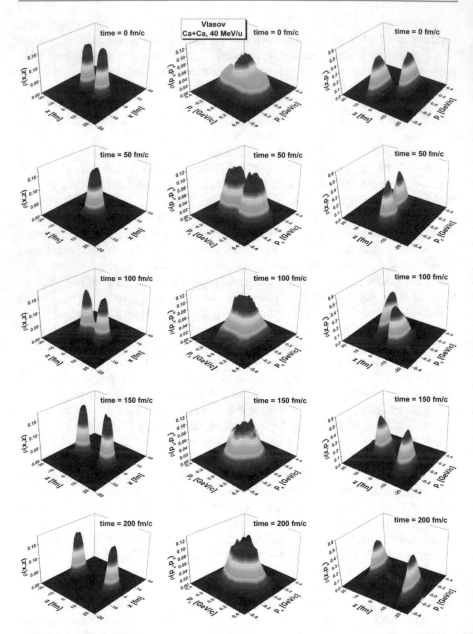

Fig. 2.3 (Left column) The density distribution $\rho(x, y = 0, z; t)$ for a central collision of $^{40}Ca +^{40} Ca$ at a beam energy of 40 MeV/u in the semiclassical limit according to the Vlasov equation (2.93) within the testparticle method (2.96). (Middle column) The momentum distribution $\rho(p_x, p_y = 0, p_z; t)$ for the same reaction in the semiclassical limit according to the Vlasov equation (2.93) within the testparticle method (2.96). (Right column) The phase-space distribution $\bar{f}(z, p_z; t)$—integrated over perpendicular degrees of freedom—for the same reaction in the semiclassical limit according to the Vlasov equation (2.93)

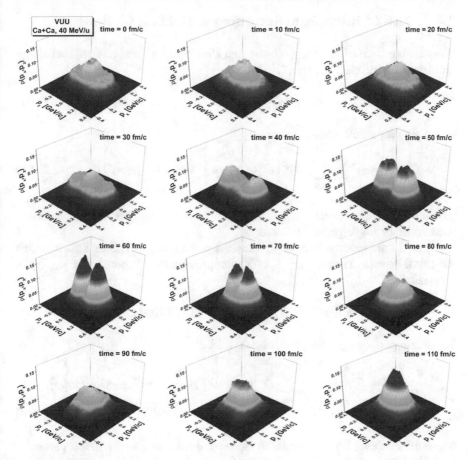

Fig. 2.4 The momentum distribution $f(p_x, p_y = 0, p_z; t)$ for a central collision of $^{40}Ca + ^{40}Ca$ at a beam energy of 40 MeV/u in the semiclassical limit according to the Vlasov equation (2.93) within the testparticle method (2.96) for different times

One may thus conclude that the solution of the Vlasov equation in the semi-classical limit well reproduces the dynamics from THDF in the energy range considered, however, relaxation or stopping phenomena are not well described in the context of Eq. (2.78) since two-body correlations have been discarded. The two-body correlations of lowest order are elastic two-body collisions as known phenomenologically from the dynamics of classical gas particles in a finite volume. Actually we will find out in the next section that the lowest order two-body correlations lead to such collisions although with some medium corrections due to the antisymmetry of the many-body fermion problem.

2.4   The Collision Term According to Uehling-Uhlenbeck

Whereas the derivation of the Vlasov equation could be realized without major problems, the collision term in (2.45)

$$I(11'; t) := -i Tr_{(2=2')}[v(12), c_2(12, 1'2'; t)] \tag{2.97}$$

or in a single-particle basis

$$I_{\alpha\alpha'}(t) = -i \sum_{\beta} \sum_{\lambda\gamma} \{ \langle \alpha\beta|v|\lambda\gamma \rangle C_{\lambda\gamma\alpha'\beta} - C_{\alpha\beta\lambda\gamma} \langle \lambda\gamma|v|\alpha'\beta \rangle \} \tag{2.98}$$

requires some more efforts and the explicit knowledge of the two-body correlations in an arbitrary single-particle basis $|\alpha >$.

We will start with the Born approximation and employ a discrete basis in which the single-particle Hamiltonian $h_{\alpha\lambda}(t)$ is approximately diagonal and $\rho_{\alpha\alpha'}(t)$ is diagonal, i.e.

$$h_{\alpha\lambda}(t) \approx \epsilon_\alpha(t) \delta_{\alpha\lambda}; \qquad \rho_{\alpha\alpha'}(t) = n_\alpha(t) \delta_{\alpha\alpha'}. \tag{2.99}$$

The single-particle equation of motion then reads

$$\frac{\partial}{\partial t} n_\alpha(t) = I_{\alpha\alpha}(t) = -i \sum_{\beta} \sum_{\lambda\gamma} \{ \langle \alpha\beta|v|\lambda\gamma \rangle C_{\lambda\gamma\alpha\beta} - C_{\alpha\beta\lambda\gamma} \langle \lambda\gamma|v|\alpha\beta \rangle \},$$
$$\tag{2.100}$$

where the sum over β is the trace over the second collision partner. In this particular basis the time evolution of the coefficients $C_{\alpha\beta\alpha'\beta'}(t)$ according to (2.54) reads in lowest order Born approximation:

$$\left\{ i \frac{\partial}{\partial t} - [\epsilon_\alpha + \epsilon_\beta - \epsilon_{\alpha'} - \epsilon_{\beta'}] \right\} C_{\alpha\beta\alpha'\beta'}(t) \tag{2.101}$$

$$= \sum_{\lambda\gamma} \{ \langle \alpha\beta|Q^= v|\lambda\gamma \rangle (\rho_{20})_{\lambda\gamma\alpha'\beta'} - (\rho_{20})_{\alpha\beta\lambda\gamma} \langle \lambda\gamma|v Q^=|\alpha'\beta' \rangle \}$$

$$= \langle \alpha\beta|v|\alpha'\beta' \rangle_A [n_{\alpha'} n_{\beta'}(1 - n_\alpha - n_\beta) - n_\alpha n_\beta(1 - n_{\alpha'} - n_{\beta'})](t)$$

$$=: \langle \alpha\beta|V_B(t)|\alpha'\beta' \rangle,$$

where we have used that $Q^=$ also becomes diagonal,

$$Q^=_{\alpha\beta\lambda\gamma}(t) = \delta_{\alpha\lambda} \delta_{\beta\gamma}[1 - n_\alpha(t) - n_\beta(t)]. \tag{2.102}$$

Equation (2.101) is a differential equation of first order in time which can directly be integrated. In the context of the further discussion about energy conservation in two-body collisions we assume that the single-particle energies $\epsilon_\alpha(t) \approx \epsilon_\alpha$ are smoothly varying functions of time with momentum distributions that might be out-of equilibrium. This approximation is well fulfilled for stationary states in a "large" volume. In case of vanishing homogenous solution of (2.101) the coefficients $C_{\alpha\beta\alpha'\beta'}(t)$ are given by

$$C_{\alpha\beta\alpha'\beta'}(t) = -i \int_{t_0}^t dt' \exp\{-i[\epsilon_\alpha + \epsilon_\beta - \epsilon_{\alpha'} - \epsilon_{\beta'}](t - t')\} \langle \alpha\beta|V_B(t')|\alpha'\beta'\rangle,$$

(2.103)

which is easy to verify by insertion in (2.101). In Eq. (2.103) t_0 denotes some arbitrary initial time. Note that the occupation numbers in (2.103) in general depend on t' and

$$\langle \alpha\beta|V_B(t)|\alpha'\beta'\rangle = -\langle \alpha'\beta'|V_B(t)|\alpha\beta\rangle.$$

(2.104)

Comment The formal solution of Eq. (2.101) may also be written as

$$C_{\alpha\beta\alpha'\beta'}(t) = -\frac{i}{2} \int_{t_0}^\infty dt' \int_{-\infty}^\infty \frac{d\omega}{2\pi} \int_{-\infty}^\infty \frac{d\omega'}{2\pi} \exp(-i(\omega - \omega')(t - t'))$$

(2.105)

$$\times \ \langle \alpha\beta|g_{20}^+(\omega)V_B(t')g_{20}^-(\omega')|\alpha'\beta'\rangle$$

with the propagators $g_{20}^\pm(\omega)$ defined by

$$g_{20}^\pm(\omega) = \frac{1}{\omega - h(1) - h(2) \pm i\gamma}$$

(2.106)

with infinitesimal $\gamma > 0$ and matrix elements

$$\langle \alpha'\beta'|g_{20}^\pm(\omega)|\alpha\beta\rangle = \delta_{\alpha\alpha'}\delta_{\beta\beta'}g_{\alpha\beta}^\pm(\omega) = \delta_{\alpha\alpha'}\delta_{\beta\beta'} \frac{1}{\omega - \epsilon_\alpha - \epsilon_\beta \pm i\gamma}.$$

(2.107)

The solution (2.105) then reads alternatively

$$C_{\alpha\beta\alpha'\beta'}(t) = -\frac{i}{2} \int_{-\infty}^\infty dt' \int_{-\infty}^\infty \frac{d\omega}{2\pi} \int_{-\infty}^\infty \frac{d\omega'}{2\pi} \exp(-i(\omega - \omega')(t - t'))$$

(2.108)

$$\times \ \frac{1}{\omega - \epsilon_\alpha - \epsilon_\beta + i\gamma} \langle \alpha\beta|V_B(t')|\alpha'\beta'\rangle \frac{1}{\omega' - \epsilon_{\alpha'} - \epsilon_{\beta'} - i\gamma}.$$

When inserting (2.108) in Eq. (2.101) we obtain

$$\left(i\frac{\partial}{\partial t} - (\epsilon_\alpha + \epsilon_\beta - \epsilon_{\alpha'} - \epsilon_{\beta'})\right) C_{\alpha\beta\alpha'\beta'}(t) \tag{2.109}$$

$$= -\frac{i}{2} \int_{-\infty}^{\infty} dt' \int_{-\infty}^{\infty} \frac{d\omega}{2\pi} \int_{-\infty}^{\infty} \frac{d\omega'}{2\pi} \exp(-i(\omega - \omega')(t - t'))$$

$$\times (\omega - \omega' - \epsilon_\alpha - \epsilon_\beta + \epsilon_{\alpha'} + \epsilon_{\beta'}) \frac{1}{\omega - \epsilon_\alpha - \epsilon_\beta + i\gamma} \langle\alpha\beta|V_B(t')|\alpha'\beta'\rangle$$

$$\times \frac{1}{\omega' - \epsilon_{\alpha'} - \epsilon_{\beta'} - i\gamma}$$

$$= -\frac{i}{2} \int_{-\infty}^{\infty} dt' \int_{-\infty}^{\infty} \frac{d\omega}{2\pi} \int_{-\infty}^{\infty} \frac{d\omega'}{2\pi} \exp(-i(\omega - \omega')(t - t'))$$

$$\times \left(\langle\alpha\beta|V_B(t')|\alpha'\beta'\rangle \frac{1}{\omega' - \epsilon_{\alpha'} - \epsilon_{\beta'} - i\gamma} \right.$$

$$\left. - \frac{1}{\omega - \epsilon_\alpha - \epsilon_\beta + i\gamma} \langle\alpha\beta|V_B(t')|\alpha'\beta'\rangle \right).$$

The integrations over $d\omega'$ (in the first term) and $d\omega$ (in the second term) can be carried out by means of the residue theorem to provide

$$\int_{-\infty}^{\infty} \frac{d\omega'}{2\pi} \exp(+i\omega'(t - t')) \frac{1}{\omega' - \epsilon_{\alpha'} - \epsilon_{\beta'} - i\gamma} = i \exp(i(\epsilon_{\alpha'} + \epsilon_{\beta'})(t - t'))\Theta(t - t') \tag{2.110}$$

(integration over the upper plane) and

$$\int_{-\infty}^{\infty} \frac{d\omega}{2\pi} \exp(-i\omega(t - t')) \frac{1}{\omega - \epsilon_\alpha - \epsilon_\beta + i\gamma} = -i \exp(-i(\epsilon_\alpha + \epsilon_\beta)(t - t'))\Theta(t - t') \tag{2.111}$$

(integration over the lower plane), respectively. Now integrating the remaining first term over $d\omega$ and the second over $d\omega'$ we obtain $\delta(t - t')$ each. Due to the factor $\delta(t - t')$ the exponential factors become unity and the final integrations over dt' give $\langle\alpha\beta|V_B(t)|\alpha'\beta'\rangle/2$ from the first and second term. This completes the proof. We will come back to this formal result when exploring higher order interaction terms in the equations of motion for $C_{\alpha\beta\alpha'\beta'}(t)$ in Sect. 2.4.

Exercise 2.6: Show that for real ω_0 and $\epsilon > 0$

$$\int_{-\infty}^{\infty} \frac{d\omega}{2\pi} \frac{1}{\omega - \omega_0 \mp i\epsilon} = \pm i. \tag{2.112}$$

For the diagonal elements of the collision term (2.98), which are of relevance in the limit (2.99), we obtain with (2.103)

$$I_{\alpha\alpha}(t) = -i \sum_{\beta} \sum_{\lambda\gamma} \{\langle\alpha\beta|v|\lambda\gamma\rangle C_{\lambda\gamma\alpha\beta} - C_{\alpha\beta\lambda\gamma}\langle\lambda\gamma|v|\alpha\beta\rangle\} \tag{2.113}$$

$$= -\sum_{\beta} \sum_{\lambda\gamma} \int_{t_0}^{t} dt' \{\exp\{-i[\epsilon_\lambda + \epsilon_\gamma - \epsilon_\alpha - \epsilon_\beta](t - t')\}$$

$$\times \langle\alpha\beta|v|\lambda\gamma\rangle\langle\lambda\gamma|V_B(t')|\alpha\beta\rangle$$

$$- \exp\{-i[\epsilon_\alpha + \epsilon_\beta - \epsilon_\lambda - \epsilon_\gamma](t - t')\}\langle\alpha\beta|V_B(t')|\lambda\gamma\rangle\langle\lambda\gamma|v|\alpha\beta\rangle\}.$$

Inserting the matrix elements of V_B from (2.101) gives

$$I_{\alpha\alpha}(t) = \sum_{\beta} \sum_{\lambda\gamma} \int_{t_0}^{t} dt' \, 2 \, \cos\{[\epsilon_\alpha + \epsilon_\beta - \epsilon_\lambda - \epsilon_\gamma](t - t')\} \tag{2.114}$$

$$\cdot \langle\alpha\beta|v|\lambda\gamma\rangle\langle\lambda\gamma|v|\alpha\beta\rangle_A[n_\lambda(t')n_\gamma(t')\bar{n}_\alpha(t')\bar{n}_\beta(t') - n_\alpha(t')n_\beta(t')\bar{n}_\lambda(t')\bar{n}_\gamma(t')]$$

using $\bar{n}_\alpha(t') = 1 - n_\alpha(t')$ and $V_B(t')$ from (2.101). Furthermore, we have added and subtracted the term $\sim n_\alpha(t')n_\beta(t')n_\lambda(t')n_\gamma(t')$.

The last expression cannot be further evaluated analytically in case of rapidly changing occupation numbers. However, assuming $n_\alpha(t') \approx n_\alpha(t)$ (in case of weakly interacting systems) one can work out the time integration in (2.113) and obtains (with $t_0 = 0$)

$$\lim_{t\to\infty} \int_0^t dt' \, 2 \, \cos([\epsilon_\alpha + \epsilon_\beta - \epsilon_\lambda - \epsilon_\gamma](t - t'))$$

$$= \lim_{t\to\infty} \frac{2}{[\epsilon_\alpha + \epsilon_\beta - \epsilon_\lambda - \epsilon_\gamma]} \sin([\epsilon_\alpha + \epsilon_\beta - \epsilon_\lambda - \epsilon_\gamma]t) \tag{2.115}$$

$$= 2\pi \, \delta(\epsilon_\alpha + \epsilon_\beta - \epsilon_\lambda - \epsilon_\gamma) =: 2\pi \, \delta(\Delta\omega),$$

the energy conservation for individual two-body interactions with $\Delta\omega = \epsilon_\alpha + \epsilon_\beta - \epsilon_\lambda - \epsilon_\gamma$. This implies that the time in between subsequent two-body collisions τ_s is large compared to the microscopic collision time τ_c, such that the energy uncertainty—associated with the collision—$\Delta \epsilon \approx 2\pi \hbar / \tau_s = h/\tau_s$ becomes sufficiently small. Note that the integral of (2.115) over $\Delta \omega$ gives 2π also for finite t.

For the diagonal elements of the collision term we then obtain

$$I_{\alpha\alpha}(t) \approx 2\pi \sum_{\beta} \sum_{\lambda\gamma} \delta(\epsilon_\alpha + \epsilon_\beta - \epsilon_\lambda - \epsilon_\gamma) \langle \alpha\beta|v|\lambda\gamma \rangle \langle \lambda\gamma|v|\alpha\beta \rangle_{\mathcal{A}} \qquad (2.116)$$

$$\times [n_\lambda n_\gamma \bar{n}_\alpha \bar{n}_\beta - n_\alpha n_\beta \bar{n}_\lambda \bar{n}_\gamma](t)$$

in the basis $|\alpha >$, which diagonalizes the one-body density matrix $\rho_{\alpha\alpha'}$. In quantum mechanics Planck's constant $h = 2\pi\hbar$ is actually a large number in case of few-body systems and the approximation (2.115) not valid, however, in case of sufficiently large systems the sum over β, λ, γ in (2.114) with oscillating contributions in time t' dominantly gives contributions close to the energy shell.

In Appendix D the issue of on-shell collisions and interactions off the energy shell is investigated in a finite box with periodic boundary conditions for the damping of the quadrupole moment in momentum space for nuclear matter slightly above saturation density. As it is found there the off-shell and on-shell scattering results give very similar decay rates for the quadrupole moment since the contributions from off-shell matrix elements in (2.114) give oscillating contributions in time and cancel out to a large extent after summing over the states for all collision partners β and the final states λ, γ. Accordingly, the on-shell collision limit holds well in this case with discrete energy differences, however, there might be physical examples—with a low number of basis states involved—that may show sizeable off-shell scattering effects!

The further evaluation of (2.116) will be carried out for sufficiently extended and homogenous systems in the basis of plane waves

$$\langle \mathbf{r}|\alpha \rangle = \frac{1}{(2\pi)^{3/2}} \exp\{i\mathbf{p}_\alpha \cdot \mathbf{r}\}, \qquad (2.117)$$

such that the density matrix ρ becomes diagonal in momentum $\mathbf{p}$:

$$\rho(\mathbf{p}, \mathbf{p}') = \int d^3r \, \exp\{i(\mathbf{p} - \mathbf{p}') \cdot \mathbf{r}\} \, \varphi(\mathbf{p})\varphi^*(\mathbf{p}') = \delta(\mathbf{p} - \mathbf{p}')n(\mathbf{p}). \qquad (2.118)$$

In (2.118) $n(\mathbf{p})$ then has the physical meaning of an occupation probability for a state with momentum $\mathbf{p}$.

To simplify notation let us assume that the matrix elements of the interaction v in (2.116) do not depend on spin and isospin and are spin and isospin conserving, i.e. its space representation will read:

$$\langle \mathbf{r}_1\mathbf{r}_2|v|\mathbf{r}_{1'}\mathbf{r}_{2'} \rangle = \delta(\mathbf{r}_1 - \mathbf{r}_{1'})\delta(\mathbf{r}_2 - \mathbf{r}_{2'})v(\mathbf{r}_1 - \mathbf{r}_2), \qquad (2.119)$$

or in momentum-space representation[12]

$$\langle \mathbf{p}_1 \mathbf{p}_2 | v | \mathbf{p}_{1'} \mathbf{p}_{2'} \rangle = (2\pi)^3 \delta^3 (\mathbf{p}_1 + \mathbf{p}_2 - \mathbf{p}_{1'} - \mathbf{p}_{2'}) v(\mathbf{p}_2 - \mathbf{p}_{2'}) \qquad (2.120)$$

with the Fourier transform

$$v(\mathbf{p}_2 - \mathbf{p}_{2'}) = \int d^3 s \, \exp(-i (\mathbf{p}_2 - \mathbf{p}_{2'}) \cdot \mathbf{s}) \, v(\mathbf{s}).$$

In Eq. (2.120) $\delta^3 (\mathbf{p}_1 + \mathbf{p}_2 - \mathbf{p}_{1'} - \mathbf{p}_{2'})$ expresses the momentum conservation in a two-body reaction.

Exercise 2.7: Calculate the Fourier transform of the Yukawa interaction (with $\mu > 0$)

$$v(\mathbf{r}) = V_0 \frac{\exp(-\mu |\mathbf{r}|)}{|\mathbf{r}|}. \qquad (2.121)$$

What is the range of the interaction and the scattering amplitude in Born approximation?

Within these assumptions one can carry out the further calculations analytically and the expression (2.116) reads with $\epsilon(\mathbf{p}) = \mathbf{p}^2 / 2m$,

$$I(\mathbf{p}_1, \mathbf{p}_1; t) = (2s + 1)(2\tau + 1) \int \frac{d^3 p_2}{(2\pi)^3} \frac{d^3 p_3}{(2\pi)^3} \frac{d^3 p_4}{(2\pi)^3} \qquad (2.122)$$

$$\times 2\pi \delta \left(\frac{1}{2m} [p_1^2 + p_2^2 - p_3^2 - p_4^2] \right) (2\pi)^3 \delta^3 (\mathbf{p}_1 + \mathbf{p}_2 - \mathbf{p}_3 - \mathbf{p}_4) \, v(\mathbf{p}_2 - \mathbf{p}_4)$$

$$\times v_A (\mathbf{p}_4 - \mathbf{p}_2)$$

$$\times (n(\mathbf{p}_3; t) n(\mathbf{p}_4; t) \bar{n}(\mathbf{p}_1; t) \bar{n}(\mathbf{p}_2; t) - n(\mathbf{p}_1; t) n(\mathbf{p}_2; t) \bar{n}(\mathbf{p}_3; t) \bar{n}(\mathbf{p}_4; t)).$$

Equation (2.122) describes elastic scattering processes $\mathbf{p}_1 + \mathbf{p}_2 \to \mathbf{p}_3 + \mathbf{p}_4$ ("loss"-term) as well as $\mathbf{p}_3 + \mathbf{p}_4 \to \mathbf{p}_1 + \mathbf{p}_2$ ("gain"-term) respecting energy-momentum conservation (cf. Fig. 2.5). Furthermore, the states with momenta $\mathbf{p}_3$, $\mathbf{p}_4$ (in the "loss"-term) or $\mathbf{p}_1$, $\mathbf{p}_2$ (in the "gain"-term) should not be fully occupied due to the factors $\bar{n}(\mathbf{p}_i; t)$, which represent the Pauli principle for fermions in the final state. The factors $(2s + 1)$ for the summation over the spin of particle 2 and $(2\tau + 1)$ for the summation over isospin (or further internal degrees of freedom) of particle 2 can

---

[12] In the basis of plane waves this gives an additional factor $(2\pi)^{-6}$.

Fig. 2.5 Example for elastic
two-body scattering
$\mathbf{p}_1 + \mathbf{p}_2 \leftrightarrow \mathbf{p}_3 + \mathbf{p}_4$ in the
center-of-mass system with
the scattering angle Θ

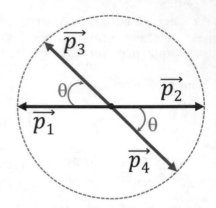

be comprised in a factor d; in case of electrons we have $(s = 1/2, \tau = 0)\ d = 2$, while for nucleons $(s = 1/2, \tau = 1/2)$ in spin and isospin symmetric systems we get $d = 4$.

Comment The collision integral (2.122) formally is an integral over all collision partners $(\int d^3 p_2)$ and all final states $(\int d^3 p_3 d^3 p_4)$ accounting for the final-state blocking factors $\bar{n}(\mathbf{p}_3; t)\ \bar{n}(\mathbf{p}_4; t)$ in the loss term, where the strength of the interaction is controlled by the two-body matrix element (squared) $W_{2,2} := v v_\mathcal{A}$ and the energy-momentum conserving δ-functions. The gain term is the corresponding time-reversed channel. We will come back to this formal result in Sect. 2.4.

The connection to two-body scattering becomes more transparent when relating the product $v \cdot v_\mathcal{A}$ to the differential cross section $d\sigma/d\Omega$ in first-order Born approximation,

$$\frac{d\sigma}{d\Omega}(\mathbf{p}_1 + \mathbf{p}_2, \mathbf{p}_2 - \mathbf{p}_4) = \frac{m^2}{16\pi^2} v(\mathbf{p}_2 - \mathbf{p}_4) v_\mathcal{A}(\mathbf{p}_4 - \mathbf{p}_2), \tag{2.123}$$

which leads to

$$I(\mathbf{p}_1, \mathbf{p}_1; t) = \frac{16\pi^2}{m^2} \frac{d}{(2\pi)^5} \int d^3 p_2 d^3 p_3 d^3 p_4 \tag{2.124}$$

$$\times \delta\left(\frac{1}{2m}[p_1^2 + p_2^2 - p_3^2 - p_4^2]\right) \delta(\mathbf{p}_1 + \mathbf{p}_2 - \mathbf{p}_3 - \mathbf{p}_4) \frac{d\sigma}{d\Omega}(\mathbf{p}_1 + \mathbf{p}_2, \mathbf{p}_2 - \mathbf{p}_4)$$

$$\times \{n(\mathbf{p}_3; t) n(\mathbf{p}_4; t) \bar{n}(\mathbf{p}_1; t) \bar{n}(\mathbf{p}_2; t) - n(\mathbf{p}_1; t) n(\mathbf{p}_2; t) \bar{n}(\mathbf{p}_3; t) \bar{n}(\mathbf{p}_4; t)\}.$$

Equation (2.124) can be further simplified by a transformation to relative and center-of-mass coordinates and integration over the δ-functions in (2.124). With the nonrelativistic relative velocity defined by

$$v_{12} = \frac{1}{m}|\mathbf{p}_1 - \mathbf{p}_2| \tag{2.125}$$

one finally obtains

$$I(\mathbf{p}_1, \mathbf{p}_1; t) = d \int \frac{d^3 p_2}{(2\pi)^3} \int d\Omega \, v_{12} \frac{d\sigma}{d\Omega} (\mathbf{p}_1 + \mathbf{p}_2, \mathbf{p}_2 - \mathbf{p}_4) \qquad (2.126)$$

$$\times \{ n(\mathbf{p}_3; t) n(\mathbf{p}_4; t) \bar{n}(\mathbf{p}_1; t) \bar{n}(\mathbf{p}_2; t) - n(\mathbf{p}_1; t) n(\mathbf{p}_2; t) \bar{n}(\mathbf{p}_3; t) \bar{n}(\mathbf{p}_4; t) \},$$

where $\Omega = (\cos \theta, \phi)$ denotes the scattering angle in the center-of-mass system of the colliding particles. Note that here the final momenta $\mathbf{p}_3$ and $\mathbf{p}_4$ are connected to $\mathbf{p}_1$ and $\mathbf{p}_2$ by energy-momentum conservation (cf. Fig. 2.5). This form of the collision integral has been proposed early by Nordheim [11] and Uehling-Uhlenbeck [12].

Exercise 2.8: Derive Eq. (2.126) starting from (2.122) using (2.123).

Comment In case of thermodynamic equilibrium (for $t \to \infty$) the collision term vanishes, i.e. $I(\mathbf{p}; t) \to 0$ for all $\mathbf{p}$, the Fermi distribution

$$n(\epsilon) = \frac{1}{\exp(\beta(\epsilon - \mu)) + 1} \qquad (2.127)$$

(with Lagrange parameters β, μ for the (inverse) temperature and chemical potential) fulfills the necessary condition

$$\{ n(\mathbf{p}_3; t) n(\mathbf{p}_4; t) \bar{n}(\mathbf{p}_1; t) \bar{n}(\mathbf{p}_2; t) - n(\mathbf{p}_1; t) n(\mathbf{p}_2; t) \bar{n}(\mathbf{p}_3; t) \bar{n}(\mathbf{p}_4; t) \} = 0 \qquad (2.128)$$

and thus is a solution of (2.126) for $I(\mathbf{p}_1) = 0$ independent from the strength of the interaction.

Exercise 2.9: Show that the Fermi distribution (2.127) is a stationary solution of the collision term, i.e. fulfills Eq. (2.128).

2.4.1   The Vlasov–Uehling–Uhlenbeck (VUU) Equation

The Vlasov–Uehling–Uhlenbeck (VUU) equation, also known as **Vlasov-Nordheim** or **Boltzmann–Uehling–Uhlenbeck (BUU)** equation is obtained by combining the Vlasov equation for the particle phase-space distribution $\rho(\mathbf{r}, \mathbf{p}; t)$ with the collision term (2.126) in phase-space representation, i.e. replacing the occupation numbers $n(\mathbf{p}_i; t)$ by the "local" phase-space density $\rho(\mathbf{r}, \mathbf{p}_i; t)$ in (2.126):

$$\left\{\frac{\partial}{\partial t} + \frac{\mathbf{p}_1}{m} \cdot \nabla_{\mathbf{r}} - \nabla_{\mathbf{r}} U(\mathbf{r}, t) \cdot \nabla_{\mathbf{p}_1}\right\} \rho(\mathbf{r}, \mathbf{p}_1; t) = I(\mathbf{r}, \mathbf{p}_1; t) \tag{2.129}$$

$$= d \int \frac{d^3 p_2}{(2\pi)^3} \int d\Omega \, v_{12} \frac{d\sigma}{d\Omega}(\mathbf{p}_1 + \mathbf{p}_2, \mathbf{p}_2 - \mathbf{p}_4)$$

$$\times \{\rho(\mathbf{r}, \mathbf{p}_3; t)\rho(\mathbf{r}, \mathbf{p}_4; t)\bar{\rho}(\mathbf{r}, \mathbf{p}_1; t)\bar{\rho}(\mathbf{r}, \mathbf{p}_2; t)$$

$$- \rho(\mathbf{r}, \mathbf{p}_1; t)\rho(\mathbf{r}, \mathbf{p}_2; t)\bar{\rho}(\mathbf{r}, \mathbf{p}_3; t)\bar{\rho}(\mathbf{r}, \mathbf{p}_4; t)\}.$$

In Eq. (2.129) the "local" phase-space distribution has to be understood as an average phase-space distribution in a local volume ΔV and finite time interval Δt of sufficient size such that the assumptions taken in the derivation become approximately valid. Accordingly, numerical solutions of Eq. (2.129) have to be performed on an appropriate space-time grid.

Classical Limits

In case of classical particles (at low phase-space densities or high temperatures) Eq. (2.129) simplifies since the Pauli-blocking factors $\bar{\rho}(\mathbf{r}, \mathbf{p}_i; t) \approx 1$. The further approximation $\nabla_{\mathbf{r}} U(\mathbf{r}; t) = 0$ then gives the **classical Boltzmann equation** [13,14]

$$\left\{\frac{\partial}{\partial t} + \frac{\mathbf{p}_1}{m} \cdot \nabla_{\mathbf{r}}\right\} \rho(\mathbf{r}, \mathbf{p}_1; t) = d \int \frac{d^3 p_2}{(2\pi)^3} \int d\Omega \, v_{12} \frac{d\sigma}{d\Omega}(\mathbf{p}_1 + \mathbf{p}_2) \tag{2.130}$$

$$\times \{\rho(\mathbf{r}, \mathbf{p}_3; t)\rho(\mathbf{r}, \mathbf{p}_4; t) - \rho(\mathbf{r}, \mathbf{p}_1; t)\rho(\mathbf{r}, \mathbf{p}_2; t)\},$$

which is used in the description of classical gases. In case of isotropic scattering of the particles, i.e. $d\sigma/d\Omega = \sigma/(4\pi)$ the classical Boltzmann equation simplifies to

$$\left\{\frac{\partial}{\partial t} + \frac{\mathbf{p}_1}{m} \cdot \nabla_{\mathbf{r}}\right\} \rho(\mathbf{r}, \mathbf{p}_1; t) = d \int \frac{d^3 p_2}{(2\pi)^3} \, v_{12} \, \sigma(\mathbf{p}_1 + \mathbf{p}_2) \tag{2.131}$$

$$\times \{\rho(\mathbf{r}, \mathbf{p}_3; t)\rho(\mathbf{r}, \mathbf{p}_4; t) - \rho(\mathbf{r}, \mathbf{p}_1; t)\rho(\mathbf{r}, \mathbf{p}_2; t)\}.$$

Furthermore, in case of homogenous media the $\mathbf{r}$-dependence drops out and one obtains

$$\frac{\partial}{\partial t}\rho(\mathbf{p}_1; t) = d \int \frac{d^3 p_2}{(2\pi)^3} \, v_{12} \, \sigma(\mathbf{p}_1 + \mathbf{p}_2) \{\rho(\mathbf{p}_3; t)\rho(\mathbf{p}_4; t) - \rho(\mathbf{p}_1; t)\rho(\mathbf{p}_2; t)\}.$$

$$\tag{2.132}$$

The collisional width of a particle with momentum $\mathbf{p}_1$ then is determined by the loss term and reads

$$\Gamma(\mathbf{p}_1) = d \int \frac{d^3 p_2}{(2\pi)^3} \, v_{12} \, \sigma(\mathbf{p}_1 + \mathbf{p}_2) \, \rho(\mathbf{p}_2). \tag{2.133}$$

In case of approximately constant cross section $\sigma(\mathbf{p}_1 + \mathbf{p}_2)$ in center-of-mass energy the latter can be taken out of the integral and the collisional width further simplifies to

$$\Gamma(\mathbf{p}_1) = \langle v_{12} \rangle \sigma \bar{\rho} \tag{2.134}$$

with the density

$$\bar{\rho} = d \int \frac{d^3 p_2}{(2\pi)^3} \, \rho(\mathbf{p}_2) \tag{2.135}$$

and the average relative velocity $\langle v_{12} \rangle$.

2.5 Including Higher Order Interactions

So far we have considered the solution for the two-body correlations in lowest order Born approximation, however, this limit does not hold in the context of nuclear physics where a resummation of the bare interaction v in the sense of a $\mathcal{G}$-matrix is mandatory for nonrelativistic energies.[13] Accordingly we take into account further interactions in the equations of motion for $C_{\alpha\beta\alpha'\beta'}(t)$ (2.63), i.e.

$$i \frac{\partial}{\partial t} C_{\alpha\beta\alpha'\beta'} = \sum_{\lambda} \{ h_{\alpha\lambda} C_{\lambda\beta\alpha'\beta'} + h_{\beta\lambda} C_{\alpha\lambda\alpha'\beta'} - C_{\alpha\beta\lambda\beta'} h_{\lambda\alpha'} - C_{\alpha\beta\alpha'\lambda} h_{\lambda\beta'} \}$$

$$\tag{2.136}$$

$$+ \sum_{\lambda\lambda'\gamma\gamma'} \{ Q^=_{\alpha\beta\lambda'\gamma'} \langle \lambda'\gamma'|v|\lambda\gamma \rangle (\rho_{20})_{\lambda\gamma\alpha'\beta'} - (\rho_{20})_{\alpha\beta\lambda'\gamma'} \langle \lambda'\gamma'|v|\lambda\gamma \rangle Q^=_{\lambda\gamma\alpha'\beta'} \}$$

$$+ \sum_{\lambda\lambda'\gamma\gamma'} \{ Q^=_{\alpha\beta\lambda'\gamma'} \langle \lambda'\gamma'|v|\lambda\gamma \rangle C_{\lambda\gamma\alpha'\beta'} - C_{\alpha\beta\lambda'\gamma'} \langle \lambda'\gamma'|v|\lambda\gamma \rangle Q^=_{\lambda\gamma\alpha'\beta'} \},$$

where all terms with $Q^{\perp}$ (particle-hole channels) have been discarded. This limit also conserves particle number, total momentum, angular momentum, and energy.

---

[13] In Appendix C we recall the two-body scattering in vacuum and demonstrate that the Born approximation gives inadequate results for low energy nucleon-nucleon scattering such that a resummation of the interaction in terms of a T-matrix is mandatory in this case.

For the further formulation it is of advantage to rewrite (2.136) as

$$\langle\alpha\beta|(i\frac{\partial}{\partial t}C_2 - (h(1) + h(2) + Q^=v)C_2(t) + C_2(t)(h(1') + h(2') + vQ^{=\dagger})|\alpha'\beta'\rangle \tag{2.137}$$
$$= \langle\alpha\beta|Q^=v\rho_{20} - \rho_{20}vQ^{=\dagger}|\alpha'\beta'\rangle(t) = \langle\alpha\beta|V_B(t)|\alpha'\beta'\rangle.$$

This equation is of similar structure as the Born limit (2.101), however, with extra terms $Q^=vC_2$ and $C_2vQ^=$ on the l.h.s. , while the inhomogenous terms on the r.h.s. are identical.

Instead of the propagators $g_{20}^\pm(\omega)$ (2.106) we now define the in-medium two-body propagators

$$g_2^\pm(\omega) = \frac{1}{\omega - h(1) - h(2) - Q^=v \pm i\gamma} \tag{2.138}$$

with infinitesimal $\gamma > 0$. The relation between $g_2(\omega)$ and $g_{20}(\omega)$ is given by the Dyson equation[14]

$$g_2^\pm(\omega) = g_{20}^\pm(\omega) + g_{20}^\pm(\omega)Q^=vg_2^\pm(\omega) = g_{20}^\pm(\omega) + g_{20}^\pm(\omega)Q^=vg_{20}^\pm(\omega) \tag{2.139}$$
$$+ g_{20}^\pm(\omega)Q^=vg_{20}^\pm(\omega)Q^=vg_{20}^\pm(\omega) + \cdots$$
$$= g_{20}^\pm(\omega)(\sum_{n=0}^{\infty}(Q^=vg_{20}^\pm(\omega))^n = \hat{\Omega}^\pm(\omega)g_{20}^\pm(\omega)$$

with the in-medium Moeller operator defined by

$$\hat{\Omega}^\pm(\omega) = \sum_{n=0}^{\infty}\left(g_{20}^\pm(\omega)Q^=v\right)^n \tag{2.140}$$

that includes the in-medium interactions v as well as the blocking operator $Q^=$ in infinite order. Here we have assumed that the continuum spectra of $g_2^\pm(\omega)$ and $g_{20}^\pm(\omega)$ are the same and complete. The Moeller operator follows

$$\hat{\Omega}^\pm(\omega) = 1 + g_{20}^\pm(\omega)Q^=v\hat{\Omega}^\pm(\omega) = \frac{1}{1 - g_{20}^\pm(\omega)Q^=v}, \tag{2.141}$$

which gives

$$\hat{\Omega}^\pm(\omega) - 1 = g_{20}^\pm(\omega)Q^=v\hat{\Omega}^\pm(\omega) = \hat{\Omega}^\pm(\omega)g_{20}^\pm(\omega)Q^=v. \tag{2.142}$$

---

[14] Here we use the operator identity $[A + B]^{-1} = A^{-1}(1 - B[A + B]^{-1})$ and employ $A = (\omega - h(1) - h(2) \pm +i\gamma)$ and $B = -Q^=v$.

Furthermore, note that $(\hat{\Omega}^+(\omega) - 1)^\dagger = \hat{\Omega}^-(\omega) - 1$. We mention in passing that the in-medium Moeller operator differs from that in vacuum[15] only by the additional Pauli-blocking operator $Q^=$.

Exercise 2.10: Prove the operator identity $[A+B]^{-1} = A^{-1}(1-B[A+B]^{-1})$.

The formal solution to Eq. (2.136) is given (in analogy to (2.105)) by

$$\langle\alpha\beta|C_2|\alpha'\beta'\rangle(t) = -\frac{i}{2}\int_{t_0}^{\infty} dt' \int_{-\infty}^{\infty}\frac{d\omega}{2\pi}\int_{-\infty}^{\infty}\frac{d\omega'}{2\pi}\,\exp(-i(\omega-\omega')(t-t'))$$

$$(2.143)$$

$$\times\;\langle\alpha\beta|g_2^+(\omega)[Q^=v\rho_{20}(t') - \rho_{20}(t')vQ^{=\dagger}]g_2^-(\omega')|\alpha'\beta'\rangle$$

with the propagators $g_2^\pm(\omega)$ defined by Eq. (2.138). Note that $Q^=$ and ρ_{20} in general depend on time t' explicitly such that the formal solution cannot be worked out analytically in the general case. In order to get some idea about the physics incorporated we consider again the basis of plane waves (in a sufficiently large volume ΔV), approximately constant occupation numbers (in some finite time interval Δt) and use (2.139) to rewrite (2.143) as

$$\langle\alpha\beta|C_2|\alpha'\beta'\rangle(t) = -\frac{i}{2}\int_{t_0}^{\infty} dt' \int_{-\infty}^{\infty}\frac{d\omega}{2\pi}\int_{-\infty}^{\infty}\frac{d\omega'}{2\pi}\,\exp(-i(\omega-\omega')(t-t'))$$

$$(2.144)$$

$$\times\;\langle\alpha\beta|\hat{\Omega}^+(\omega)g_{20}^+(\omega)[Q^=v\rho_{20}(t') - \rho_{20}(t')vQ^{=\dagger}]g_{20}^-(\omega')\hat{\Omega}^-(\omega')|\alpha'\beta'\rangle.$$

In this case the matrix elements of $g_{20}(\omega)$ are well known and $Q^=$ as well as ρ_{20} become diagonal. With the identity (2.142) the solution may be rewritten as

$$\langle\alpha\beta|C_2|\alpha'\beta'\rangle(t) = -\frac{i}{2}\int_{t_0}^{\infty} dt' \int_{-\infty}^{\infty}\frac{d\omega}{2\pi}\int_{-\infty}^{\infty}\frac{d\omega'}{2\pi}\,\exp(-i(\omega-\omega')(t-t'))$$

$$(2.145)$$

$$\times\left(\langle\alpha\beta|(\hat{\Omega}^+(\omega) - 1)\rho_{20}(t')g_{20}^-(\omega')\hat{\Omega}^-(\omega')|\alpha'\beta'\rangle\right.$$

$$\left. -\langle\alpha\beta|\hat{\Omega}^+(\omega)g_{20}^+(\omega)\rho_{20}(t')(\hat{\Omega}^-(\omega') - 1)|\alpha'\beta'\rangle\right).$$

---

[15] cf. Appendix C.

The next step is to evaluate $\hat{\Omega}^+(\omega)$ at the energy $(\epsilon_\alpha + \epsilon_\beta)$ and $\hat{\Omega}^-(\omega')$ at $(\epsilon_{\alpha'} + \epsilon_{\beta'})$ which gives

$$\langle \alpha\beta | C_2 | \alpha'\beta' \rangle(t) = -\frac{i}{2} \int_{t_0}^\infty dt' \int_{-\infty}^\infty \frac{d\omega}{2\pi} \int_{-\infty}^\infty \frac{d\omega'}{2\pi} \, \exp(-i(\omega - \omega')(t - t')$$

(2.146)

$$\times \, (\langle \alpha\beta | \hat{\Omega}^+(\epsilon_\alpha + \epsilon_\beta)\rho_{20}(t')g_{20}^-(\omega')\hat{\Omega}^-(\epsilon_{\alpha'} + \epsilon_{\beta'})|\alpha'\beta'\rangle$$
$$- \langle \alpha\beta | \hat{\Omega}^+(\epsilon_\alpha + \epsilon_\beta)g_{20}^+(\omega)\rho_{20}(t')\hat{\Omega}^-(\epsilon_{\alpha'} + \epsilon_{\beta'})|\alpha'\beta'\rangle$$
$$- \langle \alpha\beta | \rho_{20}(t')g_2^-(\omega')|\alpha'\beta'\rangle + \langle \alpha\beta | g_2^+(\omega)\rho_{20}(t')|\alpha'\beta'\rangle).$$

The integrations over $d\omega$ (or $d\omega'$) give a δ-function in $(t - t')$ as in the beginning of Sect. 2.4. Then integrating the propagators over $d\omega'/(2\pi)$ (or $d\omega/(2\pi)$) gives factors of i and $-i$, respectively. The final integration over dt' then leads to

$$\langle \alpha\beta | C_2 | \alpha'\beta' \rangle(t) = \langle \alpha\beta | \hat{\Omega}^+(\epsilon_\alpha + \epsilon_\beta)\rho_{20}(t)\hat{\Omega}^-(\epsilon_{\alpha'} + \epsilon_{\beta'})|\alpha'\beta'\rangle - \langle \alpha\beta | \rho_{20}(t)|\alpha'\beta'\rangle$$

(2.147)

which gives the result for the two-body density matrix

$$\langle \alpha\beta | \rho_2(t) | \alpha'\beta' \rangle = \langle \alpha\beta | \hat{\Omega}(\epsilon_\alpha + \epsilon_\beta)\rho_{20}(t)\hat{\Omega}^\dagger(\epsilon_{\alpha'} + \epsilon_{\beta'})|\alpha'\beta'\rangle$$

(2.148)

when identifying $\hat{\Omega} = \hat{\Omega}^+$ and $\hat{\Omega}^\dagger = \hat{\Omega}^-$.

2.5.1 Definition of Selfenergies

Since the states $|\alpha\beta\rangle$ are complete we may define an interacting two-body density operator $\rho_2(t)$ in ladder resummation by

$$\rho_2(t) = \hat{\Omega}(\omega)(t)\rho_{20}(t)\tilde{\Omega}(\omega)^\dagger(t)$$

(2.149)

with the in-medium Moeller operator defined by (2.140). Note that this result only is obtained within finite time intervals Δt and spatial volumes ΔV of sufficient size. Furthermore, the time dependence of $\rho_{20}(t)$ should not be too rapid in the four-volume $\Delta t \Delta V$. These conditions are fulfilled in vacuum and Eq. (2.149) is exact (and time-independent) as demonstrated for two-body scattering in the vacuum in Appendix C. Under these constraints we can define the resummed complex interaction $\mathcal{G}(\omega)$ (or $\mathcal{G}$-matrix) as

$$\mathcal{G}(\omega) = v\hat{\Omega}(\omega).$$

(2.150)

This leads to the identities ($Q^{=\dagger} = Q^=$)

$$\hat{\Omega}(\omega) = 1 + g_{20}^+(\omega)Q^= v\hat{\Omega}(\omega) = 1 + g_{20}^+(\omega)Q^= \mathcal{G}(\omega) \tag{2.151}$$

$$\hat{\Omega}(\omega)^\dagger = 1 + \mathcal{G}(\omega)^\dagger Q^= g_{20}^-(\omega)$$

which will be exploited in different versions (using e.g. $\Im(\mathcal{G}\mathcal{G}^\dagger) = 0$). This complex (and energy-dependent) two-body interaction in lowest order Born approximation is given by the bare interaction v. Furthermore, in a low density medium (i.e. $Q^= \rightarrow$ 1) we regain the free T-matrix $T(\omega)$ as recalled in Appendix C. The effect of the Pauli-blocking operator $Q^=$ will be more pronounced for low energy scattering with kinetic energies below (or of the order of) the local Fermi energy for the system leading to a substantial reduction of the interaction cross section for low invariant energies above threshold.

Having established this specific (nonperturbative) limit of the many-body problem we may rewrite the matrix elements

$$Tr_{(2=2')}\langle 1'2'|[v, \rho_2]|12\rangle = Tr_{(2=2')}\langle 1'2'|(v\rho_2 - \rho_2 v)|12\rangle \tag{2.152}$$

$$= Tr_{(2=2')}\langle 1'2'|(\mathcal{G}(\omega)\rho_{20}\hat{\Omega}(\omega)^\dagger - \hat{\Omega}(\omega)\rho_{20}\mathcal{G}(\omega)^\dagger)|12\rangle$$

and obtain (using (2.151))

$$Tr_{(2=2')}\langle 1'2'|[v, \rho_2]|12\rangle = Tr_{(2=2')}\langle 1'2'|(\mathcal{G}(\omega)\rho_{20} - \rho_{20}\mathcal{G}(\omega)^\dagger)|12\rangle \tag{2.153}$$

$$+ Tr_{(2=2')}\langle 1'2'|(\mathcal{G}(\omega)\rho_{20}\mathcal{G}(\omega)^\dagger Q^= g_{20}^-(\omega) - g_{20}^+(\omega)Q^= \mathcal{G}(\omega)\rho_{20}\mathcal{G}(\omega)^\dagger)|12\rangle.$$

This allows to define the real part of the selfenergy $\Re(\Sigma)$ as:

$$\langle 1'|[\Re(\Sigma), \rho]|1\rangle = Tr_{(2=2')}\langle 1'2'|[\Re(\mathcal{G}), \rho_{20}]|12\rangle \tag{2.154}$$

or

$$\langle 1'|\Re(\Sigma)|1\rangle = Tr_{(2=2')}\langle 1'2'|\Re(\mathcal{G})\mathcal{A}\rho|12\rangle. \tag{2.155}$$

The imaginary part of the selfenergy follows accordingly from the imaginary part of $\mathcal{G}$ as:

$$\langle 1'|[\Im(\Sigma), \rho]|1\rangle = Tr_{(2=2')}\langle 1'2'|[\Im(\mathcal{G}), \rho_{20}]|12\rangle \tag{2.156}$$

or

$$\langle 1'|\Im(\Sigma)|1\rangle = Tr_{(2=2')}\langle 1'2'|\Im(\mathcal{G})\mathcal{A}\rho|12\rangle, \tag{2.157}$$

which will turn out to give half the loss part in the collision term (see below). In lowest order in the interaction v we regain the selfconsistent Hartree–Fock potential U^s in (2.42) while the series in (2.49) provides a controlled higher order expansion for the selfenergy $\Re(\Sigma)$.

We now separate the real and imaginary parts of the commutator (2.152) employing additionally the identity

$$\Im(\mathcal{G}) = \mathcal{G}^\dagger Q^= \mathcal{G} \Im(g_{20}^+) \tag{2.158}$$

since $\mathcal{G}^\dagger Q^= \mathcal{G}$, $Q^=$ and v are Hermitian operators.

For the remaining parts of the commutator we have

$$\mathrm{Tr}_{(2=2')} \langle 1'2' | (\Im(\mathcal{G}(\omega)) \rho_{20} - \rho_{20} \Im(\mathcal{G}(\omega)^\dagger)) | 12 \rangle \tag{2.159}$$

$$+ \mathrm{Tr}_{(2=2')} \langle 1'2' | (\mathcal{G}(\omega) \rho_{20} \mathcal{G}(\omega)^\dagger Q^= g_{20}^-(\omega) - g_{20}^+(\omega) Q^= \mathcal{G}(\omega) \rho_{20} \mathcal{G}(\omega)^\dagger) | 12 \rangle$$

since ρ_{20} is Hermitian. The further steps in the evaluation of the collision term follow the same procedure as in the beginning of Sect. 2.4 within a basis of plane waves. Accordingly, (2.159) is identical to an Uehling-Uhlenbeck on-shell collision term for fermions where the imaginary part of g_{20}^+ gives an energy conserving δ-function, the matrix element squared is given by $\mathcal{G}\mathcal{G}^\dagger \mathcal{A}$ and $Q^=$ introduces the Pauli-blocking factors. Recall that in this basis the one-body density matrix in momentum space and $Q^=$ are diagonal, i.e.

$$\langle \mathbf{p}' | \rho | \mathbf{p} \rangle = \delta^3(\mathbf{p}' - \mathbf{p}) f(\mathbf{p}) \tag{2.160}$$

$$\langle \mathbf{p}_1' \mathbf{p}_2' | Q^= | \mathbf{p}_1 \mathbf{p}_2 \rangle = \delta^3(\mathbf{p}_1' - \mathbf{p}_1) \delta^3(\mathbf{p}_2' - \mathbf{p}_2) \left(1 - f(\mathbf{p}_1') - f(\mathbf{p}_2')\right)$$

$$\langle \mathbf{p}_1' \mathbf{p}_2' | \mathcal{G} | \mathbf{p}_1 \mathbf{p}_2 \rangle = \delta^3(\mathbf{p}_1 + \mathbf{p}_2 - \mathbf{p}_1' - \mathbf{p}_2') \mathcal{G}(\mathbf{p}_2 - \mathbf{p}_2').$$

Here we have dropped again the discrete quantum numbers in the matrix elements (spin, isospin, etc.). Note that

$$\langle \mathbf{p}_1' \mathbf{p}_2' | \Im(g_{20}^+(\omega)) | \mathbf{p}_1 \mathbf{p}_2 \rangle = \delta^3(\mathbf{p}_1' - \mathbf{p}_1) \delta^3(\mathbf{p}_2' - \mathbf{p}_2) \frac{-i\epsilon}{(\omega - h(\mathbf{p}_1) - h(\mathbf{p}_2))^2 + \epsilon^2} \tag{2.161}$$

$$= -\pi \delta^3(\mathbf{p}_1' - \mathbf{p}_1) \delta^3(\mathbf{p}_1' - \mathbf{p}_1) \delta(\omega - h(\mathbf{p}_1) - h(\mathbf{p}_2)).$$

In lowest order we regain the Born transition matrix element squared $vv_\mathcal{A}^\dagger$ in the collision term, where the antisymmetrization $\mathcal{A}$ only works for the interaction of fermions with the same spin (and isospin). The relation to the differential cross section $d\sigma/d\Omega$ for nonidentical particles is given by

$$\frac{d\sigma}{d\Omega} = \frac{m^2}{16\pi^2} \mathcal{G}(\mathbf{q}) \mathcal{G}^\dagger(\mathbf{q}) \tag{2.162}$$

where $\mathbf{q}$ is the momentum transfer in the elastic scattering event. Defining the (resummed) single-particle Hamiltonian by

$$h_{eff}(\mathbf{r}, \mathbf{p}; t) = \frac{\mathbf{p}^2}{2M} + \Re(\Sigma(\mathbf{r}, \mathbf{p}; t)) \qquad (2.163)$$

the nonrelativistic (and nonperturbative) transport equation for Fermions finally reads:

$$\left(\frac{\partial}{\partial t} + \nabla_p h_{eff}(\mathbf{r}, \mathbf{p}; t) \cdot \nabla_r - \nabla_r h_{eff}(\mathbf{r}, \mathbf{p}; t) \cdot \nabla_p \right) \rho(\mathbf{r}, \mathbf{p}; t) \qquad (2.164)$$

$$= d \int \frac{d^3 p_2}{(2\pi)^3} \int \frac{d^3 p'_1}{(2\pi)^3} \int \frac{d^3 p'_2}{(2\pi)^3} |\mathcal{G}\mathcal{G}^\dagger A(\mathbf{p}_1 + \mathbf{p}_2, \mathbf{p}'_2 - \mathbf{p}_2)| \qquad (2.165)$$

$$\times (2\pi)^4 \delta(h(\mathbf{p}_1) + h(\mathbf{p}_2) - h(\mathbf{p}'_1) - h(\mathbf{p}'_2)) \, \delta^3(\mathbf{p}_1 + \mathbf{p}_2 - \mathbf{p}'_1 - \mathbf{p}'_2)$$

$$\times \left(\rho(\mathbf{r}, \mathbf{p}'_1; t)\rho(\mathbf{r}, \mathbf{p}'_2; t)(1 - \rho(\mathbf{r}, \mathbf{p}; t))(1 - \rho(\mathbf{r}, \mathbf{p}_2; t)) \right.$$

$$\left. - \rho(\mathbf{r}, \mathbf{p}; t)\rho(\mathbf{r}, \mathbf{p}_2; t)(1 - \rho(\mathbf{r}, \mathbf{p}'_1; t))(1 - \rho(\mathbf{r}, \mathbf{p}'_2 s; t)) \right),$$

where (2.164) represents the selfconsistent Vlasov term while (2.165) describes the on-shell collision term with a nonperturbative transition probability. We recall again that the space-time coordinates $(\mathbf{r}, t)$ in Eq. (2.164) should be considered on a finite grid in space-time $\Delta t \Delta V$ that is sufficiently large to fulfill the assumptions and approximations made in the derivation of the collision term.

2.5.2 Effective Parametrizations for the $\mathcal{G}$-matrix

In practice the real part of the $\mathcal{G}$-matrix in the nuclear physics context is parametrized by some functional of the nuclear density ρ, e.g.

$$\Re(\mathcal{G}(\mathbf{r}_1 - \mathbf{r}_2) \approx -A\delta(\mathbf{r}_1 - \mathbf{r}_2) + B\delta(\mathbf{r}_1 - \mathbf{r}_2)\rho((\mathbf{r}_1 + \mathbf{r}_2)/2))^\gamma, \qquad (2.166)$$

where A and B denote the strength of the attractive and repulsive interaction and $0.3 \leq \gamma \leq 1$ some density dependence of the repulsive interaction that controls the incompressibility of nuclear matter. Spin and isospin dependencies have been discarded as well as finite-range (Yukawa) interactions. A nonrelativistic mean field (or selfenergy) is obtained (as a function of the density ρ) by integration over ρ,

$$\Sigma_{eff}(\rho) = U_{eff}(\rho) = \frac{3}{4}\left(-A\rho + \frac{B}{1 + \gamma}\rho^{1+\gamma} \right), \qquad (2.167)$$

where the prefactor 3/4 stems from subtracting the Fock part of the interaction from the direct Hartree part in case of two spin and isospin degrees of freedom since the local interaction (2.166) forbids interactions of particles with identical quantum numbers at the same position due to antisymmetry. The potential energy density $\mathcal{V}$ is obtained by another integration over ρ:

$$\mathcal{V}(\rho) = \frac{3}{4} \left(-\frac{A}{2}\rho^2 + \frac{B}{(1+\gamma)(2+\gamma)}\rho^{2+\gamma} \right). \tag{2.168}$$

Alternatively, having fixed the potential energy density $\mathcal{V}$ e.g. by some effective Lagrangian, the selfenergy is obtained by a (functional) derivative with respect to the density ρ and the "local" effective interaction by another (functional) derivative.

> Exercise 2.11: Calculate the nonrelativistic energy per nucleon for symmetric nuclear matter at zero temperature as a function of the density ρ, i.e. the nuclear equation of state (EoS), using the effective interaction (2.166) or potential energy density (2.168), respectively.

2.5.3 Coupled-Channel Transport Equations

So far only elastic scattering between two fermions has been considered in the transport equation (2.164) which holds well for energies below the first inelastic threshold. In case of nucleons, however, the particles have some internal structure— being a bound state of quarks and gluons as well as quark-antiquark pairs—and have excited states of finite lifetime such as the $\Delta(1232)$ resonance or excited states of even higher mass. Furthermore, with increasing energy of the collision strange-antistrange quark pairs can be created from the nonperturbative vacuum of Quantum Chromo Dynamics (QCD) which become bound in baryons of finite strangeness such as the Λ's, Σ's, and Ξ's. The lowest energy states build up the baryon octet (of spin $1/2\ \hbar$), while first excited states are described by the baryon decuplet (of spin $3/2\ \hbar$). Additional resonances of higher mass may be included without problems if their experimental identification is sufficiently solid. On the other hand the mesonic decays of the higher resonances to pions, kaons, etc. require to incorporate also the pseudoscalar (0^-) meson nonet as well as the vector meson nonet (1^-).[16]

The form of the resulting transport equation is fixed by Eq. (2.98) where now one has to include the discrete quantum numbers explicitly, i.e. the hadron type h, spin projection σ, isospin τ, etc.; in short $c := (h, \sigma, \tau, \ldots)$. Accordingly, the phase-space distribution $\rho(\mathbf{r}, \mathbf{p}; t)$ carries a discrete index c as well as the effective single-particle Hamiltonian $h_{eff}(\mathbf{r}, \mathbf{p}; t)$. The collision term, furthermore, becomes a matrix in the channels c and c' which describes the change of the phase-space distribution ρ_c due to elastic ($c \leftrightarrow c$) or inelastic channels ($c \leftrightarrow c' \neq c$) obeying

---

[16] Here the notation J^p specifies the spin J and parity p of the state.

detailed balance. The summation over a specific state then translates to

$$\sum_\beta \rightarrow \sum_{\sigma,\tau,\cdots} \int \frac{d^3 p_\beta}{(2\pi)^3} \tag{2.169}$$

which replaces the single momentum integrals used so far. The matrix element squared $|\mathcal{G}\mathcal{G}^\dagger \mathcal{A}|$, which should be evaluated in a coupled-channel G-matrix approach, then carries additional indices for the individual channels $c_1 + c_2 \leftrightarrow c_3 + c_4$. At least in case of particle production channels relativistic expressions for the particle energies have to be incorporated in the energy conserving δ-function which in case of vanishing selfenergies imply

$$E_{kin} = \frac{\mathbf{p}^2}{2m} \rightarrow E = \sqrt{\mathbf{p}^2 + m^2}. \tag{2.170}$$

Furthermore, the antisymmetrization operator $\mathcal{A}$ for fermions has to be replaced by a symmetrization operator $\mathcal{S}$ in case of mesons (bosons) and the Pauli-blocking terms by Bose-enhancement factors, i.e.

$$(1 - \rho_j(\mathbf{r}, \mathbf{p}; t)) \rightarrow (1 + \eta \rho_j(\mathbf{r}, \mathbf{p}; t)) \tag{2.171}$$

with $\eta = 1$ for bosons and $\eta = -1$ for fermions. This leads to the coupled-channel VUU (CVUU) equations:

$$\left(\frac{\partial}{\partial t} + \nabla_p h_{eff}^c(\mathbf{r}, \mathbf{p}_1; t) \cdot \nabla_r - \nabla_r h_{eff}^c(\mathbf{r}, \mathbf{p}_1; t) \cdot \nabla_p \right) \rho_c(\mathbf{r}, \mathbf{p}_1; t) \tag{2.172}$$

$$= \sum_{c'} I_{cc'}(\mathbf{r}, \mathbf{p}_1; t) = \sum_{c'} \sum_{c3,c4} \int \frac{d^3 p_2}{(2\pi)^3} \frac{d^3 p_3}{(2\pi)^3} \frac{d^3 p_4}{(2\pi)^3} \tag{2.173}$$

$$\times (2\pi)^4 \delta^4(p_1 + p_2 - p_3 - p_4)\, \mathcal{G}\mathcal{G}^\dagger_{A,S}(p_1 + p_2, p_2 - p_4; c_1 + c_2 \leftrightarrow c_3 + c_4)$$

$$\times \left(\rho_{c3}(\mathbf{r}, \mathbf{p}_3; t) \rho_{c4}(\mathbf{r}, \mathbf{p}_4; t) \bar{\rho}_c(\mathbf{r}, \mathbf{p}_1; t) \bar{\rho}_{c'}(\mathbf{r}, \mathbf{p}_2; t) \right.$$

$$\left. - \rho_c(\mathbf{r}, \mathbf{p}_1; t) \rho_{c'}(\mathbf{r}, \mathbf{p}_2; t) \bar{\rho}_{c3}(\mathbf{r}, \mathbf{p}_3; t) \bar{\rho}_{c4}(\mathbf{r}, \mathbf{p}_4; t) \right)$$

with the Pauli-blocking or Bose-enhancement factors

$$\bar{\rho}_c(\mathbf{r}, \mathbf{p}; t) = 1 + \eta \rho_c(\mathbf{r}, \mathbf{p}; t). \tag{2.174}$$

Here the relativistic four-momenta p_j have to be incorporated in the δ^4-function for energy-momentum conservation in the transition process. Since coupled-channel G-matrix calculations practically are restricted to a limited (small) number of particles (or channels) the CVUU equations (2.172) in practice are limited to the low-lying baryon resonances including part of the meson (0^-) nonet. Furthermore, these equations lack Lorentz invariance and accordingly can only be applied to nonrelativistic many-particle systems.

2.6 Numerical Solutions

The numerical solution of the VUU equation (2.164) consists in defining the local cell size ΔV and time step Δt which in case of low energy heavy-ion collisions is usually taken to be $\Delta V = 1\,\text{fm}^3$ and $\Delta t = 0.5\,\text{fm/c}$. The phase-space distribution is taken in line with (2.96) as

$$\rho_t(\mathbf{r}, \mathbf{p}; t) = \frac{1}{N_t} \sum_{i=1}^{N \cdot N_t} \delta(\mathbf{r} - \mathbf{r}_i(t))\, \delta(\mathbf{p} - \mathbf{p}_i(t)), \tag{2.175}$$

where N denotes the total number of nucleons and N_t the number of parallel ensembles which should be of the order 300–1000. The propagation of the particles in time follows the classical equations of motion for $\dot{\mathbf{r}}_i$ and $\dot{\mathbf{p}}_i$, i.e.

$$\dot{\mathbf{r}}_i = \nabla_{\mathbf{p}} h_{eff}(\mathbf{r}_i, \mathbf{p}_i; t); \qquad \dot{\mathbf{p}}_i = -\nabla_{\mathbf{r}} h_{eff}(\mathbf{r}_i, \mathbf{p}_i; t), \tag{2.176}$$

with $h_{eff}(\mathbf{r}, \mathbf{p}; t) = \mathbf{p}^2/(2m) + U_{eff}(\mathbf{r}, \mathbf{p}; t)$ while $U_{eff}(\mathbf{r}, \mathbf{p}; t)$ is evaluated in each space-time cell according to Eq. (2.167).

The solution of the equations of motion (2.176) is performed by a predictor-corrector or Runge-Kutta method while different schemes exist for the calculation of the gradients $\nabla_{\mathbf{r}} U_{eff}(\mathbf{r}, \mathbf{p}; t)$ and $\nabla_{\mathbf{p}} U_{eff}(\mathbf{r}, \mathbf{p}; t)$. For further details we refer the reader to the original literature [15–23]. In order to avoid large numerical fluctuations some Gaussian smoothing is often employed and adjusted to achieve approximately stable nuclei in their semiclassical ground states taken in the local Thomas-Fermi approximation (for spherical nuclei):

$$\rho_{TF}(\mathbf{r}, \mathbf{p}) = \Theta(p_F(r)^2 - p^2) \tag{2.177}$$

for neutrons and protons (with two spin projections each). In (2.177) $p_F(r)$ denotes the local Fermi momentum defined by

$$p_F(r)^2 = 2m(E_F - U_{eff}(r)) \tag{2.178}$$

in the classically allowed regime with the Fermi energy $E_F (\approx 40\,\text{MeV})$.

It is useful to replace the attractive δ-force in Eq. (2.166) by some finite-range Yukawa interaction of the same integrated strength

$$A = \int d^3r \, \frac{\hat{A}}{\mu |\mathbf{r}|} \exp(-\mu|\mathbf{r}|) = \frac{4\pi \hat{A}}{\mu^3}. \tag{2.179}$$

This complicates the calculation of the mean field $U_{eff}(\mathbf{r}; t)$ slightly but helps in achieving rather stable nuclei with low fluctuations in the density during time integration.

In order two simulate heavy-ion reactions the target and projectile nuclei are shifted by some distance R in z-direction such that they do not touch. Furthermore, both nuclei are shifted in x-direction by $\pm b/2$ where b denotes the impact parameter of the collision. Then both nuclei are boosted towards each other (in z-direction) in line with the bombarding energy of interest, however, including the Coulomb repulsion at distance R as well as the Coulomb deflection in the $x - z$ plane. This completes the initialization.

Binary collisions in the time interval $[t, t + \Delta t]$ can be evaluated by Monte Carlo via the collision criterion

$$b_{max} := \sqrt{\frac{\sigma(\sqrt{s})}{\pi}} \tag{2.180}$$

if the distance of closest approach of two particles $d_c^{i,j}$ is less than b_{max} and reached in the time interval $[t, t + \Delta t]$. In Eq. (2.180) $\sqrt{s}$ denotes the invariant energy of the collision, i.e. $s = (E_1 + E_2)^2 - (\mathbf{p}_1 + \mathbf{p}_2)^2$. If a collision may occur between particles i and j according to the collision criterion ($d_c^{i,j} \leq b_{max}$) the possible final states (allowed by energy-momentum conservation) are selected by Monte Carlo in the common center-of-mass system and weighted by the angular distribution $d\sigma(\sqrt{s})/d\Omega$ and the final blocking factors. If a final state is selected (and not blocked) the new particle momenta are boosted back to the calculational frame. Some note of caution has to be added here since one has to take care that the collision probability for each particle i—summed over all particles j in the same ensemble—in the time step Δt is significantly lower than 1 in order not to miss collisions. Furthermore, the total energy-momentum conservation has to be recorded and controlled in time. Deviations by less than 1% are commonly accepted.

2.6.1 Application to Low Energy Heavy-Ion Reactions

In order to illustrate the effect of the collision term in the VUU equation (2.164) Fig. 2.6 shows the density distribution $\rho(x, y = 0, z; t)$ for a central collision of $^{40}Ca +^{40}Ca$ at a beam energy of 40 MeV/u in the semiclassical limit according to the VUU/BUU equation (2.164). Contrary to the corresponding solution of the Vlasov equation (cf. Fig. 2.3) the nuclei do not pass through each other but merge for longer times t emitting some nucleons to the continuum. Accordingly the stopping power of the collision is essentially controlled by the strength of the binary scattering that is missing in the TDHF or Vlasov solutions.

Furthermore, the middle column of Fig. 2.6 shows the momentum distribution $\rho(p_x, p_y = 0, p_z; t)$ for the same system which at $t = 0$ is elongated in p_z-direction due to the initial boosts but becomes isotropic in time due to the two-body collisions

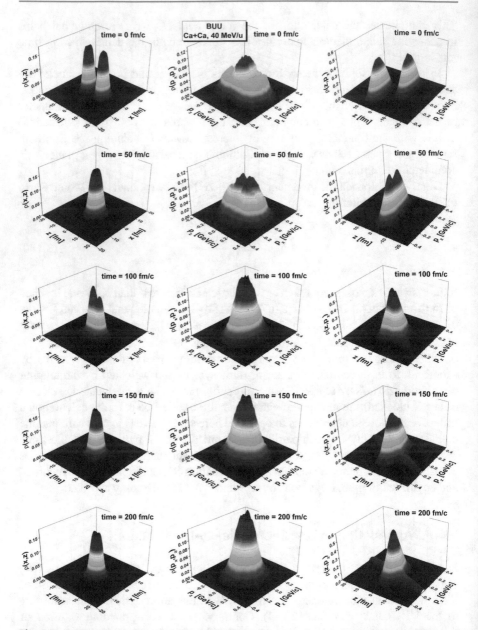

Fig. 2.6 (Left column) The density distribution $\rho(x, y = 0, z; t)$ for a central collision of $^{40}Ca + ^{40}Ca$ at a beam energy of 40 MeV/u in the semiclassical limit according to the BUU equation (2.93) within the testparticle method (2.96). (Middle column) The momentum distribution $\rho(p_x, p_y = 0, p_z; t)$ for the same reaction. (Right column) The phase-space distribution $\bar{f}(z, p_z; t)$—integrated over perpendicular degrees of freedom—for the same reaction

since the systems fuses to a single system of mass $A < 80$ while nucleons are emitted to the continuum during the equilibration process.

2.6.2 Summary

The VUU/BUU transport equation for fermions has been derived from the BBGKY hierarchy in the perturbative and nonperturbative limits for systems of sufficient size and/or particle number such that averages over space-time volumes still provide a reasonable resolution in space and time. Furthermore, the testparticle method provides a convenient scheme to solve the transport equations in the limit of large ensemble numbers N_t. Apart from elastic scattering of particles also inelastic reactions can be treated on the same footing by a Monte Carlo evaluation of the respective collision integrals.

As mentioned before the VUU/BUU equation (2.164) is nonrelativistic and studies at higher bombarding energies require a covariant formulation such that the results of actual calculations (for Lorentz-invariant observables) do not depend on the reference frame adopted, being either the laboratory frame, the nucleus-nucleus center-of-mass or the nucleon-nucleon center-of-mass. It is straight forward to employ relativistic kinematics which means to replace the velocities by their relativistic counterparts,

$$\frac{\mathbf{p}}{M} \rightarrow \frac{\mathbf{p}}{E}, \qquad \frac{|\mathbf{p}_1 - \mathbf{p}_2|}{M} \rightarrow \left| \frac{\mathbf{p}_1}{E_1} - \frac{\mathbf{p}_2}{E_2} \right|, \qquad (2.181)$$

in the nucleon-nucleon center-of-mass frame, however, the effective mean-field potential essentially depends on the nuclear density ρ, which is not a Lorentz-invariant quantity but the zeroth component of a four-vector. Accordingly, a covariant formulation of the dynamics is required with a clean specification of the transformation properties for the selfenergies as well as matrix elements in the collision term.

Solution of Exercises

Exercise 2.1: Prove Eq. (2.26) starting from (2.23).

Equation (2.23) defines the normalization

$$Tr_{(1,...,N)} \, \rho_N = N! \qquad (2.182)$$

for arbitrary N, while the recursion relation reads

$$\rho_n = \frac{1}{(N-n)!} Tr_{(n+1,\dots,N)}\rho_N. \tag{2.183}$$

Taking the trace $Tr_{(1,\dots,n)}\rho_n$ we have

$$Tr_{(1,\dots,n)}\rho_n = \frac{1}{(N-n)!}Tr_{(1,\dots,N)}\rho_N = \frac{N!}{(N-n)!} = N(N-1)(N-2)\dots(N-n+1),$$

which proves Eq. (2.26).

Exercise 2.2: Show that solutions of the TDHF equation

$$i\frac{\partial}{\partial t}\psi_\alpha(\mathbf{r};t)$$

$$= \left(-\frac{1}{2m}\nabla_r \cdot \nabla_r + \sum_\beta \int d^3r_2\, \psi_\beta^*(\mathbf{r}_2;t)v(\mathbf{r}-\mathbf{r}_2)\psi_\beta(\mathbf{r}_2;t)n_\beta\right)\psi_\alpha(\mathbf{r};t)$$

$$- \sum_\beta \int d^3r_2\, \psi_\beta^*(\mathbf{r}_2;t)v(\mathbf{r}-\mathbf{r}_2)\psi_\beta(\mathbf{r};t)\psi_\alpha(\mathbf{r}_2;t)n_\beta,$$

where n_β denote time-independent occupation numbers with $N = \sum_\beta n_\beta$, provide a solution for Eq. (2.35) with $\rho(1,1';t)$ given by

$$\rho(1,1';t) = \rho(\mathbf{r},\mathbf{r}';t) = \sum_\alpha n_\alpha \psi_\alpha^*(\mathbf{r}';t)\psi_\alpha(\mathbf{r};t).$$

The Hermitian conjugate TDHF equation reads

$$-i\frac{\partial}{\partial t}\psi_\alpha^*(\mathbf{r};t) = \left(-\frac{1}{2m}\nabla_r \cdot \nabla_r + \sum_\beta \int d^3r_2\, \psi_\beta(\mathbf{r}_2;t)v(\mathbf{r}-\mathbf{r}_2)\psi_\beta^*(\mathbf{r}_2;t)n_\beta\right)$$

$$\times \psi_\alpha^*(\mathbf{r};t)$$

$$- \sum_\beta \int d^3r_2\, \psi_\beta(\mathbf{r}_2;t)v(\mathbf{r}-\mathbf{r}_2)\psi_\beta^*(\mathbf{r};t)\psi_\alpha^*(\mathbf{r}_2;t)n_\beta = h^{HF}\psi_\alpha^*(\mathbf{r};t)$$

$$\tag{2.184}$$

and the time derivative of $\rho(\mathbf{r}, \mathbf{r}'; t)$ then gives

$$i\frac{\partial}{\partial t}\rho(\mathbf{r}, \mathbf{r}'; t) = i\frac{\partial}{\partial t}\sum_\alpha n_\alpha \psi_\alpha^*(\mathbf{r}'; t)\psi_\alpha(\mathbf{r}; t)$$

$$= i\sum_\alpha n_\alpha\left[\left(\frac{\partial}{\partial t}\psi_\alpha^*(\mathbf{r}'; t)\right)\psi_\alpha(\mathbf{r}; t) + \psi_\alpha^*(\mathbf{r}'; t)\left(\frac{\partial}{\partial t}\psi_\alpha(\mathbf{r}; t)\right)\right] \qquad (2.185)$$

$$= -\sum_\alpha n_\alpha\psi_\alpha(\mathbf{r}; t)$$

$$\times\left[\left(-\frac{1}{2m}\nabla_{r'}\cdot\nabla_{r'} + \sum_\beta n_\beta\int d^3r_2\ \psi_\beta(\mathbf{r}_2; t)v(\mathbf{r}' - \mathbf{r}_2)\psi_\beta^*(\mathbf{r}_2; t)\right)\psi_\alpha^*(\mathbf{r}'; t)\right.$$

$$\left. + \sum_\beta n_\beta\int d^3r_2\ \psi_\beta(\mathbf{r}_2; t)v(\mathbf{r}' - \mathbf{r}_2)\psi_\beta^*(\mathbf{r}'; t)\psi_\alpha^*(\mathbf{r}_2; t)\right]$$

$$+ \sum_\alpha n_\alpha\psi_\alpha^*(\mathbf{r}'; t)$$

$$\times\left[\left(-\frac{1}{2m}\nabla_r\cdot\nabla_r + \sum_\beta n_\beta\int d^3r_2\ \psi_\beta(\mathbf{r}_2; t)v(\mathbf{r} - \mathbf{r}_2)\psi_\beta^*(\mathbf{r}_2; t)\right)\psi_\alpha^*(\mathbf{r}; t)\right.$$

$$\left. - \sum_\beta n_\beta\psi_\alpha^*(\mathbf{r}_2; t)\int d^3r_2\ \psi_\beta(\mathbf{r}_2; t)v(\mathbf{r} - \mathbf{r}_2)\psi_\beta^*(\mathbf{r}; t)\psi_\alpha^*(\mathbf{r}_2; t)\right] = 0$$

since h^{HF} is Hermitian.

Exercise 2.3: Prove Eq. (2.53) for the expansions (2.51) and (2.52).

We have to show that (2.45)

$$i\frac{\partial}{\partial t}\rho(11'; t) = [h(1) - h(1')]\rho(11'; t) + \text{Tr}_{(2=2')}[v(12) - v(1'2')]c_2(12, 1'2'; t)$$

$$(2.186)$$

gives (2.53). Inserting the expansions (2.51) and (2.52) we obtain

$$i\frac{\partial}{\partial t} \sum_{\lambda\lambda'} \rho_{\lambda\lambda'}(t)\, \varphi_\lambda(\mathbf{r}_1)\varphi_{\lambda'}^*(\mathbf{r}_1') = [h(\mathbf{r}_1) - h(\mathbf{r}_1')]\sum_{\lambda\lambda'} \rho_{\lambda\lambda'}(t)\, \varphi_\lambda(\mathbf{r})\varphi_{\lambda'}^*(\mathbf{r}')$$

$$\tag{2.187}$$

$$+ \int d^3r_2\, [v(\mathbf{r}_1 - \mathbf{r}_2) - v(\mathbf{r}_1' - \mathbf{r}_2)] \sum_{\lambda\gamma\lambda'\gamma'} C_{\lambda\gamma\lambda'\gamma'}(t)\, \varphi_\lambda(\mathbf{r}_1)\varphi_\gamma(\mathbf{r}_2)\varphi_{\lambda'}^*(\mathbf{r}_1')\varphi_{\gamma'}^*(\mathbf{r}_2).$$

We now multiply from the left with $\varphi_\alpha^*(\mathbf{r}_1)\varphi_{\alpha'}(\mathbf{r}_{1'})$ and integrate over $d^3r_1 d^3r_{1'}$,

$$i\frac{\partial}{\partial t} \sum_{\lambda\lambda'} \int d^3r_1 \int d^3r_{1'}\, \varphi_\alpha^*(\mathbf{r}_1)\varphi_{\alpha'}(\mathbf{r}_{1'})\rho_{\lambda\lambda'}(t)\, \varphi_\lambda(\mathbf{r}_1)\varphi_{\lambda'}^*(\mathbf{r}_1')$$

$$= i\frac{\partial}{\partial t} \sum_{\lambda\lambda'} \rho_{\lambda\lambda'}(t)\delta_{\alpha\lambda}\,\delta_{\alpha'\lambda'} \tag{2.188}$$

$$= i\frac{\partial}{\partial t} \rho_{\alpha\alpha'}(t) = \int d^3r_1 \int d^3r_{1'}\, \varphi_\alpha^*(\mathbf{r}_1)\varphi_{\alpha'}(\mathbf{r}_{1'})[h(\mathbf{r}_1; t) - h(\mathbf{r}_1'; t)]$$

$$\times \sum_{\lambda\lambda'} \rho_{\lambda\lambda'}(t)\, \varphi_\lambda(\mathbf{r}_1)\varphi_{\lambda'}^*(\mathbf{r}_1')$$

$$+ \sum_{\lambda\gamma\lambda'\gamma'} \int d^3r_1 \int d^3r_{1'}\, \varphi_\alpha^*(\mathbf{r}_1)\varphi_{\alpha'}(\mathbf{r}_{1'}) \int d^3r_2 [v(\mathbf{r}_1 - \mathbf{r}_2) - v(\mathbf{r}_1' - \mathbf{r}_2)]C_{\lambda\gamma\lambda'\gamma'}(t)$$

$$\times \varphi_\lambda(\mathbf{r}_1)\varphi_\gamma(\mathbf{r}_2)\varphi_{\lambda'}^*(\mathbf{r}_1')\varphi_{\gamma'}^*(\mathbf{r}_2).$$

Defining

$$h_{\alpha\lambda} = \langle \alpha | h | \lambda \rangle = \int d^3r_1\, \varphi_\alpha^*(\mathbf{r}_1)\, h(\mathbf{r}_1)\, \varphi_\lambda(\mathbf{r}_1) \tag{2.189}$$

and

$$\langle \alpha\beta | v | \lambda\gamma \rangle = \int d^3r_1 \int d^3r_2\, \varphi_\alpha^*(\mathbf{r}_1)\varphi_\beta^*(\mathbf{r}_2)\, v(\mathbf{r}_1 - \mathbf{r}_2)\, \varphi_\lambda(\mathbf{r}_1)\varphi_\gamma(\mathbf{r}_2) \tag{2.190}$$

we arrive at

$$i\frac{\partial}{\partial t} \rho_{\alpha\alpha'}(t) = \sum_{\lambda\lambda'} \delta_{\lambda'\alpha'}h_{\alpha\lambda}\rho_{\lambda\lambda'}(t) - \sum_{\lambda\lambda'} \delta_{\alpha\lambda}h_{\lambda'\alpha'}\rho_{\lambda\lambda'}(t) \tag{2.191}$$

$$+ \sum_{\lambda\gamma\lambda'\gamma'} \left(\delta_{\alpha'\lambda'}\langle \alpha\gamma' | v | \lambda\gamma \rangle C_{\lambda\gamma\lambda'\gamma'}(t)\ - \delta_{\alpha\lambda}\langle \lambda'\gamma' | v | \alpha'\gamma \rangle C_{\lambda\gamma\lambda'\gamma'}(t)\right)$$

$$= \sum_{\lambda} [h_{\alpha\lambda} \rho_{\lambda\alpha'}(t) - \rho_{\alpha\lambda} h_{\lambda\alpha'}]$$

$$+ \sum_{\beta} \sum_{\lambda\gamma} \{ \langle \alpha\beta | v | \lambda\gamma \rangle C_{\lambda\gamma\alpha'\beta}(t) - C_{\alpha\beta\lambda\gamma}(t) \langle \lambda\gamma | v | \alpha'\beta \rangle \}$$

after redefining the summation indices. In shorthand form we thus may also write

$$i \frac{\partial}{\partial t} \rho_{\alpha\alpha'}(t) = [h, \rho]_{\alpha\alpha'} + \sum_{\beta} \{ \langle \alpha\beta | vC - Cv | \alpha'\beta \rangle \}. \tag{2.192}$$

Exercise 2.4: The ground state wavefunction of a three-dimensional oscillator is given by a Gaussian

$$\psi(\mathbf{r}) = N_0 \exp(-\mathbf{r} \cdot \mathbf{r}/(2b^2)).$$

Calculate the normalization factor N_0 and the Wigner transform of $\psi^*(\mathbf{r}')\psi(\mathbf{r})$. What is the value at the origin? Does the interpretation of a classical phase-space distribution hold?

The three-dimensional gaussian wavefunction in cartesian coordinates can be considered as a product of three gaussians in x, y, z-direction each with normalization factor $N_0^{1/3}$. The normalization in each direction is obtained from

$$N_0^{2/3} \int_{-\infty}^{\infty} dx \, \exp(-x^2/b^2) = N_0^{2/3} \sqrt{\pi} b = 1, \tag{2.193}$$

i.e. $N_0 = 1/(\sqrt{\pi}b)^{3/2}$.

For the Wigner transformation we have to compute the following one-dimensional integral (including $\hbar$ explicitly for the exercise)

$$f(r, p) = \int_{-\infty}^{\infty} ds \, \rho(r + s/2, r - s/2) \exp\left(-\frac{i}{\hbar} ps\right) \tag{2.194}$$

$$= \frac{1}{\sqrt{\pi}b} \int_{-\infty}^{\infty} ds \, \exp\left(-\frac{x^2 + x'^2}{2b^2}\right) \exp\left(-\frac{i}{\hbar} ps\right),$$

since the problem reduces to a product of three independent one-dimensional integrals. First we evaluate

$$x^2 + x'^2 = (r + s/2)^2 + (r - s/2)^2 = r^2 + rs + s^2/4 + r^2 - rs + s^2/4 = 2r^2 + s^2/2. \tag{2.195}$$

This leads to

$$f(r, p) = \frac{1}{\sqrt{\pi} b} \exp\left(-\frac{r^2}{b^2}\right) \int_{-\infty}^{\infty} ds \, \exp\left(-\frac{s^2}{4b^2}\right) \exp\left(-\frac{i}{\hbar} ps\right). \qquad (2.196)$$

We rewrite the (negative) argument of the exponent by introducing $\xi = \frac{i}{\hbar} p 2 b^2$, i.e.

$$\frac{s^2}{4b^2} + \frac{i}{\hbar} ps = \frac{s^2 + 2s\xi + \xi^2 - \xi^2}{4b^2} = \frac{(s + \xi)^2 - \xi^2}{4b^2}, \qquad (2.197)$$

which gives

$$f(r, p) = \frac{1}{\sqrt{\pi} b} \exp\left(-\frac{r^2}{b^2}\right) \exp\left(\frac{\xi^2}{4b^2}\right) \int_{-\infty}^{\infty} ds \, \exp\left(-\frac{(s + \xi)^2}{4b^2}\right). \qquad (2.198)$$

With $\xi^2/(4b^2) = -p^2 b^2/\hbar^2$ and $\eta = (s + \xi)/(2b)$ we arrive at

$$f(r, p) = \frac{1}{\sqrt{\pi} b} \exp\left(-\frac{r^2}{b^2}\right) \exp\left(-\frac{p^2 b^2}{\hbar^2}\right) 2b \int_{-\infty}^{\infty} d\eta \, \exp(-\eta^2). \qquad (2.199)$$

The integral over $d\eta$ gives $\sqrt{\pi}$ and we finally obtain

$$f(r, p) = \frac{1}{\sqrt{\pi} b} \exp\left(-\frac{r^2}{b^2}\right) \exp\left(-\frac{p^2 b^2}{\hbar^2}\right) 2b \sqrt{\pi} = 2 \exp\left(-\frac{r^2}{b^2}\right) \exp\left(-\frac{p^2 b^2}{\hbar^2}\right). \qquad (2.200)$$

In three dimensions we thus obtain the product

$$f(\mathbf{r}, \mathbf{p}) = 2^3 \exp\left(-\frac{\mathbf{r} \cdot \mathbf{r}}{b^2}\right) \exp\left(-\mathbf{p} \cdot \mathbf{p} \frac{b^2}{\hbar^2}\right) = 8 \exp\left(-\frac{\mathbf{r}^2}{b^2} - \frac{\mathbf{p}^2 b^2}{\hbar^2}\right). \qquad (2.201)$$

For $\mathbf{r} = 0, \mathbf{p} = 0$ we get

$$f(\mathbf{r} = 0, \mathbf{p} = 0) = 8,$$

which indicates that the Wigner transform of a wavefunction cannot be identified with the local occupation probability in phase space, i.e. the probability to find a particle with momentum $\mathbf{p}$ at position $\mathbf{r}$. Only when averaging over a phase-space volume $\Delta r \, \Delta p \geq \hbar/2$ in all three dimensions a classical "occupation probability" emerges.

Exercise 2.5: Show that

$$\vec{\nabla}_x^2 - \vec{\nabla}_{x'}^2 = \vec{\nabla}_{r+s/2}^2 - \vec{\nabla}_{r-s/2}^2 = 2\vec{\nabla}_s \cdot \vec{\nabla}_r,$$

where $\mathbf{x} = \mathbf{r} + \mathbf{s}/2$, $\qquad \mathbf{x}' = \mathbf{r} - \mathbf{s}/2$.
What is the Wigner transform of $\vec{\nabla}_s \cdot \vec{\nabla}_r \, \rho(\mathbf{r} + \mathbf{s}/2, \mathbf{r} - \mathbf{s}/2)$ when assuming that $\rho(\mathbf{r}, \mathbf{s})$ vanishes at $s_i \to \pm\infty$ for $i = x, y, z$?

We have $(i = 1, 2, 3)$

$$\frac{\partial}{\partial x_i} = \frac{\partial}{\partial r_i}\frac{\partial r_i}{\partial x_i} + \frac{\partial}{\partial s_i}\frac{\partial s_i}{\partial x_i} = \frac{1}{2}\frac{\partial}{\partial r_i} + \frac{\partial}{\partial s_i}, \qquad (2.202)$$

$$\frac{\partial}{\partial x_i'} = \frac{\partial}{\partial r_i}\frac{\partial r_i}{\partial x_i'} + \frac{\partial}{\partial s_i}\frac{\partial s_i}{\partial x_i'} = \frac{1}{2}\frac{\partial}{\partial r_i} - \frac{\partial}{\partial s_i}.$$

This leads to

$$\vec{\nabla}_x^2 - \vec{\nabla}_{x'}^2 = \vec{\nabla}_{r+s/2}^2 - \vec{\nabla}_{r-s/2}^2 = \sum_{i=1}^{3}\left(\left(\frac{1}{2}\frac{\partial}{\partial r_i} + \frac{\partial}{\partial s_i}\right)^2 - \left(\frac{1}{2}\frac{\partial}{\partial r_i} - \frac{\partial}{\partial s_i}\right)^2\right)$$

$$(2.203)$$

$$= \sum_{i=1}^{3}\left(\frac{1}{4}\frac{\partial^2}{\partial r_i^2} + \frac{\partial}{\partial r_i}\frac{\partial}{\partial s_i} + \frac{\partial^2}{\partial s_i^2} - \frac{1}{4}\frac{\partial^2}{\partial r_i^2} + \frac{\partial}{\partial r_i}\frac{\partial}{\partial s_i} - \frac{\partial^2}{\partial s_i^2}\right) = 2\vec{\nabla}_r \cdot \vec{\nabla}_s.$$

The Wigner transform of $\vec{\nabla}_s \cdot \vec{\nabla}_r \rho(\mathbf{r} + \mathbf{s}/2, \mathbf{r} - \mathbf{s}/2)$ is given by

$$I_W(\mathbf{p}) = \int d^3s \, \exp\left(-\frac{i}{\hbar}\mathbf{p}\cdot\mathbf{s}\right)\vec{\nabla}_s \cdot \vec{\nabla}_r \, \rho(\mathbf{r} + \mathbf{s}/2, \mathbf{r} - \mathbf{s}/2) \qquad (2.204)$$

$$= \vec{\nabla}_r \cdot \int d^3s \, \exp\left(-\frac{i}{\hbar}\mathbf{p}\cdot\mathbf{s}\right)\vec{\nabla}_s\rho(\mathbf{r} + \mathbf{s}/2, \mathbf{r} - \mathbf{s}/2).$$

Use that

$$\frac{d}{dx}[f(x)g(x)] = f(x)\frac{dg(x)}{dx} + g(x)\frac{df(x)}{dx}.$$

Then

$$\vec{\nabla}_s \left(\exp\left(-\frac{i}{\hbar}\mathbf{p}\cdot\mathbf{s}\right) \rho(\mathbf{r}+\mathbf{s}/2,\mathbf{r}-\mathbf{s}/2)\right) \tag{2.205}$$

$$= \exp\left(-\frac{i}{\hbar}\mathbf{p}\cdot\mathbf{s}\right) \vec{\nabla}_s \, \rho(\mathbf{r}+\mathbf{s}/2,\mathbf{r}-\mathbf{s}/2)$$

$$+ \rho(\mathbf{r}+\mathbf{s}/2,\mathbf{r}-\mathbf{s}/2)\vec{\nabla}_s \exp\left(-\frac{i}{\hbar}\mathbf{p}\cdot\mathbf{s}\right).$$

Since

$$\vec{\nabla}_s \exp\left(-\frac{i}{\hbar}\mathbf{p}\cdot\mathbf{s}\right) = -\frac{i}{\hbar}\mathbf{p}\exp\left(-\frac{i}{\hbar}\mathbf{p}\cdot\mathbf{s}\right), \tag{2.206}$$

we obtain that

$$\exp\left(-\frac{i}{\hbar}\mathbf{p}\cdot\mathbf{s}\right) \vec{\nabla}_s \, \rho(\mathbf{r}+\mathbf{s}/2,\mathbf{r}-\mathbf{s}/2) \tag{2.207}$$

$$= \vec{\nabla}_s \left(\exp\left(-\frac{i}{\hbar}\mathbf{p}\cdot\mathbf{s}\right) \rho(\mathbf{r}+\mathbf{s}/2,\mathbf{r}-\mathbf{s}/2)\right)$$

$$+ \frac{i}{\hbar}\mathbf{p}\exp\left(-\frac{i}{\hbar}\mathbf{p}\cdot\mathbf{s}\right) \rho(\mathbf{r}+\mathbf{s}/2,\mathbf{r}-\mathbf{s}/2).$$

Substitute (2.207) to Eq. (2.204):

$$I_W(\mathbf{q}) = \vec{\nabla}_\mathbf{r}\cdot\int d^3s \, \vec{\nabla}_s \left(\exp\left(-\frac{i}{\hbar}\mathbf{p}\cdot\mathbf{s}\right) \rho(\mathbf{r}+\mathbf{s}/2,\mathbf{r}-\mathbf{s}/2)\right) \tag{2.208}$$

$$- \vec{\nabla}_\mathbf{r}\cdot\int d^3s \, \left(\vec{\nabla}_s \exp\left(-\frac{i}{\hbar}\mathbf{p}\cdot\mathbf{s}\right)\right) \rho(\mathbf{r}+\mathbf{s}/2,\mathbf{r}-\mathbf{s}/2).$$

The first term in (2.208) is vanishing in the limits of partial integration since $\rho(\mathbf{r},\mathbf{s}) \to 0$ when $s_i \to \pm\infty$ for all components $i = 1, 2, 3$:

$$\int d^3s \, \vec{\nabla}_s \left(\exp\left(-\frac{i}{\hbar}\mathbf{p}\cdot\mathbf{s}\right) \rho(\mathbf{r}+\mathbf{s}/2,\mathbf{r}-\mathbf{s}/2)\right) \to 0 \tag{2.209}$$

since

$$\int_{-\infty}^{\infty} ds_i \frac{\partial}{\partial s_i} \left(\exp\left(-\frac{i}{\hbar}\mathbf{p}\cdot\mathbf{s}\right) \rho(\mathbf{r}+\mathbf{s}/2,\mathbf{r}-\mathbf{s}/2)\right)$$

$$\to \exp\left(-\frac{i}{\hbar}\mathbf{p}\cdot\mathbf{s}\right) \rho(\mathbf{r}+\mathbf{s}/2,\mathbf{r}-\mathbf{s}/2)\Big|_{s_i\to-\infty}^{s_i\to\infty} \to 0. \tag{2.210}$$

Thus,

$$
I_W(\mathbf{q}) = -\vec{\nabla}_{\mathbf{r}} \cdot \int d^3 s \left(-\frac{i}{\hbar} \mathbf{p} \right) \exp\left(-\frac{i}{\hbar} \mathbf{p} \cdot \mathbf{s} \right) \rho(\mathbf{r} + \mathbf{s}/2, \mathbf{r} - \mathbf{s}/2) \qquad (2.211)
$$

$$
= \frac{i}{\hbar} \mathbf{p} \cdot \vec{\nabla}_{\mathbf{r}} \int d^3 s \, \exp\left(-\frac{i}{\hbar} \mathbf{p} \cdot \mathbf{s} \right) \rho(\mathbf{r} + \mathbf{s}/2, \mathbf{r} - \mathbf{s}/2)
$$

$$
= \frac{i}{\hbar} \mathbf{p} \cdot \vec{\nabla}_{\mathbf{r}} f(\mathbf{r}, \mathbf{p}).
$$

Exercise 2.6: Show that for real ω_0 and $\epsilon > 0$

$$
\int_{-\infty}^{\infty} \frac{d\omega}{2\pi} \frac{1}{\omega - \omega_0 \mp i\epsilon} = \pm i.
$$

For $\epsilon > 0$ the integrand has a pole in the upper half plane for the minus sign in front of the $i\epsilon$ at $\omega_0 + i\epsilon$. In this case we can close the integration contour by a semicircle in the upper half plane since the integrand vanishes on the semicircle. Now applying the residue theorem we have

$$
\int_{-\infty}^{\infty} \frac{d\omega}{2\pi} \frac{1}{\omega - \omega_0 - i\epsilon} = \oint \frac{d\omega}{2\pi} \frac{1}{\omega - \omega_0 - i\epsilon} = 2\pi i \frac{1}{2\pi} = i. \qquad (2.212)
$$

In case of a plus sign in front of the $i\epsilon$ the pole is in the lower half plane and we close the integration contour by a semicircle in the lower half plane which gives a relative minus sign due to the different orientation.

Another important relation is

$$
i \int_{-\infty}^{\infty} \frac{d\omega}{2\pi} \frac{\exp(-i\omega t)}{\omega + i\epsilon} = i \oint \frac{d\omega}{2\pi} \frac{\exp(-i\omega t)}{\omega + i\epsilon} = \Theta(t), \qquad (2.213)
$$

i.e. the Heaviside unit step-function (after integration over the lower plane).

Exercise 2.7: Calculate the Fourier transform of the Yukawa interaction (with $\mu > 0$)

$$
v(\mathbf{r}) = V_0 \frac{\exp(-\mu|\mathbf{r}|)}{|\mathbf{r}|}.
$$

What is the range of the interaction and the scattering amplitude in Born approximation?

The Fourier transform is given by

$$
\begin{aligned}
v(\mathbf{p}) &= V_0 \int d^3r \, \exp(-i\mathbf{p} \cdot \mathbf{r}) \, \frac{\exp(-\mu|\mathbf{r}|)}{|\mathbf{r}|} \\
&= 2\pi V_0 \int_0^\infty dr \int_{-1}^1 d\xi \, r \exp(-ip\xi r) \exp(-\mu r) \\
&= 2\pi V_0 \int_0^\infty dr \, \frac{1}{ip}(\exp(ipr) - \exp(-ipr)) \exp(-\mu r) \\
&= \frac{V_0 2\pi}{ip} \int_0^\infty dr \, (\exp((ip - \mu)r) - \exp(-(ip + \mu)r)) \\
&= \frac{2\pi V_0}{ip} \left(\frac{-1}{ip - \mu} + \frac{-1}{ip + \mu} \right) = 2\pi V_0 \frac{2ip}{ip} \frac{1}{p^2 + \mu^2} = \frac{4\pi V_0}{p^2 + \mu^2}.
\end{aligned}
\tag{2.214}
$$

The range of the Yukawa interaction is $1/\mu$. In Born approximation the scattering amplitude is

$$
f(\mathbf{p}) = -\frac{1}{4\pi} v(\mathbf{p}) = -\frac{V_0}{p^2 + \mu^2},
\tag{2.215}
$$

where $\mathbf{p}$ denotes the momentum transfer in the reaction.

Exercise 2.8: Derive Eq. (2.126) starting from (2.122) using (2.123).

We start with the collision integral in the form

$$
\begin{aligned}
I(\mathbf{p}_1, \mathbf{p}_1; t) = d \int & \frac{d^3 p_2}{(2\pi)^3} \frac{d^3 p_3}{(2\pi)^3} \frac{d^3 p_4}{(2\pi)^3} \\
& \times 2\pi \delta \left(\frac{1}{2m} [p_1^2 + p_2^2 - p_3^2 - p_4^2] \right) \\
& \times (2\pi)^3 \delta^3(\mathbf{p}_1 + \mathbf{p}_2 - \mathbf{p}_3 - \mathbf{p}_4) \, v(\mathbf{p}_2 - \mathbf{p}_4) v_A(\mathbf{p}_4 - \mathbf{p}_2) \\
& \times (n(\mathbf{p}_3; t)n(\mathbf{p}_4; t)\bar{n}(\mathbf{p}_1; t)\bar{n}(\mathbf{p}_2; t) - n(\mathbf{p}_1; t)n(\mathbf{p}_2; t)\bar{n}(\mathbf{p}_3; t)\bar{n}(\mathbf{p}_4; t))
\end{aligned}
\tag{2.216}
$$

and introduce total and relative momenta in the initial and final channels as

$$
\mathbf{P}_{in} = \mathbf{p}_1 + \mathbf{p}_2, \quad \mathbf{P}_{out} = \mathbf{p}_3 + \mathbf{p}_4,
\tag{2.217}
$$

$$
\mathbf{q}_{in} = \frac{\mathbf{p}_1 - \mathbf{p}_2}{2}, \quad \mathbf{q}_{out} = \frac{\mathbf{p}_3 - \mathbf{p}_4}{2}.
$$

Now we can rewrite

$$I_{out} := \int \frac{d^3 p_3}{(2\pi)^3} \int d^3 p_4 \, 2\pi \delta \left(\frac{1}{2m} [p_1^2 + p_2^2 - p_3^2 - p_4^2] \right)$$

$$\times \delta^3(\mathbf{p}_1 + \mathbf{p}_2 - \mathbf{p}_3 - \mathbf{p}_4) \, v(\mathbf{p}_2 - \mathbf{p}_4) v_A(\mathbf{p}_4 - \mathbf{p}_2)$$

$$= \int d^3 P_{out} \int \frac{d^3 q_{out}}{(2\pi)^3} 2\pi \delta \left(\frac{1}{2m} [P_{in}^2/2 + 2q_{in}^2 - P_{out}^2/2 - 2q_{out}^2] \right)$$

$$\times \delta^3(\mathbf{P}_{in} - \mathbf{P}_{out}) \, v(\mathbf{p}_2 - \mathbf{p}_4) v_A(\mathbf{p}_4 - \mathbf{p}_2)$$

$$= \int \frac{d^3 q_{out}}{(2\pi)^2} \, \delta \left(\frac{1}{m} [q_{in}^2 - q_{out}^2] \right) \, v(\mathbf{p}_2 - \mathbf{p}_4) v_A(\mathbf{p}_4 - \mathbf{p}_2). \qquad (2.218)$$

Changing to spherical coordinates for $\mathbf{q}_{out}$ we get

$$I_{out} = \int d\Omega \int \frac{dq_{out}}{(2\pi)^2} \, q_{out}^2 \left(\frac{1}{m} [q_{in}^2 - q_{out}^2] \right) \, v(\mathbf{p}_2 - \mathbf{p}_4) v_A(\mathbf{p}_4 - \mathbf{p}_2)$$

$$= \frac{1}{4\pi^2} \int d\Omega \, \frac{m q_{in}}{2} \, v(\mathbf{p}_2 - \mathbf{p}_4) v_A(\mathbf{p}_4 - \mathbf{p}_2). \qquad (2.219)$$

Now using (in Born approximation) relation (2.123), i.e.

$$\frac{16\pi^2}{m^2} \frac{d\sigma}{d\Omega} (\mathbf{p}_1 + \mathbf{p}_2, \mathbf{p}_2 - \mathbf{p}_4) = v(\mathbf{p}_2 - \mathbf{p}_4) v_A(\mathbf{p}_4 - \mathbf{p}_2), \qquad (2.220)$$

we obtain

$$I_{out} = \int d\Omega \, \frac{1}{4\pi^2} \frac{m q_{in}}{2} \frac{16\pi^2}{m^2} \frac{d\sigma}{d\Omega} (\mathbf{p}_1 + \mathbf{p}_2, \mathbf{p}_2 - \mathbf{p}_4) \qquad (2.221)$$

$$= \int d\Omega \, \frac{2q_{in}}{m} \frac{d\sigma}{d\Omega} (\mathbf{p}_1 + \mathbf{p}_2, \mathbf{p}_2 - \mathbf{p}_4) = \int d\Omega \, v_{12} \frac{d\sigma}{d\Omega} (\mathbf{p}_1 + \mathbf{p}_2, \mathbf{p}_2 - \mathbf{p}_4)$$

with the relative velocity

$$v_{12} = \frac{2q_{in}}{m} = \frac{|\mathbf{p}_1 - \mathbf{p}_2|}{m}, \qquad (2.222)$$

which gives Eq. (2.126), i.e.

$$I(\mathbf{p}_1, \mathbf{p}_1; t) = d \int \frac{d^3 p_2}{(2\pi)^3} \int d\Omega \, v_{12} \frac{d\sigma}{d\Omega} (\mathbf{p}_1 + \mathbf{p}_2, \mathbf{p}_2 - \mathbf{p}_4) \qquad (2.223)$$

$$\times \{ n(\mathbf{p}_3; t) n(\mathbf{p}_4; t) \bar{n}(\mathbf{p}_1; t) \bar{n}(\mathbf{p}_2; t) - n(\mathbf{p}_1; t) n(\mathbf{p}_2; t) \bar{n}(\mathbf{p}_3; t) \bar{n}(\mathbf{p}_4; t) \}.$$

Exercise 2.9: Show that the Fermi distribution (2.127) is a stationary solution of the collision term, i.e. fulfills Eq. (2.128).

We have to show that $I(\mathbf{p}_1, \mathbf{p}_1; t) = 0$ in case of $n = n_F(\epsilon) = n_F(\mathbf{p}^2/(2m))$. It is sufficient to show that

$$n(\mathbf{p}_3)n(\mathbf{p}_4)\bar{n}(\mathbf{p}_1)\bar{n}(\mathbf{p}_2) = n(\mathbf{p}_1)n(\mathbf{p}_2)\bar{n}(\mathbf{p}_3)\bar{n}(\mathbf{p}_4), \qquad (2.224)$$

if $n = n_F(\epsilon)$. To this aim we write

$$1 - n_F(\epsilon) = 1 - \frac{1}{1 + \exp((\epsilon - \mu)/T)} = \frac{1 - (1 + \exp((\epsilon - \mu)/T))}{1 + \exp((\epsilon - \mu)/T)}$$

$$= -\frac{\exp((\epsilon - \mu)/T)}{1 + \exp((\epsilon - \mu)/T)}. \qquad (2.225)$$

We then have

$$\frac{\exp((\epsilon_1 - \mu)/T)}{(1 + \exp((\epsilon_1 - \mu)/T))} \frac{\exp((\epsilon_2 - \mu)/T)}{(1 + \exp((\epsilon_2 - \mu)/T))}$$

$$\times \frac{1}{(1 + \exp((\epsilon_3 - \mu)/T))} \frac{1}{(1 + \exp((\epsilon_4 - \mu)/T))}$$

$$= \frac{\exp((\epsilon_3 - \mu)/T)}{(1 + \exp((\epsilon_3 - \mu)/T))} \frac{\exp((\epsilon_4 - \mu)/T)}{(1 + \exp((\epsilon_4 - \mu)/T))}$$

$$\times \frac{1}{(1 + \exp((\epsilon_1 - \mu)/T))} \frac{1}{(1 + \exp((\epsilon_2 - \mu)/T))}$$

Since the denominators on both sides are the same one has to proof only

$$\exp((\epsilon_1 - \mu)/T) \exp((\epsilon_2 - \mu)/T) = \exp((\epsilon_3 - \mu)/T) \ \exp((\epsilon_4 - \mu)/T). \qquad (2.226)$$

The common factor $\exp(-2\mu/T)$ on both sides cancels out and we have

$$\exp(\epsilon_1/T) \exp(\epsilon_2/T) = \exp(\epsilon_3/T) \ \exp(\epsilon_4/T) \qquad (2.227)$$

or

$$\exp((\epsilon_1 + \epsilon_2)/T) = \exp((\epsilon_3 + \epsilon_4)/T). \qquad (2.228)$$

Since due to energy conservation $\delta(\epsilon_1 + \epsilon_2 - \epsilon_3 - \epsilon_4)$ we have $\epsilon_1 + \epsilon_2 = \epsilon_3 + \epsilon_4$; this proves that $n_F(\epsilon)$ is a stationary solution.

Exercise 2.10: Prove the operator identity $[A + B]^{-1} = A^{-1}(1 - B[A + B]^{-1})$.

By multiplication of the identity by $[A + B]$ from the right on both sides we obtain

$$
\begin{aligned}
1 = [A + B]^{-1}[A + B] &= A^{-1}(1 - B[A + B]^{-1})[A + B] \\
&= A^{-1}[A + B] - A^{-1}B[A + B]^{-1}[A + B] \\
&= A^{-1}A + A^{-1}B - A^{-1}B \cdot 1 = A^{-1}A = 1,
\end{aligned}
\tag{2.229}
$$

which proves the identity. The familiar Dyson equation emerges by setting $A = (\omega - h(1) - h(2) \pm i\gamma)$ and $B = -v$.

Exercise 2.11: Calculate the nonrelativistic energy per nucleon for symmetric nuclear matter at zero temperature as a function of the density, i.e. the nuclear equation of state, using the effective interaction (2.166) or potential energy density (2.168), respectively.

For symmetric nuclear matter at zero temperature the nuclear density is given by

$$
\rho = \frac{d}{(2\pi)^3} 4\pi \int_0^{p_F} dp \, p^2 = \frac{d}{6\pi^2} p_F^3
\tag{2.230}
$$

with the degeneracy $d = 4$ for given Fermi momentum p_F. Alternatively we may replace p_F by

$$
p_F = \left(\frac{6\pi^2}{d} \rho \right)^{1/3} = \left(\frac{3\pi^2}{2} \rho \right)^{1/3}.
\tag{2.231}
$$

The kinetic energy density is given by

$$
\begin{aligned}
\mathcal{E}_{kin} &= \frac{d}{(2\pi)^3} 4\pi \int_0^{p_F} dp \, p^2 \frac{p^2}{2m} = \frac{d}{20m\pi^2} p_F^5 = \frac{1}{5m\pi^2} p_F^5 = \frac{1}{5m\pi^2} \left(\frac{3\pi^2}{2} \rho \right)^{5/3} \\
&= \frac{3}{10m} \left(\frac{3}{2} \right)^{2/3} \pi^{4/3} \rho^{5/3} =: C \, \rho^{5/3}.
\end{aligned}
\tag{2.232}
$$

The total energy density then is given by adding the potential energy density (2.168),

$$\mathcal{E}(\rho) = C \, \rho^{5/3} + \frac{3}{4} \left(-\frac{A}{2}\rho^2 + \frac{B}{(1+\gamma)(2+\gamma)}\rho^{2+\gamma} \right), \qquad (2.233)$$

and the energy per nucleon is obtained (dividing by ρ) as,

$$\frac{E}{N}(\rho) = C \, \rho^{2/3} + \frac{3}{4} \left(-\frac{A}{2}\rho + \frac{B}{(1+\gamma)(2+\gamma)}\rho^{1+\gamma} \right). \qquad (2.234)$$

The parameters A, B, γ have to be fixed by the boundary conditions

$$\frac{d}{d\rho}\frac{E}{N}(\rho)|_{\rho_0} = \frac{2C}{3}\rho_0^{-1/3} + \frac{3}{4}\left(-\frac{A}{2} + \frac{B}{(1+\gamma)}\rho_0^{\gamma} \right) = 0 \qquad (2.235)$$

for a minimum at saturation density $\rho = \rho_0$, and

$$\frac{E}{N}(\rho_0) = C \, \rho_0^{2/3} + \frac{3}{4}\left(-\frac{A}{2}\rho_0 + \frac{B}{(1+\gamma)(2+\gamma)}\rho_0^{1+\gamma} \right) = -16\,\mathrm{MeV} \qquad (2.236)$$

at $\rho_0 \approx 1/6\,\mathrm{fm}^{-3}$ as well as by the incompressibility

$$K = 9\rho_0^2 \frac{d^2}{d\rho^2}\frac{E}{N}|_{\rho_0} = -9\rho_0^2\frac{2C}{9}\rho_0^{-4/3} + 9\rho_0^2\frac{3B}{4}\frac{\gamma}{2+\gamma}\rho_0^{\gamma-1}$$

$$= -2C\rho_0^{2/3} + \frac{27B}{4}\frac{\gamma}{2+\gamma}\rho_0^{1+\gamma}, \qquad (2.237)$$

which fixes the parameter γ. The incompressibility K is not so well known but in the range of 200–400 MeV. In practice one chooses some value for γ in the interval $[0.3, 1]$ and solves for the parameters A and B which gives a unique incompressibility $K(\gamma)$.

References

1. S.J. Wang, W. Cassing, Ann. Phys. **159**, 328 (1985)
2. K.A. Brueckner, Phys. Rev. **97**, 1353 (1955)
3. H.A. Bethe, Phys. Rev. **103**, 1353 (1956)
4. H.A. Bethe, J. Goldstone, Proc. Roy. Soc. Lond. Ser. A **238**, 551 (1957)
5. A. Peter, W. Cassing, J. Häuser, A. Pfitzner, Nucl. Phys. **A573**, 93 (1994)
6. M. Tohyama, Phys. Rev. C **38**, 553 (1988)
7. M. Tohyama, Nucl. Phys. A **563**, 494 (1993)
8. M. Tohyama, A.S. Umar, Phys. Rev. C **65**, 037601 (2002)
9. M. Tohyama, P. Schuck, Eur. Phys. J. A **50**, 77 (2014)
10. W. Cassing, Z. Phys. A **327**, 87 (1987)
11. L.W. Nordheim, Proc. R. Soc. Lond. Ser. A **119**, 689 (1928)

12. E.A. Uehling, G.E. Uhlenbeck, Phys. Rev. **43**, 552 (1933)
13. L. Boltzmann, Phil. Mag. **35**, 161 (1893)
14. L. Boltzmann, Phil. Mag. **51**, 414 (1895)
15. W. Bauer, G.F. Bertsch, W. Cassing, U. Mosel, Phys. Rev. **C34**, 2127 (1986)
16. W. Cassing, V. Metag, U. Mosel, K. Niita, Phys. Rep. **188**, 363 (1990)
17. W. Cassing, E. Bratkovskaya, Phys. Rept. **308**, 65 (1999)
18. S.A. Bass, M. Belkacem, M. Bleicher, M. Brandstetter, L. Bravina, C. Ernst, L. Gerland, M. Hofmann, S. Hofmann, J. Konopka, G. Mao, L. Neise, S. Soff, C. Spieles, H. Weber, L.A. Winckelmann, H. Stöcker, W. Greiner, Prog. Part. Nucl. Phys. **41**, 255 (1998)
19. B.A. Li, A.T. Sustich, B. Zhang, C.M. Ko, Int. J. Mod. Phys. E **10**, 267 (2001)
20. B.-A. Li, L.-W. Chen, C.M. Ko, Phys. Rep. **464**, 113 (2008)
21. O. Buss, T. Gaitanos, K. Gallmeister, H. van Hees, M. Kaskulov, O. Lalakulich, A.B. Larionov, T. Leitner, J. Weil, U. Mosel, Phys. Rep. **512**, 1 (2012)
22. J. Weil, V. Steinberg, J. Staudenmeier, L.G. Pang, D. Oliinychenko, J. Mohs, M. Kretz, T. Kehrenberg, A. Goldschmidt, B. Bäuchle, J. Auvinen, M. Attems, H. Petersen, Phys. Rev. C **94**, 054905 (2016)
23. P. Moreau, O. Soloveva, L. Oliva, T. Song, W. Cassing, E. Bratkovskaya, Phys. Rev. C **100**, 014911 (2019)

Relativistic On-Shell Kinetic Theories

3

As mentioned in the previous chapter a covariant transport theory requires a proper relativistic formulation of the particle dynamics which is based on the Dirac equation for fermions (of spin $1/2\hbar$) and the Klein-Gordon (or Proca) equation for the bosons. In the nuclear physics context a commonly used (and flexible) scheme is given by the Lagrangian of the isospin symmetric Quantum-Hadro-Dynamics (QHD) which consists of the free Dirac Lagrangian for the nucleons, the Lagrangian for a scalar field $\sigma(x)$ and a vector field $\omega^\mu(x)$ with self-interactions and an interaction part between bosons and fermions of Yukawa type (in the stationary limit) [1, 2]. As pointed out in Appendix H the QHD model should be considered as an effective approach on the mean-field level that approximates the selfenergies provided by relativistic Dirac-Brueckner theory [3–5]. A reminder and short survey of quantum-hadro-dynamics is given in Appendix H as well as a proof of thermodynamic consistency. This is mandatory for transport theory since a system initially out-of equilibrium asymptotically (for $t \rightarrow \infty$) has to provide the correct equilibrium distribution functions (for localized systems).

In this chapter we will thus formulate covariant transport equations for fermions and bosons, introduce an alternative method for the solution of the covariant collision terms and illustrate solutions in case of heavy-ion collisions at 1 A GeV. Since multi-particle interactions become important for high particle densities—as achieved in relativistic heavy-ion reactions—a covariant formulation of multi-particle ($n \leftrightarrow m$) collisions becomes mandatory which fulfills detailed balance. This task will be addressed in Sect. 3.2 and a suitable solution scheme be pointed out. As an illustration we consider the problem of baryon+antibaryon annihilation into a couple of mesons ($B + \bar{B} \leftrightarrow 3 - 7\pi's$).

© The Author(s), under exclusive license to Springer Nature Switzerland AG 2021
W. Cassing, *Transport Theories for Strongly-Interacting Systems*, Lecture Notes in Physics 989, https://doi.org/10.1007/978-3-030-80295-0_3

3.1   Covariant Transport Equations

We here will base the formulation on the QHD approach which is presented in detail in Appendix H. The QHD approach is an extension of the relativistic mean-field theory proposed first by Serot and Walecka [6–8] and early covariant formulations of Vlasov dynamics in Ref. [9] have been based on the Walecka model as well. Since the coupled field equations (H.7), (H.8) and (H.9) for the nucleon, σ-field and ω-field are somewhat difficult to evaluate on the quantum level, the meson fields are approximated by their expectation values and given by complex values as a function of the nucleon scalar density $\rho_s(x)$ or four-current $j^\mu(x)$. Furthermore, one applies a local-density approximation (LDA) for the meson fields which implies to neglect space-time derivatives for the meson fields $\sigma(x)$ and $\omega^\mu(x)$. As shown in Appendix H one can rearrange the equation of motion for the nucleons with spinor Ψ and write it in the form of the free Dirac equation as

$$\left(\gamma_\mu \left(i\partial^\mu - \Sigma^\mu(x)\right) - \left(M - \Sigma^s(x)\right)\right) \Psi(x) = 0. \tag{3.1}$$

In (3.1) $\Sigma^s(x)$ is the scalar selfenergy and modifies the mass while $\Sigma^\mu(x)$ is the vector selfenergy that modifies the four-momentum of the nucleons in space and time. Since the derivation of relativistic transport equations formally does not depend on the explicit self-couplings of the meson fields we will discard density-dependent couplings in the following as well as self-interactions of the vector field $\omega^\mu(x)$, i.e.

$$\Sigma^s(x) = g_s \, \sigma(x) \tag{3.2}$$

with a coupling strength g_s to the scalar field σ and

$$\Sigma^\mu(x) = g_v \, \omega^\mu(x) = \frac{g_v^2}{m_\omega^2} \, j^\mu(x), \tag{3.3}$$

where $j^\mu(x)$ denotes the baryon four-current, g_v some effective vector coupling and $m_\omega \approx 0.785$ GeV the mass of the ω meson. Note that the four-current $j^\mu(x)$ fulfills a continuity equation

$$\sum_{\mu=0}^{3} \partial_\mu j^\mu(x) = 0 \tag{3.4}$$

and that the spatial integral of j^0

$$B = \int d^3x \, j^0(x) \tag{3.5}$$

gives the conserved baryon number B. This implies that excited states of the nucleons have to be included in the calculation of the four-current.

Eq. (3.1) can be rewritten (in momentum space) as

$$\left(\gamma_\mu \Pi^\mu - M^*\right) \Psi(x) = 0, \tag{3.6}$$

with the effective four-momentum Π^μ and effective mass M^* defined by

$$\Pi_\mu = p_\mu - \Sigma_\mu, \qquad M^* = M - \Sigma^s. \tag{3.7}$$

The equations for positive and negative energy eigenstates read

$$\left(\gamma^\mu \Pi_\mu - M^*\right) u_r^*(\Pi, M^*) = 0, \qquad \left(\gamma^\mu \Pi_\mu + M^*\right) v_r^*(\Pi, M^*) = 0 \tag{3.8}$$

with $u_r^*(\Pi, M^*)$ as the effective spinor for particles and $v_r^*(\Pi, M^*)$ as the effective spinor for antiparticles (with spin $r = \pm$ or $r=1,2$). The Dirac equation then leads to the mass-shell condition

$$\Pi^\mu \Pi_\mu - M^{*2} = 0. \tag{3.9}$$

The effective spinors are obtained by replacing the mass and the energy with their effective values in the free Dirac spinors $u_r(p)$ and $v_r(p)$ and fulfill the relations,[1]

$$\bar{u}_r^*(\Pi) u_s^*(\Pi) = \delta_{rs} = -\bar{v}_r^*(\Pi) v_s^*(\Pi),$$

$$\bar{u}_s^*(\Pi) \gamma^\mu u_s^*(\Pi) = \bar{v}_s^*(\Pi) \gamma^\mu v_s^*(\Pi) = \frac{\Pi^\mu}{M^*}. \tag{3.10}$$

The normalization (3.10) implies that $u_s^{*\dagger}(\Pi) u_s^*(\Pi) = \bar{u}_s^*(\Pi) \gamma^0 u_s^*(\Pi) = \Pi^0/M^*$, which enters the density, carries an additional factor Π^0/M^* as compared to the nonrelativistic case.

The single-particle energies for positive and negative frequencies read in line with (3.7)

$$\epsilon^+(\mathbf{p}) = \sqrt{\Pi^2 + M^{*2}} + \Sigma^0, \tag{3.11}$$

$$\epsilon^-(\mathbf{p}) = -\sqrt{\Pi^2 + M^{*2}} + \Sigma^0. \tag{3.12}$$

---

[1] cf. Appendix G—in the normalization of Bjorken and Drell [10].

In general the field operator $\Psi(x)$ and the Pauli conjugated operator $\bar{\Psi}(x)$ can be expanded in terms of plane waves and effective spinors as[2]

$$\Psi(x) = \int \frac{d^3\Pi}{(2\pi)^{3/2}} \frac{M^*}{E_p^*} \times \sum_{r=1}^{2} [c_r(\Pi) u_r(\Pi, M^*) \exp(-i(\varepsilon^+(\mathbf{p})t - \mathbf{p} \cdot \mathbf{x}))$$

$$+ d_r^\dagger(\Pi) v_r(\Pi, M^*) \exp(i(\varepsilon^-(\mathbf{p})t - \mathbf{p} \cdot \mathbf{x}))], \tag{3.13}$$

$$\bar{\Psi}(x) = \int \frac{d^3\Pi}{(2\pi)^{3/2}} \frac{M^*}{E_p^*} \times \sum_{r=1}^{2} [c_r^\dagger(\Pi) \bar{u}_r(\Pi, M^*) \exp(+i(\varepsilon^+(\mathbf{p})t - \mathbf{p} \cdot \mathbf{x}))$$

$$+ d_r(\Pi) \bar{v}_r(\Pi, M^*) \exp(-i(\varepsilon^-(\mathbf{p})t - \mathbf{p} \cdot \mathbf{x}))], \tag{3.14}$$

with $E_p^* = \sqrt{\Pi^2 + M^{*2}} = \sqrt{(\mathbf{p} - \Sigma)^2 + M^{*2}}$. In Eqs. (3.13), (3.14) the operators $c_r^\dagger, c_r$ and $d_r^\dagger, d_r$ denote creation and annihilation operators for particles and antiparticles, respectively, following the anti-commutator relations

$$\{c_r(\Pi), c_s^\dagger(\Pi')\} = \{d_r(\Pi), d_s^\dagger(\Pi')\} = \frac{E_p^*}{M^*} \delta_{rs} \delta^3(\Pi - \Pi'), \tag{3.15}$$

while all other anti-commutators vanish,

$$\{c_r(\Pi), c_s(\Pi')\} = \{d_r(\Pi), d_s(\Pi')\} = \{c_r^\dagger(\Pi), c_s^\dagger(\Pi')\} = \{d_r^\dagger(\Pi), d_s^\dagger(\Pi')\} = 0, \tag{3.16}$$

as well as the mixed anti-commutators between $c, c^\dagger-$ and $d, d^\dagger-$operators. We recall that the particle number operator for quasiparticles with momentum Π and spin projection r is given by

$$N_r(\Pi) = \frac{M^*}{E_p^*} c_r^\dagger(\Pi) c_r(\Pi). \tag{3.17}$$

Eqs. (3.11), (3.13), (3.14), and (3.17) allow to calculate the quasiparticle density and the scalar density as expectation values of normal ordered operators which read

---

[2] Assuming homogenous media in a local volume of sufficient size.

for spin and isospin symmetric media,[3]

$$\rho_N(x) = 2\langle : \bar{\Psi}\gamma^0\Psi : \rangle = d \int \frac{d^3\Pi}{(2\pi)^3} \left(f_p^*(x, \Pi) - f_a^*(x, \Pi) \right), \qquad (3.18)$$

$$\rho_s(x) = 2\langle : \bar{\Psi}\Psi : \rangle = d \int \frac{d^3\Pi}{(2\pi)^3} \frac{M^*(x)}{E_p^*} \left(f_p^*(x, \Pi) + f_a^*(x, \Pi) \right), \qquad (3.19)$$

with the degeneracy factor $d = 4$ for nucleons. The functions f_p and f_a are the "space-time local" distribution functions for fermionic quasiparticles and antiparticles, respectively.

Exercise 3.1: Prove the expressions (3.18) and (3.19).

The equation for the σ field,

$$\frac{\partial U(\sigma(x))}{\partial \sigma} = g_s\, \rho_s(x), \qquad (3.20)$$

has to be solved selfconsistently with Eq. (3.19) since ρ_s depends again on M^*. The quantity $U(\sigma)$ contains the mass term for the σ field as well as self-interactions of third and fourth order,

$$U(\sigma) = \frac{1}{2}m_\sigma^2\sigma^2 + \frac{1}{3}B\sigma^3 + \frac{1}{4}C\sigma^4 \qquad (3.21)$$

with $m_\sigma \approx 0.55$ GeV denoting the mass of the σ field. Eq. (3.20) then turns to

$$m_\sigma^2\sigma(x) + B\sigma^2(x) + C\sigma^3(x) = g_s\rho_s(x) \qquad (3.22)$$

that has to be solved by iteration on a space-time grid for each space-time cell. The solution of the "gap equation" (3.22) is the extra price to pay for covariant dynamics as compared to the nonrelativistic dynamics.

The further derivation of an on-shell transport approach in principle proceeds as in case of nonrelativistic dynamics when replacing the single-particle Schrödinger equation (multiplied by twice the bare mass $2M$) with the Dirac equation:

$$i2M\frac{\partial}{\partial t}\Psi(x) - \left(\mathbf{p}^2 + 2MU_{eff}(x)\right)\psi(x) = 0 \rightarrow \left(\gamma^\mu\left(i\partial_\mu - \Sigma_\mu(x)\right) - \left(M - \Sigma^s(x)\right)\right)\hat{\Psi}(x) = 0. \qquad (3.23)$$

---

[3] With a factor of 2 for the sum over isospin.

Whereas $\psi(x)$ in the nonrelativistic case is a scalar wavefunction (or two-component Pauli-spinor) we have a spinor-field operator $\hat{\Psi}(x)$ (with four components) in the relativistic case. In the following we will neglect antiparticles, i.e. we drop the terms $\sim d_r, d_r^\dagger$ in (3.13) and (3.14). This is legitimate for heavy-ion collisions up to at least 2 A GeV since the production of nucleon-antinucleon pairs is below threshold.

3.1.1 Wigner Transformation and Gradient Expansion

A one-body (4×4) vector density matrix then is defined by

$$\rho(x, x')^\mu_{\alpha\beta} = \sum_\lambda \langle : \bar{\Psi}_\beta(x') \gamma^\mu_{\alpha\lambda} \Psi_\lambda(x) : \rangle, \tag{3.24}$$

where α, β denote Dirac indices. When taking the trace over the Dirac indices and using (3.10) the diagonal matrix elements $(x = x')$ give the four-current of quasiparticles summed over spin projection r. On the other hand, when replacing γ^μ in (3.24) by the unit matrix, i.e.

$$\langle : \bar{\Psi}_\beta(x') \Psi_\alpha(x) : \rangle = \rho^s(x, x')_{\alpha\beta}, \tag{3.25}$$

and taking the trace over the Dirac indices, the diagonal elements $(x = x')$ give the scalar density of quasiparticles summed over the spin projections r. While (3.24) transforms as a Lorentz four-vector (3.25) transforms as a Lorentz-scalar. In order to keep the further formulation transparent we consider spin-saturated systems and consequently may average over the spin degrees of freedom such that the spin indices may be dropped as in the nonrelativistic case discussed in Chap. 2.

The next step is to perform a four-dimensional Wigner transformation of (3.24), e.g. for ρ^0:

$$\tilde{F}(x, p)_{\alpha\beta} = \int d^4s \, \exp(ip_\mu s^\mu) \, \rho^0 \left(x + \frac{s}{2}, x - \frac{s}{2}\right)_{\alpha\beta}, \tag{3.26}$$

where the Dirac indices $\alpha\beta$ stem from the indices of the spinors as in (3.24). The Wigner transform of convolution integrals is most conveniently written in terms of the relativistic generalization of the Poisson bracket:

$$\sum_{\mu=0}^{3} \left(\frac{\partial F_1}{\partial X_\mu} \frac{\partial F_2}{\partial P^\mu} - \frac{\partial F_1}{\partial P_\mu} \frac{\partial F_2}{\partial X^\mu}\right) =: \{F_1, F_2\}_P. \tag{3.27}$$

This gives for convolution integrals

$$\left(F_1 \tilde{\odot} F_2\right)(x, p) = \int d^4 s \, F_1 \left(x + \frac{s}{2}, y'\right) \odot F_2 \left(y', x - \frac{s}{2}\right) \exp(ip_\mu s^\mu)$$

$$= \exp\left[\frac{i}{2}(\partial_p^{(1)}\partial_x^{(2)} - \partial_x^{(1)}\partial_p^{(2)})\right] \tilde{F}_1\tilde{F}_2, \tag{3.28}$$

which is of infinite order in the phase-space derivatives. Here the operators $\partial_x^{(j)}, \partial_p^{(j)})$ act only on F_j ($j = 1, 2$).

For many-body systems far away from the ground state and close to "classical" systems one may neglect the second and higher order derivatives (as in the nonrelativistic case) and employ the approximation:

$$\left(F_1 \tilde{\odot} F_2\right)_{\alpha\beta}(x, p) \approx \sum_\gamma \tilde{F}_{1\alpha\gamma}(x, p)\tilde{F}_{2\gamma\beta}(x, p)$$

$$+ \frac{i}{2}\sum_\gamma \{F_{1\alpha\gamma}(x, p), F_{2\gamma\beta}(x, p)\}_P. \tag{3.29}$$

The equation of motion for the spinor $\Psi(x)$ (3.1) and $\Psi^\dagger(x') = \bar{\Psi}(x')\gamma^0$ now can be employed for the time evolution of $\rho^0(x, x')$ and subsequently be Wigner transformed (as in the nonrelativistic case). The relativistic BUU (RBUU) equation then reads (for $X = (x + x')/2$)) after taking the trace over the Dirac indices $\tilde{F} = \sum_\alpha \tilde{F}_{\alpha\alpha}$,

$$\sum_\mu \left(\Pi_\mu \partial_X^\mu - [\sum_\nu \Pi_\nu(\partial_\mu^X \Sigma^\nu(X) + M^*(\partial_\mu^x \Sigma^s(X)]\partial_\Pi^\mu\right) \tilde{F}(X, \Pi)$$

$$= M^*(X)\tilde{I}_{coll}(X, \Pi), \tag{3.30}$$

where $\tilde{I}_{coll}(X, \Pi)$ describes the interactions between a quasiparticle at space-time X with effective momentum Π with other particles in the environment. Eq. (3.30) is written in covariant form and differs from the nonrelativistic case by an additional factor Π^0 on the l.h.s. and a factor M^* on the r.h.s. in front of the collision term.

The function $\tilde{F}(X, \Pi)$ is the relativistic phase-space distribution function which is nonvanishing only for $\Pi^2 = M^{*2}$, i.e. on the quasiparticle energy-shell. The usual Wigner function $\tilde{f}(X, \Pi)$ then is obtained from

$$\tilde{F}(X, \Pi) = 2\tilde{f}(X, \Pi)\Theta(\Pi^0)2\pi\delta(\Pi^0 - \epsilon^+(\mathbf{p})), \tag{3.31}$$

where the step function $\Theta(\Pi^0)$ restricts the phase space to positive energy states and the factor of 2 reflects the summation over the 2 spin projections. The function $\tilde{f}(X, \Pi)$ is the analogue to the nonrelativistic phase-space distribution in the VUU equation (2.129). In order to establish the close relation of Eq. (3.30) to the nonrelativistic limit we divide (3.30) by $\Pi^0 = E_p^*$ and use $X = (t, \mathbf{r})$ to obtain

$$
\left(\frac{\partial}{\partial t} + \frac{\Pi}{E_p^*} \cdot \nabla_r - \sum_\mu [\sum_\nu \frac{\Pi_\nu}{E_p^*} (\partial_\mu^X \Sigma^\nu(X) + \frac{M^*}{E_p^*} (\partial_\mu^X \Sigma^s(X)] \partial_\Pi^\mu \right) \tilde{f}(\mathbf{r}, \Pi; t)
$$

$$
= \frac{M^*(X)}{E_p^*} \tilde{I}_{coll}(X, \Pi). \tag{3.32}
$$

The l.h.s. of Eq. (3.32) can be solved by the testparticle Ansatz,

$$
\tilde{f}(\mathbf{r}, \Pi; t) \sim \frac{1}{N_t} \sum_{i=1}^{N_t \cdot A} \delta^3(\mathbf{r} - \mathbf{x}_i(t)) \delta^3(\Pi - \Pi_i(t)) \tag{3.33}
$$

for the number of testparticles per nucleon $N_t \rightarrow \infty$. In (3.33) A denotes the total number of nucleons (baryons) in the system. The equations of motion for the testparticles i then read (dropping the index i):

$$
\frac{dx_k}{dt} = \frac{\Pi_k}{\Pi^0} = \frac{\Pi_k}{E_p^*} = v_k, \tag{3.34}
$$

$$
\frac{dp_k}{dt} = -\frac{\partial \Sigma^0}{\partial x_k} + \sum_{j=1}^3 \frac{\partial \Sigma^j}{\partial x_k} v_j - \frac{M^*}{E_p^*} \frac{\partial \Sigma^s}{\partial x_k} \tag{3.35}
$$

for $k = 1, 2, 3$ with $E_p^* = \sqrt{\Pi^2 + M^{*2}}$, where $\mathbf{p}$ is the generalized momentum. Alternatively, the equations of motion for the quasiparticle kinetic momenta Π_k read,

$$
\frac{d\Pi_k}{dt} = \frac{dp_k}{dt} - \frac{d\Sigma^k}{dt} = -\frac{\partial \Sigma^0}{\partial x_k} - \frac{\partial \Sigma^k}{\partial t} + \sum_{j=1}^3 \left(\frac{\partial \Sigma^j}{\partial x_k} v_j - \frac{\partial \Sigma^k}{\partial x_j} v_j \right) - \frac{M*}{E_p^*} \frac{\partial \Sigma^s}{\partial x_k}
$$

$$
= (\mathbf{E}^s + \mathbf{v} \times \mathbf{B}^s)_k - \frac{M^*}{E_p^*} \frac{\partial \Sigma^s}{\partial x_k} \tag{3.36}
$$

with

$$
\mathbf{E}^s = -\nabla \Sigma^0 - \frac{\partial \Sigma}{\partial t}, \qquad \mathbf{B}^s = \nabla \times \Sigma. \tag{3.37}
$$

In (3.36) one identifies a (repulsive) Lorentz force—stemming from the vector selfenergy—and an attractive scalar contribution that arises from the gradient of the effective mass M^* or the scalar selfenergy Σ^s, respectively. The upper index s

in (3.37) indicates that the Lorentz force is due to the strong interaction. In Sect. 4.6 we will encounter another Lorentz force due to the electromagnetic interaction.

The collision term can be worked out in leading order in the interaction by using the spinor expansion (3.13) and (3.14) and dropping the terms $\sim d_r, d_r^\dagger$. Assuming again a homogenous system in some "local" space-time cell of volume $\Delta t \Delta V$ around X—as in the nonrelativistic case—one finally ends up with ($\Pi = \Pi_1$),

$$
I_{coll}(X, \Pi_1) = \sum_j \int \frac{d^3 \Pi_j}{(2\pi)^3} \frac{M_j^*}{E_j^*} \int \frac{d^3 \Pi_3}{(2\pi)^3} \frac{M_3^*}{E_3^*} \int \frac{d^3 \Pi_4}{(2\pi)^3} \frac{M_4^*}{E_4^*}
$$

$$
\times |\mathcal{M}(\Pi_1, \Pi_2, \Pi_3, \Pi_4)|^2 (2\pi)^4 \delta^4(\Pi_1 + \Pi_2 - \Pi_3 - \Pi_4)
$$

$$
\times \left(\tilde{f}_3 \tilde{f}_4 (1 - \tilde{f}_1)(1 - \tilde{f}_j) - \tilde{f}_1 \tilde{f}_j (1 - \tilde{f}_3)(1 - \tilde{f}_4) \right), \tag{3.38}
$$

where the shorthand notation $\tilde{f}_j = \tilde{f}_j(X, \Pi_j)$ has been used and the sum over j runs over all baryon species j. For spin-isospin symmetric matter (without excited nucleon states) the sum over j just gives a degeneracy factor $d=4$.[4]

The Lorentz-invariant matrix element squared $|\mathcal{M}(\Pi_1, \Pi_2, \Pi_3, \Pi_4)|^2$ should be evaluated for the local quasiparticles in terms of a coupled-channel Dirac-Brueckner approach and be antisymmetrized in case of identical fermions, however, this is quite demanding and out-of scale presently. In practice one adopts matrix elements from nuclear reactions in vacuum as a function of the invariant energy $\sqrt{s}$ and shifts the invariant energy in line with the selfenergies of the particles involved in the reaction. When applied to nucleus-nucleus collisions in the range of 100 A MeV to about a few A GeV this strategy should work sufficiently well since the scattering is dominated by invariant energies far from threshold. The actual scattering is evaluated in the center-of-mass of the colliding nucleons where their four-momenta Π_1, Π_2 are given by a Lorentz transformation with the individual center-of-mass velocity $\mathbf{v}_{cm}$. The final state is selected by Monte Carlo following energy-momentum conservation and an "experimental" angular distribution. Then the final four-momenta Π_3, Π_4 are boosted back to the calculational frame. In this way the scattering processes do not depend on the actual calculational frame and conserve energy-momentum in any frame. Furthermore, the selfenergies Σ^s and Σ^μ transform as a Lorentz-scalar or Lorentz-vector, respectively. In this way one achieves a relativistic transport approach with well-defined transformation properties with respect to Lorentz transformations.

---

[4] For readers interested in the explicit spin dynamics we refer to Ref. [11].

It is instructive to compare the relativistic collision term (3.38) to the nonrelativistic case (2.122),

$$
I_{coll}^{nr}(X, \mathbf{p}_1; t) = (2s + 1)(2\tau + 1) \int \frac{d^3 p_2}{(2\pi)^3} \frac{d^3 p_3}{(2\pi)^3} \frac{d^3 p_4}{(2\pi)^3}
$$

$$
\times\, 2\pi \delta(\frac{1}{2m}[p_1^2 + p_2^2 - p_3^2 - p_4^2])
$$

$$
\times\, (2\pi)^3 \delta^3(\mathbf{p}_1 + \mathbf{p}_2 - \mathbf{p}_3 - \mathbf{p}_4)\, v(\mathbf{p}_2 - \mathbf{p}_4) v_A(\mathbf{p}_4 - \mathbf{p}_2)
$$

$$
\times\, (n(X, \mathbf{p}_3; t) n(X, \mathbf{p}_4; t) \bar{n}(X, \mathbf{p}_1; t) \bar{n}(X, \mathbf{p}_2; t)
$$

$$
- n(X, \mathbf{p}_1; t) n(X, \mathbf{p}_2; t) \bar{n}(X, \mathbf{p}_3; t) \bar{n}(X, \mathbf{p}_4; t)), \qquad (3.39)
$$

where $n(X, \mathbf{p}_i; t)$ denote the "local" occupation numbers of states with momentum $\mathbf{p}$ and $\bar{n}_i = 1 - n_i$ in a space-time cell located around X.[5] Furthermore, the factor $(2s + 1)(2\tau + 1)$ results from summation over spin and isospin giving a factor of $d = 4$ for spin and isospin symmetric nuclear matter.

Formally the relativistic version is obtained by the substitution

$$
\int d^3 p \to \int d^3 \Pi \frac{M^*}{E_p^*}, \qquad (3.40)
$$

which gives a Lorentz-invariant integral over the (quasi-) particle momenta via the (dimensionless) factors M^*/E_p^*. The latter are the consequence of γ^0 matrix elements with the effective spinors since $\Psi^\dagger(x)\Psi(x) = \bar{\Psi}(x)\gamma^0\Psi(x)$ in the relativistic case. The separate δ-functions for energy and momentum also merge to a single δ^4-function for the sum of the four-momenta.

When using an alternative convention for the spinor normalization, which implies to replace the Lorentz-invariant integrations $\int (d^3\Pi M^*)/E_p^*$ by $\int (d^3\Pi)/(2E_p^*)$ (as in case of bosons) one obtains

$$
I_{coll}(X, \Pi_1) = \sum_j \int \frac{d^3\Pi_j}{(2\pi)^3 2E_j^*} \int \frac{d^3\Pi_3}{(2\pi)^3 2E_3^*} \int \frac{d^3\Pi_4}{(2\pi)^3 2E_4^*}
$$

$$
\times\, |\tilde{\mathcal{M}}(\Pi_1, \Pi_2, \Pi_3, \Pi_4)|^2\, (2\pi)^4 \delta^4(\Pi_1 + \Pi_2 - \Pi_3 - \Pi_4)
$$

$$
\times \left(\tilde{f}_3 \tilde{f}_4 (1 - \tilde{f}_1)(1 - \tilde{f}_j) - \tilde{f}_1 \tilde{f}_j (1 - \tilde{f}_3)(1 - \tilde{f}_4) \right), \qquad (3.41)
$$

---

[5] Note that a local nonrelativistic potential U(X) drops out in the energy conserving δ-function.

where the matrix element $\tilde{\mathcal{M}}(\Pi_1, \Pi_2, \Pi_3, \Pi_4)$ has to be evaluated in the different normalization. Note, however, that the prefactor of the collision term in (3.32) changes, too,

$$\frac{M^*}{E_p^*} \to \frac{1}{2E_p^*}. \tag{3.42}$$

We will adopt the convention (3.41) when treating many-body interactions of fermions and bosons in Sect. 3.2 on the same footing.

3.1.2   Nuclear Equation of State in the QHD Model

In the QHD model the energy-momentum tensor,

$$T^{\mu\nu} = \frac{\partial \mathcal{L}}{\partial\left(\partial_\mu \Psi\right)} \frac{\partial \Psi}{\partial x_\nu} - g^{\mu\nu} \mathcal{L}, \tag{3.43}$$

can be evaluated in a straight forward manner, where the energy density $\mathcal{E}$ and the pressure P of a system are given as normal ordered expectation values from the diagonal elements of the tensor,

$$\mathcal{E} = \langle: T^{00} :\rangle, \tag{3.44}$$

$$P = \langle: T^{ii} :\rangle = \frac{1}{3}\sum_{i=1}^{3} \langle: T^{ii} :\rangle. \tag{3.45}$$

For systems in thermal and chemical equilibrium at rest (with $\Pi = \mathbf{p}$ and $\Sigma = 0$) the energy density in mean-field approximation is given by [12]

$$\mathcal{E} = U(\sigma) + \frac{m_\omega^2}{2g_v^2}\Sigma_0^2 + E_0(T, \mu^*, M^*). \tag{3.46}$$

In (3.46) E_0 is the energy density for a non-interacting particle system evaluated at the effective chemical potential $\mu^* = \mu - \Sigma_0$ and with the effective mass $M^* = M_0 - \Sigma^s$, i.e.

$$E_0(T, \mu^*, M^*) = d \int \frac{d^3 p}{(2\pi)^3}\, E_p^* \left(n_F(T, \mu^*, M^*) + n_{\bar{F}}(T, \mu^*, M^*) \right), \tag{3.47}$$

where n_F and $n_{\bar{F}}$ denote the equilibrium Fermi distribution functions for non-interacting particles and antiparticles, respectively.

The pressure in mean-field approximation reads

$$P = -U(\sigma) + \frac{m_\omega^2}{2g_v^2}\Sigma_0^2 + \frac{2}{3}\sum_{i=1}^{3}\langle : i\bar{\Psi}\gamma^i\partial_i\Psi : \rangle , \tag{3.48}$$

where the factor of 2 in front of the sum stems from summation over isospin. This gives

$$P = -U(\sigma) + \frac{m_\omega^2}{2g_v^2}\Sigma_0^2 + P_0(T, \mu^*, M^*), \tag{3.49}$$

where P_0 is the pressure for non-interacting particles with the effective quantities μ^* and M^* at temperature T,

$$P_0(T, \mu^*, M^*) = \frac{d}{3} \int \frac{d^3p}{(2\pi)^3} \frac{\mathbf{p}^2}{E_p^*} \left(n_F(T, \mu^*, M^*) + n_{\bar{F}}(T, \mu^*, M^*)\right). \tag{3.50}$$

> Exercise 3.2: Derive the expressions (3.46) and (3.49) for systems in thermal equilibrium.

In the following we give an example for a relativistic mean-field calculation with respect to the nuclear equation of state (EoS) for spin and isospin symmetric nuclear matter at vanishing temperature T. Then only nucleons contribute to the energy density and the baryon density is equal to the nucleon density, i.e. $\rho_B = \rho_N$. The binding energy per nucleon E_B/A is defined by

$$\frac{E_B}{A} = \frac{\mathcal{E}(\mu^*)}{\rho_N(\mu^*)} - M_0, \tag{3.51}$$

where M_0 denotes the nucleon mass in vacuum, $\rho_N(\mu^*)$ the nucleon density for the effective chemical potential and $\mathcal{E}(\mu^*)$ the energy density (3.46).

Figure 3.1 displays E_B/A as a function of the nucleon density ρ_N for the parameters $m_s = 550$ MeV, $m_v = 783$ MeV, $g_s = 9.39$, $g_v = 11$, $B_s = 2.95$ fm and $C_s = 7.83$ fm^2 showing a minimum at $\rho_N = \rho_0 \approx 0.168$ fm^{-3} of -16 MeV with an incompressibility

$$K := 9\rho_0^2 \frac{d^2 E_B/A}{d^2\rho_N}|_{\rho_0}, \tag{3.52}$$

which gives $K \approx 375$ MeV in this case and corresponds to a "hard" equation of state (denoted by NL6). By varying the parameters one can also describe an EoS with the same minimum but a lower incompressibility K. Furthermore, the effective

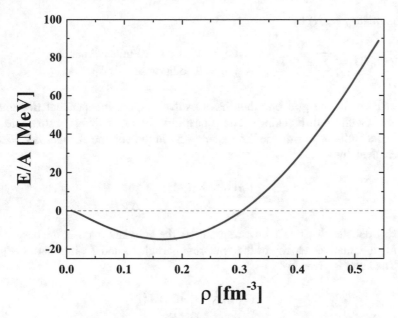

Fig. 3.1 The equation of state (EoS) for infinite nuclear matter, i.e. the binding energy per nucleon as a function of the nucleon density $\rho = \rho_N$, at vanishing temperature T for the parameter set NL6

mass at ρ_0 is $M^* \approx 0.65 \, M_0$ which is another value providing information on the momentum dependence of the Schrödinger-equivalent potential

$$U_{SEP}(\mathbf{p}) = \Sigma_s + \Sigma_0 + \frac{1}{2M_0}(\Sigma_s^2 - \Sigma_0^2) + \frac{\Sigma_0}{M_0}\epsilon_{kin}(\mathbf{p}) \tag{3.53}$$

with the kinetic energy $\epsilon_{kin}(\mathbf{p}) = \sqrt{\mathbf{p}^2 + M_0^2} - M_0$ that is controlled by experimental data on elastic proton-nucleus scattering.

3.1.3 The Local-Ensemble Method for the Solution of Binary Collision Terms

Before coming to actual results for relativistic heavy-ion collisions we report on an alternative solution of the collision integral which is denoted as the "local-ensemble" method [13]. We start with the following assumption: Let the phase-space density $f(x, \Pi)$ be a slowly varying function (on an appropriately small chosen scale) of the four-vector x. This assumption has already been made by discretizing the Vlasov-part of the relativistic transport equation when computing the fields only for discrete space-time cells. At each time-step we can therefore approximate $f(\mathbf{r}, \Pi)$ by

$$\tilde{f}(\mathbf{r}, \Pi) := \sum_i \delta_i f_i(\Pi) \qquad \begin{cases} \delta_i = 1, \text{ if } \mathbf{r} \in \text{ volume element } i \\ \delta_i = 0, \text{ otherwise} \end{cases} . \qquad (3.54)$$

At each cell of the grid one thus has to solve a space-independent Boltzmann-equation (with Pauli blocking). The probability for one pair of "testparticles" to undergo a collision during the time interval Δt in the volume element $\Delta V = \Delta^3 x$ is then given by

$$W = \frac{\sigma_B(\sqrt{s}, M_1, M_2)}{N} v_{\text{rel}} \frac{\Delta t}{\Delta V}. \qquad (3.55)$$

Here σ_B denotes the (total) cross section, $\sqrt{s}$ the invariant mass of the baryon pair and v_{rel} the relative velocity of the scattering particles 1 and 2 with masses M_1 and M_2. The latter is given by

$$v_{\text{rel}} = \frac{\lambda^{1/2}(s, M_1^2, M_2^2)}{2 E_1^* E_2^*} \qquad (3.56)$$

with $\lambda(x, y, z) = (x - y - z)^2 - 4yz$.

Out of the $n(n - 1)/2$ possible pairs (n being the number of "testparticles" in the cell $\Delta^3 x = \Delta V$) we choose at random $\lfloor n/2 \rfloor$ collision pairs. We therefore have to replace the probability W by W',

$$W' := W \frac{n(n - 1)/2}{\lfloor n/2 \rfloor} \qquad (3.57)$$

in order to obtain the correct total transition rate in the cell. In the limit $\Delta V \to 0, \Delta t \to 0, N \to \infty$ the solutions obtained by this method will converge to the exact solutions of the Boltzmann equation [14]. Furthermore, this prescription is evidently covariant. Since we are dealing with transition-rates and do not employ a geometrical interpretation of the parallel-ensemble method as described in Sect. 2.6, no problems connected with the time-ordering of the collision processes occur. Furthermore, this method can be extended to multi-particle transitions as addressed in Sect. 3.2.

3.1.4   Application of RBUU to $Au + Au$ Collisions at 1 A GeV

As an example we consider $Au + Au$ collisions at a bombarding energy of 1 A GeV at an impact parameter of $b= 7$ fm which is about the radius of a Au-nucleus. Accordingly the nuclei are initially shifted by ± 3.5 fm in x-direction—apart from the initial distance in z-direction—and boosted towards each other. In the initial

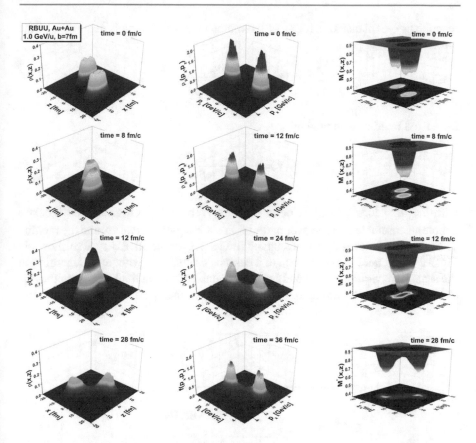

Fig. 3.2 (Left column) The baryon density distribution $\rho(x, y = 0, z; t)$ for a $Au + Au$ collision at 1 A GeV and impact parameter $b = 7$ fm at different times. (Middle column) The momentum distribution $\rho_p(p_x, p_z; t)$ for the same reaction at different times (integrated over p_y). (Right column) The effective mass $M^*(x, y = 0, z; t)$ for the same reaction at different times

state the nuclei then are Lorentz contracted in z-direction by the factor $1/\gamma_{cm} \approx 0.82$ while the initial momentum distributions are elongated in p_z-direction by the factor $\gamma_{cm} \approx 1.22$. During the course of the reaction the nuclei overlap, become compressed in the overlap region and the nucleons start scattering. The "spectators" or non-interacting nucleons propagate approximately with their original speed and separate in coordinate space from the interaction zone for large times (cf. l.h.s. of Fig. 3.2). The middle column shows the momentum distribution of the baryons— integrated over p_y—as a function of time. The initially separated Fermi distributions mix due to scattering in the overlap region and populate the momentum distribution for lower momenta in the center-of-mass system. Moreover, the momentum distribution becomes tilted in p_x-direction thus showing a momentum anisotropy in the reaction $(x - z)$ plane. This tilting of the momentum distribution is quantified by the directed flow

$$v_1(y) := \langle \frac{p_x}{\sqrt{p_x^2 + p_y^2}} \rangle_y \tag{3.58}$$

for a fixed rapidity y defined by

$$y = \frac{1}{2} \ln(\frac{1+\beta}{1-\beta}) \tag{3.59}$$

with the final longitudinal velocity $\beta = p_z / \sqrt{\mathbf{p}^2 + M_0^2}$. In (3.58) the brackets denote an ensemble average over particles in some small interval around the rapidity y. The directed flow $v_1(y)$ is displayed in Fig. 3.3 for the RBBU system shown in Fig. 3.2 (solid line) in comparison to a cascade calculation (without selfenergies) at the same bombarding energy and impact parameter (dashed line). It is seen that the directed flow close to midrapidity ($y \approx 0$) can be further quantified by the derivative

$$F_1 = \frac{dv_1(y)}{dy}|_{y=0}, \tag{3.60}$$

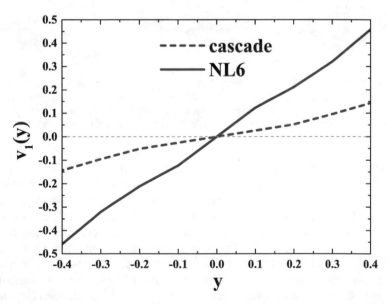

Fig. 3.3 The directed flow $v_1(y)$ for nucleons as a function of rapidity y for a $Au+Au$ collision at 1 A GeV and impact parameter $b=7$ fm. The solid line shows the result from the RBUU calculation for the parameter set NL6 while the dashed line displays the result for a cascade calculation (without selfenergies)

which is significantly larger in size for the calculation including the scalar and vector selfenergies (solid line) than without selfenergies. Accordingly experimental measurements of the proton directed flow (or 'bounce-off') provide constraints on the size of the selfenergies employed.

The sizeable directed flow $v_1(y)$ results from the interplay of the scalar and vector selfenergies which are working in opposite directions. The effect of the scalar selfenergy is seen directly in the effective mass $M^*(\mathbf{r}; t)$ which is shown for the same system in the right column of Fig. 3.2. In the center of the impinging nuclei the effective mass drops by about 1/3 due to the scalar selfenergy of about -0.33 GeV; this drop of M^* becomes larger in the overlap region at t= 10–15 fm/c and vanishes for late times because the reaction zone disintegrates and the spectator nucleons become bound again to nuclei of smaller mass.

The impact of the selfenergies on the collective expansion becomes visible also in the elliptic flow defined by

$$v_2(y) := \langle \frac{p_x^2 - p_y^2}{p_x^2 + p_y^2} \rangle_y \tag{3.61}$$

which is displayed in Fig. 3.4 for the RBBU system shown in Fig. 3.2 (solid line) in comparison to a cascade calculation at the same bombarding energy and impact parameter (dashed line). Again the effect of the selfenergies is sizeable and reflects the collective acceleration by the Lorentz and scalar forces on the particles. The

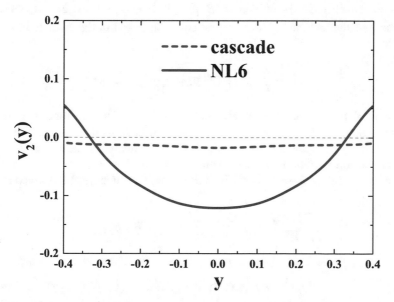

Fig. 3.4 The elliptic flow $v_2(y)$ for nucleons as a function of rapidity y for a $Au + Au$ collision at 1 A GeV and impact parameter b=7 fm. The solid line shows the result from the RBUU calculation for the parameter set NL6 while the dashed line displays the result for a cascade calculation

negative elliptic flow is also denoted as "squeeze-out" since the particles are pushed in a direction perpendicular to the reaction plane. For further applications of the relativistic BUU approach we refer the reader to the reviews [15, 16].

In summarizing this section we have achieved a covariant formulation of the transport equations on the basis of an effective Lagrangian model (QHD) and pointed out the importance of strong scalar and vector selfenergies for the buildup of collective flow in the colliding systems. However, this model does not allow to compute the matrix elements for the two-body interactions in a consistent fashion and is based on the on-shell quasiparticle picture which loses its validity due to a sizeable scattering width of the particles at higher bombarding energy (and baryon density). Furthermore, with increasing bombarding energy inelastic baryon-baryon collisions become dominant such that final channels with more than two particles $(2 \rightarrow n)$ become important. The question then arises how to treat the backward channels $(n \rightarrow 2)$ in order to maintain detailed balance and the correct equilibrium state. This issue will be addressed in the next section.

3.2 Multi-Particle Transitions

For the discussion of the general on-shell collision terms the hadron selfenergies will be discarded for transparency which implies to replace the four-momentum (E_p^*, Π) by $p = (E, \mathbf{p})$ and M^* by M for fermions with the on-shell energy $E = \sqrt{\mathbf{p}^2 + M^2}$. Furthermore, we will also neglect selfenergies for the mesons such that their on-shell four-momenta can also be written as $(E, \mathbf{p})$ with $E = \sqrt{\mathbf{p}^2 + M_h^2}$ and M_h denoting the mass of the hadron h. The transport equation (3.41) in the cascade limit then reduces to

$$(\partial_t + \frac{\vec{p}}{E} \cdot \vec{\partial}_r) f_i(X, \mathbf{p}) = \frac{1}{2E} I_{coll}^i(X, \mathbf{p}), \qquad (3.62)$$

where $f_i(X, \mathbf{p})$ is the on-shell phase-space distribution function for a hadron with quantum numbers i. It describes the free flow of hadrons that interact with each other during the propagation. In order to describe fermions and bosons on the same footing the normalization for baryon spinors is adopted as in (3.41), i.e. the on-shell collision integral for $2 \leftrightarrow 2$ interactions can be written (in case of fermions) as

$$I_{coll}^{i-on}[2 \leftrightarrow 2] = \sum_j \sum_{k,l} \int \frac{d^3 p_2}{(2\pi)^3 2E_2} \frac{d^3 p_3}{(2\pi)^3 2E_3} \frac{d^3 p_4}{(2\pi)^3 2E_4}$$

$$\times W_{2,2}(p, p_2; i, j \mid p_3, p_4; k, l) (2\pi)^4 \delta^4(p^\mu + p_2^\mu - p_3^\mu - p_4^\mu)$$

$$\times [f_k(X, \mathbf{p}_3) f_l(X, \mathbf{p}_4)(1 - f_i(X, \mathbf{p}))(1 - f_j(X, \mathbf{p}_2))$$

$$- f_i(X, \mathbf{p}) f_j(X, \mathbf{p}_2)(1 - f_k(X, \mathbf{p}_3))(1 - f_l(X, \mathbf{p}_4))], \qquad (3.63)$$

where the on-shell energy p_k^0 of particle k is rewritten as E_k. Here we have replaced the matrix element (squared) by $W_{2,2}(p, p_2; i, j \mid p_3, p_4; k, l)$ indicating explicitly the discrete quantum numbers of the initial states (i, j) and final states (k, l) to allow for a suitable formulation of n-body transitions.

The on-shell version of the collision integral for $(2 \leftrightarrow m)$ interactions then reads in straight forward generalization (for nonidentical particles):

$$
I_{coll}^{i-on}[2 \leftrightarrow m] = \sum_\nu \sum_\lambda \int \frac{d^3 p_j}{(2\pi)^3 2E_j} \prod_{k=1}^m \frac{d^3 p_k}{(2\pi)^3 2E_k}
$$

$$
\times W_{2,m}(p, p_j; i, \nu \mid p_k; \lambda) (2\pi)^4 \, \delta^4(p^\mu + p_j^\mu - \sum_{k=1}^m p_k^\mu)
$$

$$
\times [\bar{f}_i(X, \mathbf{p}) \bar{f}_j(X, \mathbf{p}_j) \prod_{k=1}^m f_k(X, \mathbf{p}_k)
$$

$$
- f_i(X, \mathbf{p}) f_j(X, \mathbf{p}_j) \prod_{k=1}^m \bar{f}_k(X, \mathbf{p}_k)]. \tag{3.64}
$$

The discrete quantum numbers for particle 2 with momentum $\mathbf{p}_j$ are denoted by ν while those for the final states are abbreviated by λ while the transition matrix element (squared) is denoted by $W_{2,m}(p, p_j; i, \nu \mid p_k; \lambda)$ which should be evaluated within a suitable many-body approach. However, for $2 \rightarrow m$ transitions these may be extracted to some extend from two-body scattering data (see below). Furthermore, we have used

$$
\bar{f}_j(X, \mathbf{p}_k) = 1 + \eta f_j(X, \mathbf{p}_k) \tag{3.65}
$$

with $\eta = -1$ for fermions and $\eta = 1$ for bosons to allow for a simultaneous description of fermions and bosons in the initial and final states.[6]

The extension to $n \leftrightarrow m$ transitions gives,

$$
I_{coll}^{i-on}[n \leftrightarrow m] = \sum_\nu \sum_\lambda \int \prod_{j=2}^n \frac{d^3 p_j}{(2\pi)^3 2E_j} \prod_{k=1}^m \frac{d^3 p_k}{(2\pi)^3 2E_k}
$$

$$
\times W_{n,m}(p, p_j; i, \nu \mid p_k; \lambda) (2\pi)^4 \, \delta^4(p^\mu + \sum_{j=2}^n p_j^\mu - \sum_{k=1}^m p_k^\mu)
$$

---

[6] Note that in case of identical bosons of type i there are additional statistical factors of $1/n_i!$ in the collision integral.

$$\times [\bar{f}_i(X, \mathbf{p}) \prod_{j=2}^{n} \bar{f}_j(X, \mathbf{p}_j) \prod_{k=1}^{m} f_k(X, \mathbf{p}_k)$$

$$-f_i(X, \mathbf{p}) \prod_{j=2}^{n} f_j(X, \mathbf{p}_j) \prod_{k=1}^{m} \bar{f}_k(X, \mathbf{p}_k)]. \tag{3.66}$$

For large times ($t \to \infty$) all collision integrals will vanish in equilibrium, which implies that "gain" and "loss" terms become equal in magnitude.

The number of reactions in the covariant 4-volume $d^3 r dt = dV dt$ is obtained by dividing the gain and loss terms in the collision integrals by twice the energy $2p_0 = 2E_1$ (according to Eq. (3.62)), integrating over $d^3 p/(2\pi)^3$ and summing over the discrete quantum numbers i. For the case of fermion two-body collisions this gives (using $\mathbf{p} = \mathbf{p}_1$) for the "loss" term

$$\frac{dN_{coll}[2 \to 2]}{dt dV} = \sum_{i,j} \sum_{k,l} \int \frac{d^3 p_1}{(2\pi)^3 2E_1} \frac{d^3 p_2}{(2\pi)^3 2E_2} \frac{d^3 p_3}{(2\pi)^3 2E_3} \frac{d^3 p_4}{(2\pi)^3 2E_4}$$

$$\times W_{2,2}(p_1, p_2; i, j \mid p_3, p_4; k, l) \, (2\pi)^4 \, \delta^4(p_1^\mu + p_2^\mu - p_3^\mu - p_4^\mu)$$

$$\times [f_i(X, \mathbf{p}_1) f_j(X, \mathbf{p}_2)(1 - f_k(X, \mathbf{p}_3))(1 - f_l(X, \mathbf{p}_4))], \tag{3.67}$$

where X is within the space-time volume $dt dV$. In case of $n \to m$ processes this leads to

$$\frac{dN_{coll}[n \to m]}{dt dV} = \sum_{i,\nu} \sum_{\lambda} \int \prod_{j=1}^{n} \frac{d^3 p_j}{(2\pi)^3 2E_j} \prod_{k=1}^{m} \frac{d^3 p_k}{(2\pi)^3 2E_k}$$

$$\times W_{n,m}(p_j; i, \nu \mid p_k; \lambda) \, (2\pi)^4 \, \delta^4(\sum_{j=1}^{n} p_j^\mu - \sum_{k=1}^{m} p_k^\mu)$$

$$\times (\prod_{j=1}^{n} f_j(X, \mathbf{p}_j) \prod_{k=1}^{m} \bar{f}_k(X, \mathbf{p}_k)) \tag{3.68}$$

and in case of $m \to n$ processes to

$$\frac{dN_{coll}[m \to n]}{dt dV} = \sum_{i,\nu} \sum_{\lambda} \int \prod_{j=1}^{n} \frac{d^3 p_j}{(2\pi)^3 2E_j} \prod_{k=1}^{m} \frac{d^3 p_k}{(2\pi)^3 2E_k}$$

$$\times W_{n,m}(p_j; i, \nu \mid p_k; \lambda) \, (2\pi)^4 \, \delta^4 \sum_{j=1}^{n} p_j^\mu - \sum_{k=1}^{m} p_k^\mu$$

$$\times \left(\prod_{k=1}^{m} f_k(X, \mathbf{p}_k) \prod_{j=1}^{n} \bar{f}_j(X, \mathbf{p}_j) \right). \tag{3.69}$$

In high energy nucleus-nucleus collisions, which are dominated by produced mesons of low phase-space density, the Pauli blocking or Bose-enhancement terms $\tilde{f}_k$ are ≈ 1, which implies to replace the quantum statistical ensembles by classical ones. In this limit the integrals over the final momenta can be carried out provided that the transition probabilities $W_{n,m}$ do not sensitively depend on the final momenta p_k and essentially only depend on the invariant energy $\sqrt{s}$. For example, for proton-antiproton annihilation, where the final state on average consists of 5 pions, this is a reasonable approximation.

Employing the definition of the n-body phase-space integrals for total 4-momentum P^μ [17],

$$R_n(P^\mu; M_1, .., M_n) = \int \prod_{k=1}^{n} \frac{d^3 p_k}{(2\pi)^3 2E_k} (2\pi)^4 \, \delta^4(P^\mu - \sum_{j=1}^{n} p_j^\mu), \qquad (3.70)$$

one obtains the recursion relation (cf. Appendix E for a detailed derivation)

$$R_n(P^\mu, M_1, .., M_n) = \int \frac{d^3 p_n}{(2\pi)^3 2E_n} R_{n-1}\left(P^\mu - p_n^\mu; M_1, .., M_{n-1}\right). \qquad (3.71)$$

Note, that the phase-space integrals are of dimension GeV^{2n-4} or $(1/\mathrm{fm})^{2n-4}$. Inserting (3.70) this gives in case of $n \to m$ processes

$$\frac{dN_{coll}[n \to m]}{dt \, dV} = \sum_{i,\nu} \sum_\lambda \int \left(\prod_{j=1}^{n} \frac{d^3 p_j}{(2\pi)^3 2E_j}\right) W_{n,m}(P)$$

$$\times R_m(P^\mu = \sum_{j=1}^{n} p_j^\mu; M_1, .., M_m) \prod_{j=1}^{n} f_j(X, \mathbf{p}_j)$$

$$= \sum_{i,\nu} \sum_\lambda \int \left(\prod_{j=1}^{n} \frac{d^3 p_j}{(2\pi)^3}\right) P(n \to m)_{i,\nu}^\lambda \left(\prod_{j=1}^{n} f_j(X, \mathbf{p}_j)\right),$$

$$(3.72)$$

where $M_1, .., M_m$ stand for the masses in the final state, and in case of $m \to n$ processes

$$\frac{dN_{coll}[m \to n]}{dt \, dV} = \sum_{i,\nu} \sum_\lambda \int \left(\prod_{k=1}^{m} \frac{d^3 p_k}{(2\pi)^3 2E_k}\right) W_{n,m}(P)$$

$$\times R_n(P^\mu = \sum_{k=1}^{m} p_k^\mu; M_1, \ldots, M_n) \left(\prod_{k=1}^{m} f_k(X, \mathbf{p}_k)\right)$$

$$= \sum_{i,v} \sum_{\lambda} \int \left(\prod_{k=1}^{m} \frac{d^3 p_k}{(2\pi)^3} \right) P(m \to n)_{\lambda}^{i,v} \left(\prod_{k=1}^{m} f_k(X, \mathbf{p}_k) \right).$$

(3.73)

For fixed sets of quantum numbers (i, v) and λ in the initial and final states this relates the integrands $P(n \to m)$ in (3.72) and $P(m \to n)$ in (3.73) for individual scatterings as (dropping the indices for quantum numbers):

$$\frac{P(m \to n)}{P(n \to m)} = \left[\prod_{k=1}^{m} \frac{1}{2E_k} \right] \left[\prod_{j=1}^{n} 2E_j \right] \frac{R_n(P^{\mu} = \sum_{k=1}^{n} p_k^{\mu}; M_1, \dots, M_n)}{R_m(P^{\mu} = \sum_{j=1}^{m} p_j^{\mu}; M_1, \dots, M_m)},$$

(3.74)

if $W_{n,m}$ essentially depends only on the invariant energy $\sqrt{s} = \sqrt{P^2}$. Note, that the r.h.s. of (3.74) is in units of $\text{GeV}^{3(n-m)}$ or $\text{fm}^{3(m-n)}$ such that a factor $(dV)^{n-m}$ is needed to interpret the quantities as relative "probabilities." Thus, once the transition probabilities $W_{n,m}$ are known as a function of $\sqrt{s}$ for a given set of quantum numbers, the integrand $P(n \to m)$ in (3.72) is determined by phase space and the backward reactions in (3.73) are fixed by Eq. (3.74).

3.2.1 Baryon-Antibaryon Annihilation and Recreation

As an example for multi-particle production we consider the processes $B\bar{B} \leftrightarrow m$ mesons, which are of relevance for annihilation of antibaryons on baryons and the recreation of $B\bar{B}$ pairs by m *meson* interactions. The 4-differential collision rate for baryon-antibaryon annihilation $(1 + 2 \to 3, \dots, m + 2)$ then is given by

$$\frac{dN_{coll}[B\bar{B} \to m \text{ mesons}]}{dt dV} = \sum_{i,j} \sum_{\lambda_m} \int \frac{d^3 p_1}{(2\pi)^3 2E_1} \frac{d^3 p_2}{(2\pi)^3 2E_2}$$
$$\times W_{2,m}(P = p_1 + p_2; i, j; \lambda_m)$$
$$\times R_m(P^{\mu}; M_3, \dots, M_{m+2}) f_i(X, \mathbf{p}_1) f_j(X, \mathbf{p}_2),$$

(3.75)

where f_i and f_j denote the baryon and antibaryon phase-space distributions, respectively. The integrand is related to the annihilation cross section $\sigma_{ann.}(\sqrt{s})$ for a baryon-antibaryon pair with quantum numbers i, j as

$$\sum_{m} \sum_{\lambda_m} W_{2,m}(p_1 + p_2; i, j; \lambda_m) R_m(p_1^{\mu} + p_2^{\mu}; M_3, \dots, M_{m+2})$$

$$= 2\sqrt{\tilde{\lambda}(s, M_1^2, M_2^2)} \, \sigma_{ann.}(\sqrt{s}) = 4E_1 E_2 \, v_{rel} \, \sigma_{ann.}(\sqrt{s})$$

(3.76)

with the relativistic relative velocity

$$v_{rel} = \frac{\sqrt{\tilde{\lambda}(s, M_1^2, M_2^2)}}{2E_1 E_2}, \tag{3.77}$$

involving

$$\tilde{\lambda}(x, y, z) = (x - y - z)^2 - 4yz. \tag{3.78}$$

In (3.76) the sum runs over the final meson multiplicity ($m \approx 3, \ldots, 9$) in the final state and over λ_m which denotes all discrete quantum numbers of the final mesons for given multiplicity m.

Note, that by summing (3.75) additionally over m, but keeping the quantum numbers i, j fixed, one arrives at

$$\frac{dN_{coll}^{i,j}}{dtdV} = \int \frac{d^3 p_1}{(2\pi)^3} \frac{d^3 p_2}{(2\pi)^3} \, v_{rel}(p_1, p_2) \, \sigma_{ann}\left(\sqrt{s}\right) \, f_i(X, \mathbf{p}_1) f_j(X, \mathbf{p}_2). \tag{3.79}$$

If the product of the relative velocity and the cross section, i.e. $v_{rel}\sigma_{ann}$, is approximately constant (see below) the integrals over the momenta in (3.79) give the classical Boltzmann limit

$$\frac{dN_{coll}^{i,j}}{dtdV} = < v_{rel} \, \sigma_{ann} > \; \rho_i(X)\rho_j(X), \tag{3.80}$$

where $\rho_i(X)$ is the density of the hadron with quantum numbers i.

The number of reactions per volume and time for the back processes is then given by ($\lambda_m = k_1, .., k_m$)

$$\frac{dN_{coll}[m \text{ mesons} \to B\bar{B}]}{dtdV} = \sum_{i,j} \sum_{\lambda_m} \int \left(\prod_{k=3}^{m+2} \frac{d^3 p_k}{(2\pi)^3 2E_k} \right)$$

$$\times W_{2,m}(\sqrt{s}; i, j, \lambda_m) \, R_2(P^\mu = \sum_{k=3}^{m+2} p_k^\mu; M_1, M_2) \left(\prod_{k=3}^{m+2} f_k(X, \mathbf{p}_k) \right), \tag{3.81}$$

assuming $W_{2,m}(\sqrt{s}; i, j, \lambda_m)$ to depend only on the available energy $\sqrt{s}$ and conserved quantum numbers.

To proceed further, some simplifying assumptions have to be invoked to lead to a tractable problem for antibaryon annihilation and production. Experimentally, the differential multiplicity in the pions from $p\bar{p}$ annihilation at low $\sqrt{s}$ above threshold can be described as

$$P(N_\pi) \approx \frac{1}{\sqrt{2\pi}D} \exp(-\frac{(N_\pi - <N_\pi>)^2}{2D^2}) \tag{3.82}$$

with an average pion multiplicity of $<N_\pi> \approx 5$ and $D^2 = 0.95$ [18]. This observation is reminiscent of flavor rearrangement processes in the $B\bar{B}$ annihilation reaction to vector mesons and pseudoscalar mesons, e.g. $\rho + \rho + \pi$ or $\omega + \omega + \pi^0$, where the ρ and ω "later" decay to 2 or 3 pions, respectively. In this picture the $\rho + \rho + \pi$ final channel in $p\bar{p}$ annihilation is the dominant process leading finally to 5 pions. Alternatively, the $\omega + \omega + \pi^0$ channel leads to 7 pions in the final channel which will appear on the scale of the ω-meson lifetime. Three pions are obtained in the direct 3 pion decay which, however, is substantially suppressed at higher $\sqrt{s}$ due to spin multiplicities.

For the actual example we employ a quark rearrangement model for $B\bar{B}$ annihilation to 3 mesons, where the final mesons M_i may be pseudoscalar or vector mesons, i.e. (π, η) or (ρ, ω), respectively, when restricting to the nonstrange baryon-antibaryon reactions. In the following, the quantum numbers denoted by λ_m will be separated into different channels c, that can be distinguished by their mass decomposition, and degenerate quantum numbers such as spin multiplicities and isospin projections. In the latter sense the sum over the final quantum numbers λ_m in (3.75) and (3.81) then includes a sum over the mass partitions $c = (M_3, M_4, M_5)$, a sum over the spins of the mesons and a sum over all isospin quantum numbers that are compatible with charge conservation in the transition. The probability for a channel $c = (M_3, M_4, M_5)$ then reads

$$P_c(\sqrt{s}; M_3, M_4, M_5) = N_3(\sqrt{s}) \ R_3(\sqrt{s}; M_3, M_4, M_5) \ N_{fin}^c, \tag{3.83}$$

where the number of "equivalent" meson final states in the channel c is given by

$$N_{fin}^c = (2s_3 + 1)(2s_4 + 1)(2s_5 + 1)\frac{F_{iso}}{N_{id}!}. \tag{3.84}$$

In (3.84) s_j denote the spins of the final mesons, F_{iso} is the number of isospin projections compatible with charge conservation while N_{id} is the number of identical mesons in the final channel (e.g. $N_{id} = 3$ for the $\pi^0\pi^0\pi^0$ final channel). This combinatorial problem for the final number of states N_{fin}^c is of finite dimension and easily tractable numerically. For each mass partition $c = (M_3, M_4, M_5)$ the decay probability then is given by the 3-body phase space $R_3(\sqrt{s}, M_3, M_4, M_5)$ in the center-of-mass system and the allowed number of final states N_{fin}^c since the absolute normalization—described by $N_3(\sqrt{s})$—is fixed by the constraint $\sum_c P_c = 1$.

In case of nucleon-antinucleon annihilation the following final meson channels contribute:

$$(1) \; \pi\pi\pi \; (2) \; \pi\pi\rho \; (3) \; \pi\pi\omega \; (4) \; \pi\rho\rho \; (5) \; \pi\rho\omega \; (6) \; \pi\omega\omega, \qquad (3.85)$$

when excluding 3 vector mesons in the final channel. According to (3.83) the distribution in the final number of pions (including the explicit vector meson decays to pions) can be evaluated as a function of $\sqrt{s}$ since it only depends on the phase space and the number of possible final states N_{fin}^c in each channel c.

For the backward reactions, i.e. the 3 meson fusion to a $B\bar{B}$ pair, the quarks and antiquarks are redistributed in a baryon and antibaryon, respectively, incorporating the baryons N and Δ as well as their antiparticles. In line with (3.83) the relative population of states (with the same quark content) is determined by phase space, i.e.

$$P_{c'}(\sqrt{s}; M_1, M_2) = N_2\left(\sqrt{s}\right) \; R_2(\sqrt{s}; c' = (M_1, M_2)) \; (2s_1 + 1)(2s_2 + 1)$$
$$= N_2\left(\sqrt{s}\right) \; R_2(\sqrt{s}; M_1, M_2) \; N_B^{c'}, \qquad (3.86)$$

where $N_B^{c'}$ now denotes the number of final states for the particular mass channel c' in the backward reaction. The absolute normalization $N_2(\sqrt{s})$ is fixed again by the constraint $\sum_{c'} P_{c'} = 1$. In this particular case the number of mass channels is 2 for baryon-antibaryon pairs while for mesons there are 6 different mass channels (3.85). Accordingly, a 2×6 mass channel matrix has to be calculated and stored (or parametrized).

As a special (and reduced) example we consider the reactions $\pi^-\pi^+\pi^-$ or $\pi^-\rho^+\pi^-$ or $\pi^-\rho^+\rho^-$ (and isospin combinations), i.e. in terms of quarks and antiquarks $\bar{u}d + \bar{d}u + \bar{u}d \rightarrow (\bar{u}\bar{u}d) + (udd)$: here the final states may be either $\bar{p} + n$, $\bar{\Delta}^- + n$, $\bar{p} + \Delta^0$ or $\bar{\Delta}^- + \Delta^0$ within the Fock space considered. Thus the transition channel mass matrix even reduces to a 2×3 matrix. Note that the final states with a Δ-resonance are favored due to the spin factors in (3.86), however, somewhat suppressed by the 2-body phase-space integral $R_2(\sqrt{s})$ for low $\sqrt{s}$.

One is thus left with the $B\bar{B}$ annihilation problem

$$\frac{dN_{coll}[B\bar{B} \rightarrow 3\,mesons]}{dt\,dV} = \sum_c \sum_{c'} \int \frac{d^3 p_1}{(2\pi)^3 2E_1} \frac{d^3 p_2}{(2\pi)^3 2E_2} W_{2,3}\left(\sqrt{s}\right)$$

$$\times N_3\left(\sqrt{s}\right) \; R_3(p_1 + p_2; c = (M_3, M_4, M_5)) \; N_{fin}^c \; f_i(X, \mathbf{p}_1) f_j(X, \mathbf{p}_2),$$

$$(3.87)$$

where (M_1, M_2) denote the baryon and antibaryon masses in the channel c' and (M_3, M_4, M_5) the final meson masses in the channel c. Equation (3.87) can be rewritten as

$$\frac{dN_{coll}[B\bar{B} \to 3\, mesons]}{dt\,dV}$$
$$= \sum_c \sum_{c'} \int \frac{d^3 p_1}{(2\pi)^3} \frac{d^3 p_2}{(2\pi)^3} \, P_{cc'}^{2,3}\left(\sqrt{s}\right) \, f_i(X, \mathbf{p}_1) f_j(X, \mathbf{p}_2) \qquad (3.88)$$

with the channel probabilities

$$P_{cc'}^{2,3}\left(\sqrt{s}\right) = \frac{1}{4E_1 E_2} \, W^{2,3}\left(\sqrt{s}\right) \, N_3\left(\sqrt{s}\right) \, R_3(p_1 + p_2; (M_3, M_4, M_5)) \, N_{fin}^c. \qquad (3.89)$$

Note, that by construction we have

$$\sum_c P_{cc'}^{2,3}\left(\sqrt{s}\right) = \frac{1}{4E_1 E_2} \, W^{2,3}\left(\sqrt{s}\right) = v_{rel}\, \sigma_{ann}\left(\sqrt{s}\right)_{c'}, \qquad (3.90)$$

where v_{rel} denotes the relative velocity (3.77) and $\sigma_{ann}(\sqrt{s})_{c'}$ is the total annihilation cross section for $B\bar{B}$ pairs of channel c'.

The backward invariant collision rate is given by

$$\frac{dN_{coll}[3\, mesons \to B\bar{B}]}{dt\,dV} = \sum_c \sum_{c'} \int \left(\prod_{k=3}^{5} \frac{d^3 p_k}{(2\pi)^3 2E_k}\right) W_{2,3}\left(\sqrt{s}\right)$$
$$\times N_2\left(\sqrt{s}\right) \, R_2(\sum_{k=3}^{5} p_k; c' = (M_1, M_2)) \, N_B^{c'} \left(\prod_{k=3}^{5} f_k(X, \mathbf{p}_k)\right). \qquad (3.91)$$

Using Eq. (3.74), the relation (3.76) for 3 mesons in the final state and (3.90) one arrives at

$$\frac{dN_{coll}[3\, mesons \to B\bar{B}]}{dt\,dV} = \sum_c \sum_{c'} \int \frac{d^3 p_3}{(2\pi)^3 2E_3} \frac{d^3 p_4}{(2\pi)^3 2E_4} \frac{d^3 p_5}{(2\pi)^3 2E_5} 4E_1 E_2$$
$$\times v_{rel}\, \sigma\left(\sqrt{s}\right)_{c'} \frac{N_2\left(\sqrt{s}\right)}{N_3\left(\sqrt{s}\right)} \frac{R_2(P^\mu; c' = (M_1, M_2))}{R_3(P^\mu; c = (M_3, M_4, M_5))} \frac{N_B^{c'}}{N_{fin}^c}$$
$$\times f_3(X, \mathbf{p}_3) f_4(X, \mathbf{p}_4) f_5(X, \mathbf{p}_5) \qquad (3.92)$$

for the backward reaction $3 + 4 + 5 \rightarrow 1 + 2$. Equation (3.92) can now be rewritten as

$$\frac{dN_{coll}[3 \ mesons \rightarrow B\bar{B}]}{dt dV}$$

$$= \sum_c \sum_{c'} \int \frac{d^3 p_3}{(2\pi)^3} \frac{d^3 p_4}{(2\pi)^3} \frac{d^3 p_5}{(2\pi)^3} \ P_{cc'}^{3,2} \left(\sqrt{s}\right)$$

$$\times f_3(X, \mathbf{p}_3) f_4(X, \mathbf{p}_4) f_5(X, \mathbf{p}_5) \tag{3.93}$$

with the 'transition integrand'

$$P_{cc'}^{3,2} \left(\sqrt{s}\right) = \frac{E_1 E_2}{2 E_3 E_4 E_5} v_{rel} \ \sigma \left(\sqrt{s}\right)_{c'} \frac{N_2 \left(\sqrt{s}\right)}{N_3 \left(\sqrt{s}\right)} \frac{R_2(P; c' = (M_1, M_2))}{R_3(P; c = (M_3, M_4, M_5))} \frac{N_B^{c'}}{N_{fin}^c}, \tag{3.94}$$

which is of dimension GeV^{-3} or fm^3.

3.2.2 Numerical Implementation

For a reformulation of the "transition integrands" (specified in (3.94)) in a testparticle representation one has to recall that the average density of a meson with quantum numbers k is obtained by integration over momentum as:

$$n_k(X) = \int \frac{d^3 p}{(2\pi)^3} \ f_k(X, \mathbf{p}), \tag{3.95}$$

where e.g. charge, strange flavor content, total spin and spin projection are specified by the discrete quantum number k. The conversion formula thus reads:

$$\int \frac{d^3 p}{(2\pi)^3} \ f_k(X, \mathbf{p}) \rightarrow \frac{1}{dV} \sum_{i \ \epsilon \ dV}, \tag{3.96}$$

where dV is a (small) finite volume and the sum runs over all test particles in the volume dV with quantum numbers k. The number of $B\bar{B}$ annihilations in the volume dV during the time dt is thus given by

$$N_{B\bar{B}} = \frac{dt}{dV} \sum_{i,j \ \epsilon \ dV} v_{rel}(i, j) \sigma_{ann}(\sqrt{s}_{i,j}) \tag{3.97}$$

with the invariant energy squared

$$s_{i,j} = (p_1 + p_2)^2, \tag{3.98}$$

where p_1, p_2 denote the 4-momenta of the colliding $B\bar{B}$ pair. The relative velocity $v_{rel}(i, j)$ is given by (3.77) while the annihilation cross section $\sigma_{ann}(\sqrt{s})$ has to be specified for all baryon-antibaryon pairs. In case of nucleon-antinucleon reactions this cross section is rather well known experimentally [19] and the product $v_{rel}\,\sigma_{ann}(\sqrt{s}) \approx 50$ mb for a wide range of invariant energies $\sqrt{s}$,

$$\sigma_{ann}\left(\sqrt{s}\right) = \frac{50[mb]}{v_{rel}}, \tag{3.99}$$

which holds well in the dynamical range of interest. We thus can adopt the Boltzmann limit (3.80) to estimate the $B\bar{B}$ annihilation time at nucleon density ρ as

$$\tau_{ann.} \approx (5fm^2\rho)^{-1} \approx 1.2\,\frac{\rho_0}{\rho}\,[fm/c]. \tag{3.100}$$

The number of backward reactions by 3 mesons in the testparticle picture in the volume dV and time dt according to (3.92) for a given mass channel c' is given by

$$N_{3meson} = \frac{dt}{dVdV}\sum_{i,j,k\,\epsilon\,dV}\frac{E_1E_2}{2E_iE_jE_k}v_{rel}(1,2)\sigma\left(\sqrt{s}\right)_{c'}$$

$$\times\,\frac{N_2\left(\sqrt{s}\right)}{N_3\left(\sqrt{s}\right)}\frac{R_2(\sqrt{s};c'=(M_1,M_2))}{R_3(\sqrt{s};c=(M_i,M_j,M_k))}\frac{N_B^{c'}}{N_{fin}^c} = \sum_{i,j,k\,\epsilon\,dV}P_{ijk}, \tag{3.101}$$

where the channel c is defined by the colliding mesons (cf. (3.81)) and the outgoing channel c' by the $B\bar{B}$ pair with masses M_1 and M_2 and energies E_1 and E_2, respectively. In (3.101) the summation over the mesons in the volume dV is restricted to $i < j < k$ in case of 3 identical mesons (e.g. 3 π^0's) and to $i < j$ in case of 2 identical mesons i, j in order to account for the statistical factor $N_{id}!$ in Eq. (3.84).

Equations (3.97) and (3.101) are well suited for a Monte Carlo decision problem, i.e. a transition in the space-time volume $dtdV$ is accepted if the probability P_{ijk} is larger than some random number in the interval [0,1]. One has to assure only, that all P_{ijk} are smaller than 1, which—for a fixed volume dV—can easily be achieved by adjusting the time-step dt. This evaluation of scattering probabilities—in the mutual center-of-mass system—is Lorentz-invariant and does not suffer from geometrical collision criteria as in the standard approaches that imply a different sequence of collisions when changing the reference frame by a Lorentz transformation (cf. Sect. 3.1.3). It is worth to point out that this numerical implementation is a promising way to treat $n \leftrightarrow m$ transitions in transport theories without violating

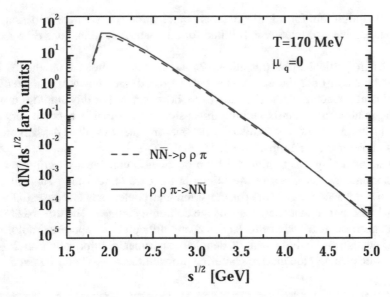

Fig. 3.5 The number of $N\bar{N} \rightarrow \rho\rho\pi$ reactions as a function of the invariant energy $\sqrt{s}$ for a system in thermal and chemical equilibrium at temperature $T = 170$ MeV and chemical potential $\mu_q = 0$. The solid line denotes the differential number in the backward $(\rho\rho\pi)$ collisions, respectively

covariance or causality. In case of infinitesimal volumes dV and time steps dt it gives the correct solution to the many-particle Boltzmann equation, however, the constraints in the actual derivation require space-time volumes of sufficient size.

As a numerical test the number of collisions in a single box of volume 10 fm^3 during the time dt =1 fm/c has been calculated with spatially uniform phase-space distributions given by a classical system of hadrons in thermal and chemical equilibrium, i.e.

$$f_k(p) = \frac{(2s+1)(2I+1)}{(2\pi)^3} \exp(-E_k(p)/T) \qquad (3.102)$$

with s and I denoting spin and isospin, respectively. The particles taken into account are N, Δ and their antiparticles and π, ρ on the meson side in the strangeness sector $S=0$. The numerical results for the number of $B\bar{B}$ annihilation collisions $(\rightarrow \pi\rho\rho)$ are shown in Fig. 3.5 in terms of the dashed line as a function of $\sqrt{s}$, which corresponds to the invariant energy in an individual collision.[7] As can be seen from Fig. 3.5 the dashed line very well coincides with the solid line that corresponds to the energy differential number of $\pi\rho\rho$ collisions for the backward reactions. Thus the numerical scheme employed well reproduces the detailed balance relation in

---

[7] This figure is taken from Ref. [20].

thermal equilibrium for a given channel combination cc'. We note in passing that the detailed balance relation is fulfilled for all channel combinations cc' specified above.

Thus in particular the transitions $2 \leftrightarrow n$ can be treated on the basis of detailed balance once the matrix elements (squared) are known or extracted from experimental reaction data. The drawback, however, is that this employs a multi-dimensional channel matrix and the dimensions of this matrix become very large when including all $J^P = 0^-$ and $J^P = 1^-$ meson states as well as the baryon octet and decuplet and their antiparticles [21].

In summarizing this section we have obtained a consistent description of the collision integral for $n \leftrightarrow m$ transitions in the case of vanishing Pauli blocking and Bose-enhancement factors for transition matrix elements that essentially only depend on the invariant energy $\sqrt{s}$ of the colliding particles. The drawback is that these transition matrix elements (squared) are difficult to compute and notoriously model dependent. Only in case of detailed experimental information on $2 \rightarrow m$ production channels the inverse scattering probabilities $m \rightarrow 2$ may be fixed.

Solution of Exercises

Exercise 3.1: Prove the expressions (3.18) and (3.19).

The field operator $\Psi(x)$ and the Pauli-adjoint operator $\bar{\Psi}(x)$ can be expanded in terms of plane waves and effective spinors as (cf. (3.13) and (3.14)) as

$$
\Psi(x) = \int \frac{d^3\Pi}{(2\pi)^{3/2}} \frac{M^*}{E_p^*} \times \sum_{r=1}^{2} [c_r(x, \Pi) u_r(\Pi, M^*) \exp(-i(\varepsilon^+(\mathbf{p})t - \mathbf{p} \cdot \mathbf{x}))
$$

$$
+ d_r^\dagger(x, \Pi) v_r(\Pi, M^*) \exp(i(\varepsilon^-(\mathbf{p})t - \mathbf{p} \cdot \mathbf{x}))], \tag{3.103}
$$

$$
\bar{\Psi}(x) = \int \frac{d^3\Pi}{(2\pi)^{3/2}} \frac{M^*}{E_p^*} \times \sum_{r=1}^{2} [c_r^\dagger(x, \Pi) \bar{u}_r(\Pi, M^*) \exp(+i(\varepsilon^+(\mathbf{p})t - \mathbf{p} \cdot \mathbf{x}))
$$

$$
+ d_r(x, \Pi) \bar{v}_r(\Pi, M^*) \exp(-i(\varepsilon^-(\mathbf{p})t - \mathbf{p} \cdot \mathbf{x}))],
$$

with $E_p^* = \sqrt{\Pi^2 + M^{*2}} = \sqrt{(\mathbf{p} - \Sigma)^2 + M^{*2}}$ and

$$
\epsilon^+(\mathbf{p}) = \sqrt{\Pi^2 + M^{*2}} + \Sigma^0, \ \epsilon^-(\mathbf{p}) = -\sqrt{\Pi^2 + M^{*2}} + \Sigma^0. \tag{3.104}
$$

The quasiparticle scalar density ρ_s in some space-time cell at x for isospin symmetric nuclear matter then is given by

$$\rho_s(x) = 2\langle: \bar{\Psi}\Psi :\rangle = 2\sum_{r=1}^{2} \int \frac{d^3\Pi}{(2\pi)^3} \left(\frac{M^*}{E_p^*}\right)^2 \tag{3.105}$$

$$\times <: [c_r^\dagger(x, \Pi)\bar{u}_r(\Pi, M^*) \exp(+i(\varepsilon^+(\mathbf{p})t - \mathbf{p}\cdot\mathbf{x}))$$

$$+ d_r(x, \Pi)\bar{v}_r(\Pi, M^*) \exp(-i(\varepsilon^-(\mathbf{p})t - \mathbf{p}\cdot\mathbf{x}))]$$

$$\times [c_r(x, \Pi)u_r(\Pi, M^*) \exp(-i(\varepsilon^+(\mathbf{p})t - \mathbf{p}\cdot\mathbf{x}))$$

$$+ d_r^\dagger(x, \Pi)v_r(\Pi, M^*) \exp(i(\varepsilon^-(\mathbf{p})t - \mathbf{p}\cdot\mathbf{x}))] :>$$

$$= 2\sum_{r=1}^{2} \int \frac{d^3\Pi}{(2\pi)^3} \left(\frac{M^*}{E_p^*}\right)^2$$

$$\times \left(<: c_r^\dagger(x, \Pi)\bar{u}_r(\Pi, M^*)c_r(x, \Pi)u_r(\Pi, M^*) :>\right.$$

$$\left. + <: d_r(x, \Pi)\bar{v}_r(\Pi, M^*)d_r^\dagger(x, \Pi)v_r(\Pi, M^*) :>\right)$$

since mixed terms of $c-$ and $d-$ operators do not contribute. The factor of 2 stems from summation over isospin. After normal ordering the $d_r d_r^\dagger$-term picks up a minus sign, however, the normalization $\bar{v}_r(\Pi)v_r(\Pi)=-1$ gives another minus sign while $\bar{u}_r(\Pi)u_r(\Pi)=1$. With the particle/antiparticle density operators (cf. Appendix G)

$$N_r^c(x, \Pi) = \frac{M^*}{E_p^*} c_r^\dagger(x, \Pi)c_r(\Pi), \quad N_r^d(x, \Pi) = \frac{M^*}{E_p^*} d_r^\dagger(x, \Pi)d_r^\dagger(x, \Pi) \tag{3.106}$$

we arrive at

$$\rho_s(x) = 2\sum_{r=1}^{2} \int \frac{d^3\Pi}{(2\pi)^3} \frac{M^*}{E_p^*} \left(N_r^c(x, \Pi) + N_r^d(x, \Pi)\right), \tag{3.107}$$

since a factor M^*/E_p^* is contained in the definition of the particle number operators. For spin and isospin symmetric nuclear matter the particle number operators do not depend on the spin index r such that the summation over spin gives another factor of 2. Thus identifying

$$f_p^*(x, \Pi) \equiv N^c(x, \Pi), \quad f_a^*(x, \Pi) \equiv N^d(x, \Pi) \tag{3.108}$$

in the local cell of average space-time position x we obtain Eq. (3.19). Here N^c and N^d denote spin averaged occupation densities.

The quasiparticle density ρ_N for isospin symmetric nuclear matter is evaluated in a similar fashion:

$$\rho_N(x) = 2\langle: \bar{\Psi}\gamma^0\Psi :\rangle = 2\sum_{r=1}^{2}\int \frac{d^3\Pi}{(2\pi)^3}\left(\frac{M^*}{E_p^*}\right)^2 \tag{3.109}$$

$$\times \left(<: c_r^\dagger(x,\Pi)\bar{u}_r(\Pi,M^*)c_r(x,\Pi)\,\gamma^0\,u_r(\Pi,M^*) :>\right.$$

$$\left.+ <: d_r(x,\Pi)\bar{v}_r(\Pi,M^*)\,\gamma^0\,d_r^\dagger(x,\Pi)v_r(\Pi,M^*) :>\right).$$

After normal ordering the $d_r d_r^\dagger$ term changes sign and

$$\bar{u}_r(\Pi,M^*)\gamma^0 u_r(\Pi,M^*) = \bar{v}_r(\Pi,M^*)\gamma^0 v_r(\Pi,M^*) = \frac{E_p^*}{M^*} \tag{3.110}$$

cancels a factor M^*/E_p^*. We thus obtain

$$\rho_N(x) = 2\sum_{r=1}^{2}\int \frac{d^3\Pi}{(2\pi)^3}\left(N_c^r(x,\Pi) - N_d^r(x,\Pi)\right)$$

$$\equiv 4\int \frac{d^3\Pi}{(2\pi)^3}\left(f_p^*(x,\Pi) - f_a^*(x,\Pi)\right), \tag{3.111}$$

which proves Eq. (3.18).

Exercise 3.2: Derive the expressions (3.46) and (3.49) for systems in thermal equilibrium.

The energy density in mean-field approximation for isospin symmetric nuclear matter in equilibrium is given by

$$\mathcal{E} = U(\sigma) - O(\omega) + 2\langle: i\bar{\Psi}\gamma^0\partial_0\Psi :\rangle - 2\langle: \bar{\Psi}\left(\gamma_\mu\left(i\partial^\mu - \Sigma^\mu\right) - \left(M - \Sigma^s\right)\right)\Psi :\rangle$$

$$= U(\sigma) - O(\omega) + 2\langle: i\bar{\Psi}\gamma^0\partial_0\Psi :\rangle - 2\underbrace{\langle: \bar{\Psi}\left(\gamma_\mu\Pi^\mu - M^*\right)\Psi :\rangle}_{=0} \tag{3.112}$$

with $\omega^\mu = (\omega, 0, 0, 0)$ and $O(\omega) = m_\omega^2 \omega^2 / 2$. With the spinor expansions for Ψ and $\bar{\Psi}$ we obtain:

$$
\mathcal{E} = U(\sigma) - O(\omega) + 2 \sum_{\lambda=1}^{2} \int \frac{d^3 p}{(2\pi)^3} \frac{M^*}{E_p^*} \left(\epsilon^+(\mathbf{p}) \langle: c_\lambda^\dagger(\mathbf{p}) c_\lambda(\mathbf{p}) :\rangle \right.
$$

$$
\left. - \epsilon^-(\mathbf{p}) \langle: d_\lambda(\mathbf{p}) d_\lambda^\dagger(\mathbf{p}) :\rangle \right)
$$

$$
= U(\sigma) - O(\omega) + d \int \frac{d^3 p}{(2\pi)^3} \left((E_p^* + \Sigma^0) \, n_F(T, \mu^*, M^*) \right.
$$

$$
\left. - (-E_p^* + \Sigma^0) \, n_{\bar{F}}(T, \mu^*, M^*) \right)
$$

$$
= U(\sigma) - O(\omega) + \Sigma^0 \rho_B + d \int \frac{d^3 p}{(2\pi)^3} E_p^* \left(n_F(T, \mu^*, M^*) + n_{\bar{F}}(T, \mu^*, M^*) \right)
$$

$$
= U(\sigma) - O(\omega) + \Sigma^0 \rho_B + E_0(T, \mu^*, M^*)
$$

$$
= U(\sigma) + \frac{m_\omega^2}{2 g_v^2} \Sigma_0^2 + E_0(T, \mu^*, M^*) \tag{3.113}
$$

with the Fermi functions n_F and $n_{\bar{F}}$ for the equilibrium occupation densities for particles and antiparticles. Here E_0 is the energy density for a non-interacting particle evaluated at the effective chemical potential μ^* and with the effective mass M^*. Furthermore, Eq. (3.3) or

$$
\Sigma^0 \rho_B = g_v \, \omega^0 \rho_B = \frac{g_v^2}{m_\omega^2} \rho_B^2
$$

has been used in the last line.

The pressure in mean-field approximation reads

$$
P = -U(\sigma) + O(\omega) + \frac{1}{3} \sum_{i=1}^{3} 2 \langle: i \bar{\Psi} \gamma^i \partial_i \Psi :\rangle. \tag{3.114}
$$

The further evaluation is similar to the energy density case above and gives

$$
P = -U(\sigma) + O(\omega) + \frac{2}{3} \sum_{\lambda=1}^{2} \sum_{i=1}^{3} \int \frac{d^3 p}{(2\pi)^3} \left(\frac{M^*}{E_p^*} \right)^2
$$

$$
\times \left(\frac{p^i}{M^*} \langle: c_\lambda^\dagger(\mathbf{p}) c_\lambda(\mathbf{p}) :\rangle p^i + \frac{p^i}{M^*} \langle: d_\lambda(\mathbf{p}) d_\lambda^\dagger(\mathbf{p}) :\rangle p^i \right)
$$

$$= -U(\sigma) + O(\omega) + \frac{d}{3} \int \frac{d^3 p}{(2\pi)^3} \frac{\mathbf{p}^2}{E_p^*} \left(n_F(T, \mu^*, M^*) + n_{\bar{F}}(T, \mu^*, M^*) \right)$$

$$= -U(\sigma) + O(\omega) + P_0(T, \mu^*, M^*) = -U(\sigma) + \frac{m_\omega^2}{2g_v^2} \Sigma_0^2 + P_0(T, \mu^*, M^*),$$

$$(3.115)$$

where P_0 is the pressure for a non-interacting particle with the effective quantities μ^* and M^*.

References

1. C. Fuchs, H. Lenske, and H. H. Wolter, Phys. Rev. C 52 (1995) 3043.
2. F. Hofmann, C. M. Keil, and H. Lenske, Phys. Rev. C 64 (2001) 034314.
3. W. Botermans and R. Malfliet, Phys. Rep. 198 (1990) 115.
4. R. Malfliet, Prog. Part. Nucl. Phys. **21**, 207 (1988)
5. A. Faessler, Prog. Part. Nucl. Phys. **30**, 229 (1993)
6. J.D. Walecka, Ann. Phys. **83**, 491 (1974)
7. B.D. Serot, J.D. Walecka, Adv. Nucl. Phys. **16**, 1 (1986)
8. B.D. Serot, J.D. Walecka, Int. J. Mod. Phys. **6**, 515 (1997)
9. C.M. Ko, Q. Li, R.-C. Wang, Phys. Rev. Lett. **59**, 1084 (1987)
10. J.D. Bjorken, S.D. Drell, *Relativistische Quantenmechanik* (Bibliographisches Institut, Mannheim, 1967)
11. S. Mrowczynski, U. Heinz, Ann. Phys. **229**, 1 (1994)
12. T. Steinert, W. Cassing, Phys. Rev. C **98**, 014908 (2018)
13. A. Lang, H. Babovsky, W. Cassing, U. Mosel, H.-G. Reusch, K. Weber, J. Comput. Phys. **106**, 391 (1993)
14. H. Babovsky, Eur. J. Mech. **B8**, 41 (1989)
15. W. Cassing, E. Bratkovskaya, Phys. Rep. **308**, 65 (1999)
16. O. Buss, T. Gaitanos, K. Gallmeister, H. van Hees, M. Kaskulov, O. Lalakulich, A.B. Larionov, T. Leitner, J. Weil, U. Mosel, Phys. Rep. **512**, 1 (2012)
17. E. Byckling, K. Kajante, *Particle Kinematics* (Wiley, London, 1973)
18. C.B. Dover, T. Gutsche, M. Maruyama, A. Faessler, Prog. Part. Nucl. Phys. **29**, 87 (1992)
19. Particle Data Group, E. Phys. J. C **3**, 1 (1998)
20. W. Cassing, Nucl. Phys. A **700**, 618 (2002)
21. E. Seifert, W. Cassing, Phys. Rev. C **97**, 024913 (2018)

Relativistic Dynamics and Off-Shell Transport

4

This chapter is devoted to the covariant dynamics of strongly-interacting systems on the level of Kadanoff–Baym equations that are solved explicitly for the scalar ϕ^4-theory in a finite box. Special attention is paid to the spectral functions of the degrees of freedom and the final equilibrium state. The quantum Boltzmann equation is derived in the on-shell limit and the solution of the Kadanoff–Baym equations is compared to those from the on-shell Boltzmann limit as a function of the coupling strength. Furthermore, covariant off-shell transport equations are derived in first-order gradient expansion of the (Wigner transformed) Kadanoff–Baym equations and an extended testparticle Ansatz is introduced that allows for a convenient solution of the transport equations. A related derivation for fermion systems is included accordingly. As an example for off-shell transport we compare the results for vector mesons in heavy-ion reactions at 2 A GeV in the on-shell and off-shell versions. Furthermore, retarded electromagnetic fields—generated by moving charges in ultra-relativistic heavy-ion reactions—are computed for noncentral collisions of $Au + Au$ and $Cu + Au$ at an invariant energy $\sqrt{s_{NN}} = 200$ GeV without introducing any additional parameter since the electromagnetic coupling $e^2/(4\pi)$ is well known.[1]

4.1 Relativistic Formulations

Relativistic formulations of the many-body problem are essentially described within covariant field theory. Since the fields themselves are distributions in space-time $x = (t, \mathbf{x})$ one changes from the Schrödinger picture discussed before to the Heisenberg picture.[2] Furthermore, the field theoretical problem in principle encounters infinitely many particles in a wavefunction such that a "top-down" scenario ($N \rightarrow n < N$)

---

[1] In this chapter we will use the Einstein convention throughout if not specified explicitly.

[2] The different pictures of quantum mechanics are recalled in Appendix A.

© The Author(s), under exclusive license to Springer Nature Switzerland AG 2021
W. Cassing, *Transport Theories for Strongly-Interacting Systems*, Lecture Notes in Physics 989, https://doi.org/10.1007/978-3-030-80295-0_4

is no longer suitable. Nevertheless, we will encounter very similar structures to the BBGKY hierarchy in the Martin–Schwinger hierarchy.

In the Heisenberg picture the time evolution of the system is described by time-dependent operators that are evolved with the help of the time evolution operator $\hat{U}(t, t_0)$ which follows

$$i \frac{\partial \hat{U}(t, t_0)}{\partial t} = \hat{H}(t) \hat{U}(t, t_0), \tag{4.1}$$

with $\hat{H}$ denoting the Hamilton operator of the system. Equation (4.1) is formally solved by

$$\hat{U}(t, t_0) = T \left(\exp \left[-i \int_{t_0}^{t} dz \, \hat{H}(z) \right] \right) = \sum_{n=0}^{\infty} \frac{1}{n!} T[-i \int_{t_0}^{t} dz \, \hat{H}(z)]^n \quad , \tag{4.2}$$

where T denotes the time-ordering operator, which is also denoted as Dyson series. Let us assume that the initial state is given by some density matrix $\hat{\rho}$, which may be a pure or mixed state, then the time evolution of any operator $\hat{O}$ in the Heisenberg picture from time t_0 to t is given by

$$O(t) = \langle \hat{O}_H(t) \rangle = \mathrm{Tr} \left(\hat{\rho} \, \hat{O}_H(t) \right) = \mathrm{Tr} \left(\hat{\rho} \, \hat{U}(t_0, t) \hat{O} \, \hat{U}(t, t_0) \right)$$

$$= \mathrm{Tr} \left(\hat{\rho} \, \hat{U}^{\dagger}(t, t_0) \hat{O} \, \hat{U}(t, t_0) \right). \tag{4.3}$$

This implies that first the system is evolved from t_0 to t and then backward from t to t_0. This may be expressed as a time integral along the (Keldysh-)Contour or Closed-Time-Path (CTP) [1–4] shown in Fig. 4.1. A time integration along the CPT thus implies

$$\int_{C} dt .. = \int_{t_0}^{t} dt^{+} .. + \int_{t}^{t_0} dt^{-} .. = \int_{t_0}^{t} dt^{+} .. - \int_{t_0}^{t} dt^{-}, \tag{4.4}$$

thus picking up a minus sign for the lower branch when considering the interval $[t_0, t]$.

Fig. 4.1 The Keldysh contour for the time integration in the Heisenberg picture

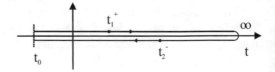

4.1.1   Two-Point Functions on the CTP

Now Green's functions on the contour may have time arguments on the same branch of the contour or on opposite branches. This gives four possibilities for the Green's functions defined—in case of a field theory with only scalar fields $\phi(x)$—by

$$i G^c(x, y) \quad = i G^{++}(x, y) = \langle\, \hat{T}^c(\phi(x)\phi(y))\, \rangle \tag{4.5}$$

$$i G^<(x, y) \qquad = i G^{+-}(x, y) = \langle \phi(y)\phi(x)\rangle \tag{4.6}$$

$$i G^>(x, y) \qquad = i G^{-+}(x, y) = \langle \phi(x)\phi(y)\rangle \tag{4.7}$$

$$i G^a(x, y) = i G^{--}(x, y) = \langle\, \hat{T}^a(\phi(x)\phi(y))\, \rangle \quad , \tag{4.8}$$

which are not independent! Here $x = (x^0, \mathbf{x})$ and $y = (y^0, \mathbf{y})$ denote space-time four-vectors. The index $+-$ means that x^0 is on the upper branch and y^0 on the lower branch while $-+$ implies that x^0 is on the lower branch and y^0 on the upper one. Time-ordering has to be fulfilled if both time arguments are on the same axis. The causal time-ordering operator T^c places fields at later times to the left while the anticausal operator T^a places fields at later times to the right. The Green's functions $G^>$ and $G^<$ are denoted as **Wightman functions** and will play the essential role in the dynamical description of the system. One may also write the Green's function on the Keldysh contour in terms of a 2x2 matrix

$$G(x, y) = \begin{array}{c} \\ + \\ - \end{array} \overset{\displaystyle + \qquad\qquad -}{\left(\begin{array}{cc} G^c(x, y) & G^<(x, y) \\ G^>(x, y) & G^a(x, y) \end{array} \right)} \quad . \tag{4.9}$$

Note that the Green's functions defined in (4.5) to (4.8) are two-point functions, i.e. they correspond to a single-particle species!

The further derivation again starts with a Dyson equation (cf. (C.7))

$$G(x, y) = G_0(x, y) + [G_0 \Sigma G](x, y) \tag{4.10}$$

which is of one-body type and instead of $G(x, y)$ we might write in shorthand notation $G(11')$. The selfenergy $\Sigma(x, y) = \Sigma(1, 1')$ has the meaning of a one-body mean-field potential (for bosons) and has the dimension [energy]^2.

The relation to the one-body density matrix ρ in the previous chapter is given by

$$\rho(\mathbf{x}, \mathbf{x}'; t) \equiv -i G^<(\mathbf{x}, \mathbf{x}'; t, t) \tag{4.11}$$

since the time diagonal Green's function can be identified with an integral over the time difference $\tau - \tau'$ (for $t = (\tau + \tau')/2$)

$$G^<(\mathbf{x}, \mathbf{x}'; t) = \int_{-\infty}^{\infty} d(\tau - \tau')\, G^<(\mathbf{x}, \mathbf{x}'; \tau, \tau'). \tag{4.12}$$

Arbitrary two-point functions F on the closed-time-path (CTP) generally can be expressed by retarded and advanced components as

$$F^R(x, y) = F^c(x, y) - F^<(x, y) = F^>(x, y) - F^a(x, y), \tag{4.13}$$

$$F^A(x, y) = F^c(x, y) - F^>(x, y) = F^<(x, y) - F^a(x, y) \tag{4.14}$$

giving in particular the relation

$$F^R(x, y) - F^A(x, y) = F^>(x, y) - F^<(x, y). \tag{4.15}$$

Note that the advanced and retarded components of the Green's functions only contain spectral and no statistical information,

$$G^{R/A}(x, y) = G_0 \, \delta(t_1 - t_2) \pm \Theta(\pm(t_1 - t_2)) \, [G^>(x, y) - G^<(x, y)]. \tag{4.16}$$

4.1.2 The Dyson–Schwinger Equation on the CTP

The Dyson–Schwinger equation (4.10) on the closed-time-path reads in matrix form:

$$
\begin{pmatrix} G^c(x, y) & G^<(x, y) \\ G^>(x, y) & G^a(x, y) \end{pmatrix} = \begin{pmatrix} G_0^c(x, y) & G_0^<(x, y) \\ G_0^>(x, y) & G_0^a(x, y) \end{pmatrix}
$$
$$
+ \begin{pmatrix} G_0^c(x, x') & G_0^<(x, x') \\ G_0^>(x, x') & G_0^a(x, x') \end{pmatrix}
$$
$$
\odot \begin{pmatrix} \Sigma^c(x', y') & -\Sigma^<(x', y') \\ -\Sigma^>(x', y') & \Sigma^a(x', y') \end{pmatrix} \odot \begin{pmatrix} G^c(y', y) & G^<(y', y) \\ G^>(y', y) & G^a(y', y) \end{pmatrix}, \tag{4.17}
$$

where the symbol $\odot$ stands for an intermediate integration over space-time on the CTP, i.e. x' or y'. The selfenergy Σ on the CPT is defined along (4.14) and incorporates interactions of higher order. In lowest order $\Sigma/2M$ is given by the Hartree mean field but it follows a nonperturbative expansion in analogy to (2.154).

An example for this formal procedure may be given by the scalar ϕ^4-theory which is a laboratory for testing theoretical approximations. Its Lagrangian reads

$$\mathcal{L}(x) = \frac{1}{2}\partial_\mu^x \phi(x)\partial_x^\mu \phi(x) - \frac{1}{2}m^2\phi(x)^2 - \frac{\lambda}{4!}\phi^4(x) \tag{4.18}$$

and incorporates a self-coupling of $4th$ order. In this (Bose) case the free propagator is defined via the negative inverse Klein–Gordon operator in space-time representation

$$\hat{G}_{0x}^{-1} = -(\partial_\mu^x \partial_x^\mu + m^2), \tag{4.19}$$

which is a solution of the Klein–Gordon equation in the following sense:

$$
\hat{G}_{0x}^{-1}
\begin{pmatrix}
G_0^c(x, y) & G_0^<(x, y) \\
G_0^>(x, y) & G_0^a(x, y)
\end{pmatrix}
= \delta(\mathbf{x} - \mathbf{y})
\begin{pmatrix}
\delta(x_0 - y_0) & 0 \\
0 & -\delta(x_0 - y_0)
\end{pmatrix}
$$

$$
= \delta(\mathbf{x} - \mathbf{y})\delta_p(x_0 - y_0) \quad ,
$$

$$
\hat{G}_{0x}^{-1} G_0^{R/A}(x, y) = \delta(x - y) \quad , \tag{4.20}
$$

with δ_p denoting the δ-function on the CTP. In (4.19) m denotes the bare mass of the scalar field.

4.1.3 Kadanoff–Baym Equations

To derive the **Kadanoff–Baym equations** one multiplies (4.17) with G_{0x}^{-1} (4.19). This gives four equations which can be cast into the form:

$$
-(\partial_\mu^x \partial_x^\mu + m^2) G^{R/A}(x, y) = \delta(x - y) + \Sigma^{R/A}(x, x') \odot G^{R/A}(x', y), \tag{4.21}
$$

$$
-(\partial_\mu^x \partial_x^\mu + m^2) G^<(x, y) = \Sigma^R(x, x') \odot G^<(x', y) + \Sigma^<(x, x') \odot G^A(x', y), \tag{4.22}
$$

$$
-(\partial_\mu^x \partial_x^\mu + m^2) G^>(x, y) = \Sigma^R(x, x') \odot G^>(x', y) + \Sigma^>(x, x') \odot G^A(x', y). \tag{4.23}
$$

The propagation of the Green's functions in the variable y is defined by the adjoint equations:

$$
-(\partial_\mu^y \partial_y^\mu + m^2) G^{R/A}(x, y) = \delta(x - y) + G^{R/A}(x, x') \odot \Sigma^{R/A}(x', y), \tag{4.24}
$$

$$
-(\partial_\mu^y \partial_y^\mu + m^2) G^<(x, y) = G^R(x, x') \odot \Sigma^<(x', y) + G^<(x, x') \odot \Sigma^A(x', y), \tag{4.25}
$$

$$
-(\partial_\mu^y \partial_y^\mu + m^2) G^>(x, y) = G^R(x, x') \odot \Sigma^>(x', y) + G^>(x, x') \odot \Sigma^A(x', y). \tag{4.26}
$$

Note again that the evolution of the retarded/advanced Green's functions only depends on retarded/advanced quantities.

> Exercise 4.1: Derive Eqs. (4.21)–(4.23) starting from (4.17).

4.1.4 Definition of Selfenergies

For the solution of the KB equations the computation/fixing of the selfenergies Σ is mandatory. In the context of field theory the latter is extracted from the effective action (for neutral scalar fields [5, 6])

$$\Gamma[G] = \Gamma^0[G_0] + \frac{i}{2}[\ln(1 - G_0\Sigma) + G\Sigma] + \Phi[G] \tag{4.27}$$

assuming a vanishing vacuum expectation value $\langle 0|\phi(x)|0\rangle$. Here $\Gamma^0[G_0]$ only depends on the free Green's function and can be considered as constant in the following. Note that all intermediate and final integrations have to be performed over the CTP. In $\Phi[G]$ all closed two-particle irreducible (2PI) diagrams are included in lowest (nontrivial) order.[3] 2PI diagrams are those that cannot be separated in two disjunct diagrams by cutting two propagator lines; formally this implies that after second order differentiation with respect to G no separate diagrams survive. The functional $\Phi[G]$ plays a similar role as the potential energy density $\mathcal{V}(\rho)$ (2.168) in the nonrelativistic case where the (nonrelativistic) selfenergy results from functional derivation of $\mathcal{V}$ with respect to ρ, i.e. $\Sigma = \delta\mathcal{V}/\delta\rho$.

For the derivation of selfenergies one now considers the variation of the action $\Gamma[G]$ with respect to G requiring $\delta\Gamma = 0$,

$$\delta\Gamma = 0 = \frac{i}{2}\Sigma\,\delta G - \frac{i}{2}\frac{G_0}{1 - G_0\,\Sigma}\delta\Sigma + \frac{i}{2}G\,\delta\Sigma + \delta\Phi$$

$$= \frac{i}{2}\Sigma\,\delta G - \frac{i}{2}\underbrace{\frac{1}{G_0^{-1} - \Sigma}}_{=G}\delta\Sigma + \frac{i}{2}G\,\delta\Sigma + \delta\Phi = \frac{i}{2}\Sigma\,\delta G + \delta\Phi \quad . \tag{4.28}$$

$$\Rightarrow \Sigma = 2i\frac{\delta\Phi}{\delta G} = 2\frac{\delta\Phi}{\delta(-iG)} \quad . \tag{4.29}$$

Note that $-iG^<$ plays the role of the one-body density matrix in nonrelativistic formulations at equal times. The selfenergies thus are obtained by opening of a propagator-line in the irreducible diagrams Φ. Note that this definition of the selfenergy preserves all conservation laws of the theory (as well as causality) and does not introduce additional conserved currents. In principle the Φ-functional includes irreducible diagrams up to infinite order, but here we will consider only the contributions up to second order in the coupling (2PI). For our present purpose this approximation is sufficient since we include the leading mean-field effects as well as the leading order scattering processes that pave the way to thermalization.

---

[3] In the previous sections this limit was denoted as Born approximation .

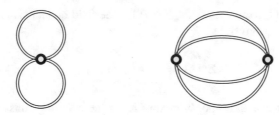

Fig. 4.2 Contributions to the Φ-functional for the Kadanoff–Baym equation: two-loop contribution (l.h.s.) giving the tadpole selfenergy and three-loop contribution (r.h.s.) generating the sunset selfenergy . The Φ-functional is built-up by full Green's functions (double lines) while open dots symbolize the integration over the inner coordinates

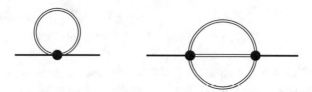

Fig. 4.3 Selfenergies of the Kadanoff–Baym equation: tadpole selfenergy (l.h.s.) and sunset selfenergy (r.h.s.) for the ϕ^4-theory. Since the lines represent full Green's functions the selfenergies are selfconsistent (see text) with the external coordinates indicated by full dots

4.1.5   Application to the Scalar ϕ^4-Theory

The contributions up to the 3-loop order for the Φ-functional (cf. Fig. 4.2) read explicitly for the ϕ^4-theory

$$i\Phi = \frac{i\lambda}{8}\int_C d^{d+1}x \ G(x,x)^2 - \frac{\lambda^2}{48}\int_C d^{d+1}x\int_C d^{d+1}y \ G(x,y)^4, \quad (4.30)$$

where d here denotes the spatial dimension of the problem.

$$\Sigma(x,y) = 2i\frac{\delta\Phi}{\delta G(y,x)} = -\frac{i\lambda}{2}G(x,x) - \frac{\lambda^2}{6}G(x,y)G(x,y)G(y,x)$$

$$= -\frac{\lambda}{2}iG(x,x) - \frac{\lambda^2}{6}[G(x,y)]^3$$

$$= \Sigma_p^\delta\delta(x_0 - y_0) + \theta_p(x_0 - y_0)\Sigma^>(x,y) + \theta_p(y_0 - x_0)\Sigma^<(x,y),$$
$$(4.31)$$

with δ_p defined in (4.20) while θ_p is the Heavyside function on the CTP (taking care about the sign on the upper (+) or lower (-) branch).

Within the 3-loop approximation for the 2PI effective action (i.e. the Φ-functional (4.30)) we get two different selfenergies: In leading order of the coupling constant only the tadpole diagram (l.h.s. of Fig. 4.3) contributes and leads to the generation of an effective mass (squared) for the field quanta. This selfenergy (in coordinate space) is given by

$$\Sigma^\delta(x) = \frac{\lambda}{2} i\, G^<(x, x)\,, \tag{4.32}$$

and is local in space and time. In next order in the coupling constant (i.e. λ^2) the nonlocal sunset selfenergy (r.h.s. of Fig. 4.3) enters the time evolution as

$$\Sigma^{\lessgtr}(x, y) = -\frac{\lambda^2}{6} G^{\lessgtr}(x, y)\, G^{\lessgtr}(x, y)\, G^{\gtrless}(y, x) \tag{4.33}$$

$$\longrightarrow \quad \Sigma^{\lessgtr}(x, y) = -\frac{\lambda^2}{6} \left[G^{\lessgtr}(x, y) \right]^3. \tag{4.34}$$

Thus the Kadanoff–Baym equation (4.22) in our case includes the influence of a mean field on the particle propagation—generated by the tadpole diagram—as well as scattering processes as inherent in the sunset diagram.

The Kadanoff–Baym equation (4.22) describes the full quantum nonequilibrium time evolution on the two-point level for a system prepared at an initial time t_0, i.e. when higher order correlations are discarded. The causal structure of this initial value problem is obvious since the time integrations are performed over the past up to the actual time x_0 (or y_0, respectively) and do not extend to the future.

4.1.6   Homogeneous Systems in Space

In the following we will consider homogeneous systems in space. To obtain a numerical solution the Kadanoff–Baym equation (4.22) is transformed to momentum space in case of the ϕ^4-theory:

$$\partial_{t_1}^2 G^<(\mathbf{p}, t_1, t_2) = -[\,\mathbf{p}^2 + m^2 + \tilde{\Sigma}^\delta(t_1)\,]\, G^<(\mathbf{p}, t_1, t_2) \tag{4.35}$$

$$-\int_{t_0}^{t_1} dt'\, \left[\Sigma^>(\mathbf{p}, t_1, t') - \Sigma^<(\mathbf{p}, t_1, t') \right]\, G^<(\mathbf{p}, t', t_2)$$

$$+\int_{t_0}^{t_2} dt'\, \Sigma^<(\mathbf{p}, t_1, t')\, \left[G^>(\mathbf{p}, t', t_2) - G^<(\mathbf{p}, t', t_2) \right]$$

$$= -[\,\mathbf{p}^2 + m^2 + \tilde{\Sigma}^\delta(t_1)\,]\, G^<(\mathbf{p}, t_1, t_2) + I_1^<(\mathbf{p}, t_1, t_2),$$

where both memory integrals are summarized in the function $I_1^<$. The equation of motion in the second time direction t_2 is given analogously. In two-time, momentum space $(\mathbf{p}, t, t')$ representation the selfenergies read

$$\tilde{\Sigma}^\delta(t) = \frac{\lambda}{2} \int \frac{d^d p}{(2\pi)^d} \; i \, G^<(\mathbf{p}, t, t) \,, \tag{4.36}$$

$$\Sigma^{\lessgtr}(\mathbf{p}, t, t') = -\frac{\lambda^2}{6} \int \frac{d^d q}{(2\pi)^d} \int \frac{d^d r}{(2\pi)^d} \; G^{\lessgtr}(\mathbf{q}, t, t')$$

$$\times G^{\lessgtr}(\mathbf{r}, t, t') \; G^{\gtrless}(\mathbf{q}+\mathbf{r}-\mathbf{p}, t', t) \,.$$

$$= -\frac{\lambda^2}{6} \int \frac{d^d q}{(2\pi)^d} \int \frac{d^d r}{(2\pi)^d} \; G^{\lessgtr}(\mathbf{q}, t, t')$$

$$\times G^{\lessgtr}(\mathbf{r}, t, t') \; G^{\lessgtr}(\mathbf{p}-\mathbf{q}-\mathbf{r}, t, t') \,.$$

By insertion of (4.36) in Eq. (4.35) we get for the collision term:

$$I_1^<(\mathbf{p}, t_1, t_2) \tag{4.37}$$

$$= +\int_{t_0}^{t_1} dt' \, \frac{\lambda^2}{6} \int \frac{d^d q}{(2\pi)^d} \int \frac{d^d r}{(2\pi)^d} \; G^>(\mathbf{q}, t_1, t') \; G^>(\mathbf{r}, t_1, t')$$

$$\times G^<(\mathbf{q}+\mathbf{r}-\mathbf{p}, t', t_1) \, G^<(\mathbf{p}, t', t_2)$$

$$-\int_{t_0}^{t_2} dt' \, \frac{\lambda^2}{6} \int \frac{d^d q}{(2\pi)^d} \int \frac{d^d r}{(2\pi)^d} \; G^<(\mathbf{q}, t_1, t') \; G^<(\mathbf{r}, t_1, t')$$

$$\times G^>(\mathbf{q}+\mathbf{r}-\mathbf{p}, t', t_1) \, G^>(\mathbf{p}, t', t_2)$$

$$+\int_{t_0}^{t_2} dt' \, \frac{\lambda^2}{6} \int \frac{d^d q}{(2\pi)^d} \int \frac{d^d r}{(2\pi)^d} \; G^<(\mathbf{q}, t_1, t') \; G^<(\mathbf{r}, t_1, t')$$

$$\times G^>(\mathbf{q}+\mathbf{r}-\mathbf{p}, t', t_1) \, G^<(\mathbf{p}, t', t_2)$$

$$-\int_{t_0}^{t_1} dt' \, \frac{\lambda^2}{6} \int \frac{d^d q}{(2\pi)^d} \int \frac{d^d r}{(2\pi)^d} \; G^<(\mathbf{q}, t_1, t') \; G^<(\mathbf{r}, t_1, t')$$

$$\times G^>(\mathbf{q}+\mathbf{r}-\mathbf{p}, t', t_1) \, G^<(\mathbf{p}, t', t_2),$$

which apart from '2 ↔ 2' processes also involves '1 ↔ 3' processes which are not allowed by energy conservation in an on-shell collision term for massive particles!

For the solution of the Kadanoff–Baym equations (4.35) a flexible and accurate algorithm works as follows: Instead of solving the second order differential equation (4.35) one can generate a set of first-order differential equations for the Green's functions in the Heisenberg picture,

$$i\, G^<_{\phi\phi}(x_1, x_2) = \langle \phi(x_2)\, \phi(x_1) \rangle = i\, G^<(x_1, x_2)\,, \tag{4.38}$$

$$i\, G^<_{\pi\phi}(x_1, x_2) = \langle \phi(x_2)\, \pi(x_1) \rangle = \partial_{t_1} i\, G^<_{\phi\phi}(x_1, x_2)\,,$$

$$i\, G^<_{\phi\pi}(x_1, x_2) = \langle \pi(x_2)\, \phi(x_1) \rangle = \partial_{t_2} i\, G^<_{\phi\phi}(x_1, x_2)\,,$$

$$i\, G^<_{\pi\pi}(x_1, x_2) = \langle \pi(x_2)\, \pi(x_1) \rangle = \partial_{t_1}\, \partial_{t_2} i\, G^<_{\phi\phi}(x_1, x_2)\,,$$

with the canonical field momentum $\pi(x) = \partial_{x_0}\phi(x)$. The first index π or ϕ is always related to the first space-time argument. Exploiting the time-reflection symmetry of the Green's functions some of the differential equations are redundant. The required equations of motion are given as [7]

$$\partial_{t_1}\, G^<_{\phi\phi}(\mathbf{p}, t_1, t_2) = G^<_{\pi\phi}(\mathbf{p}, t_1, t_2)\,, \tag{4.39}$$

$$\partial_{\tilde{t}}\, G^<_{\phi\phi}(\mathbf{p}, \tilde{t}, \tilde{t}) = 2i\, \Im\{ G^<_{\pi\phi}(\mathbf{p}, \tilde{t}, \tilde{t}) \}\,,$$

$$\partial_{t_1}\, G^<_{\pi\phi}(\mathbf{p}, t_1, t_2) = -\, \Omega^2(t_1)\, G^<_{\phi\phi}(\mathbf{p}, t_1, t_2) + I^<_1(\mathbf{p}, t_1, t_2)\,,$$

$$\partial_{t_2}\, G^<_{\pi\phi}(\mathbf{p}, t_1, t_2) = G^<_{\pi\pi}(\mathbf{p}, t_1, t_2)\,,$$

$$\partial_{\tilde{t}}\, G^<_{\pi\phi}(\mathbf{p}, \tilde{t}, \tilde{t}) = -\, \Omega^2(\tilde{t})\, G^<_{\phi\phi}(\mathbf{p}, \tilde{t}, \tilde{t}) + G^<_{\pi\pi}(\mathbf{p}, \tilde{t}, \tilde{t}) + I^<_1(\mathbf{p}, \tilde{t}, \tilde{t})\,,$$

$$\partial_{t_1}\, G^<_{\pi\pi}(\mathbf{p}, t_1, t_2) = -\, \Omega^2(t_1)\, G^<_{\phi\pi}(\mathbf{p}, t_1, t_2) + I^<_{1,2}(\mathbf{p}, t_1, t_2)\,,$$

$$\partial_{\tilde{t}}\, G^<_{\pi\pi}(\mathbf{p}, \tilde{t}, \tilde{t}) = -\, \Omega^2(\tilde{t})\, 2i\, \Im\{ G^<_{\pi\phi}(\mathbf{p}, \tilde{t}, \tilde{t}) \} + 2i\, \Im\{ I^<_{1,2}(\mathbf{p}, \tilde{t}, \tilde{t}) \}\,,$$

where $\tilde{t} = (t_1 + t_2)/2$ is the mean time variable. Thus one explicitly considers the propagation in the time diagonal. In the equations of motion (4.39) the current (renormalized) effective energy including the time dependent tadpole contribution enters,

$$\Omega^2(t) = \mathbf{p}^2 + m^2 + \delta m^2_{tad} + \delta m^2_{sun} + \bar{\Sigma}^\delta(t), \tag{4.40}$$

with the counterterms δm^2_{tad} and δm^2_{sun}.[4] The evolution in the t_2 direction has not to be taken into account for $G^<_{\phi\phi}$ and $G^<_{\pi\pi}$ since the Green's functions beyond the time diagonal ($t_2 > t_1$) are determined via the time-reflection symmetry, e.g. $G^<_{\phi\phi}(\mathbf{p}, t_1, t_2) = -[G^<_{\phi\phi}(\mathbf{p}, t_2, t_1)]^*$, from the known values for the lower time triangle in both cases. Since there is no time-reflection symmetry for the $G_{\pi\phi}$ functions, they have to be calculated (and stored) in the whole t_1, t_2 range. However, we can ignore the evolution of $G_{\phi\pi}$ since it is obtained by the relation $G^<_{\phi\pi}(\mathbf{p}, t_1, t_2) = -[G^<_{\pi\phi}(\mathbf{p}, t_2, t_1)]^*$. The correlation integrals in (4.39) are given by

$$I^<_1(\mathbf{p}, t_1, t_2) = -\int_0^{t_1} dt' \; [\Sigma^>(\mathbf{p}, t_1, t') - \Sigma^<(\mathbf{p}, t_1, t')] \; G^<_{\phi\phi}(\mathbf{p}, t', t_2) \quad (4.41)$$

$$+\int_0^{t_2} dt' \; \Sigma^<(\mathbf{p}, t_1, t') \; \left[G^<_{\phi\phi}(-\mathbf{p}, t_2, t') - G^<_{\phi\phi}(\mathbf{p}, t', t_2) \right] ,$$

$$I^<_{1,2}(\mathbf{p}, t_1, t_2) = \partial_{t_2} I^<_1(\mathbf{p}, t_1, t_2) \quad (4.42)$$

$$= -\int_0^{t_1} dt' \; [\Sigma^>(\mathbf{p}, t_1, t') - \Sigma^<(\mathbf{p}, t_1, t')] \; G^<_{\phi\pi}(\mathbf{p}, t', t_2)$$

$$+\int_0^{t_2} dt' \; \Sigma^<(\mathbf{p}, t_1, t') \; \left[G^<_{\pi\phi}(-\mathbf{p}, t_2, t') - G^<_{\phi\pi}(\mathbf{p}, t', t_2) \right] .$$

In (4.39) and (4.42) one can replace $G^<_{\phi\pi}(\mathbf{p}, t_1, t_2) = -[G^<_{\pi\phi}(\mathbf{p}, t_2, t_1)]^*$ such that the set of equations is closed in the Green's functions $G^<_{\phi\phi}$, $G^<_{\pi\phi}$ and $G^<_{\pi\pi}$.

The disadvantage, to integrate more Green's functions in time in this first-order scheme, is compensated by its good accuracy. As mentioned before, we especially take into account the propagation along the time diagonal which leads to an improved numerical precision. The set of differential equations (4.39) is solved by means of a 4th order Runge–Kutta algorithm. For the calculation of the selfenergies a Fast-Fourier transformation method is applied. The selfenergies (4.36), furthermore, are calculated in coordinate space—where they are products of coordinate-space Green's functions (that are available by Fourier transformation)—and finally transformed to momentum space.

In order to obtain a solution of the KB equations some initial conditions for $iG^<(\mathbf{p}, t = 0, t = 0)$ have to be specified. The corresponding initial distribution functions in the occupation density $n(\mathbf{p}, t = 0)$, related to $iG^<(\mathbf{p}, t = 0, t = 0)$ by

$$2\omega_{\mathbf{p}} iG^<(\mathbf{p}, t = 0, t = 0) = 2n(\mathbf{p}, t = 0) + 1 \quad (4.43)$$

---

[4] The explicit form of the counterterms are not of relevance here. They are specified in detail in Ref. [7].

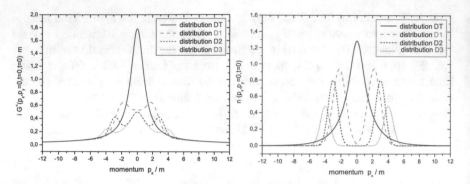

Fig. 4.4 Initial Green's functions $iG^{<}(|\vec{p}|, t = 0, t = 0)$ (l.h.s.) and corresponding initial distribution functions $n(|\vec{p}|, t = 0)$ (r.h.s.) for the distributions D1, D2, D3, and DT in momentum space (for a cut of the polar symmetric distribution in p_x direction for $p_y = 0$)

follow immediately. To this aim we consider four different initial distributions that are all characterized by the same energy density. Consequently, for large times ($\rightarrow \infty$) all initial value problems should lead to the same equilibrium final state. The initial equal-time Green's functions $iG^{<}(\mathbf{p}, t = 0, t = 0)$ adopted are displayed in Fig. 4.4 (l.h.s.) as a function of the momentum p_z. We concentrate on polar symmetric configurations due to the large numerical expense for this exploratory investigation. Since the equal-time Green's functions $G^{<}(\mathbf{p}, t, t,)$ are purely imaginary we show only the real part of $iG^{<}$ in Fig. 4.4.

Since we consider a finite volume in two dimensions ($V = a^2$) we work in a basis of momentum modes characterized by the number of nodes in each direction. The number of momentum modes is typically in the order of 40 which is found to be sufficient for numerically stable results. For times $t < 0$ we consider the systems to be non-interacting and switch on the interaction ($\sim \lambda$) for $t=0$ to explore the quantum dynamics of the interacting system for $t > 0$.

4.1.7 The Spectral Function

The spectral function of the fields ϕ is of particular interest since it follows from the field commutator at unequal times and reflects the quantization of the theory. For scalar, symmetric fields ϕ it is given by

$$A(x, y) = \langle [\phi(x), \phi(y)]_- \rangle = i[G^{>}(x, y) - G^{<}(x, y)] = i[G^{R}(x, y) - G^{A}(x, y)]$$

$$(4.44)$$

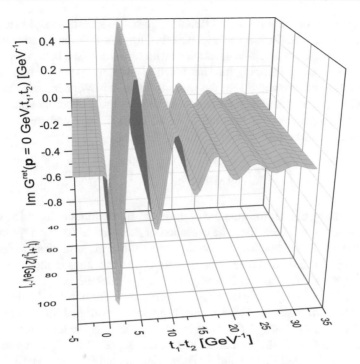

Fig. 4.5 The imaginary part of the retarded Green's function as a function of $t_1 - t_2$ and the average time $(t_1 + t_2)/2$ for ϕ^4-theory in strong coupling as emerging from the Kadanoff–Baym approach

or in momentum-time representation as

$$A(\mathbf{p}, t_1, t_2) = i[G^<(\mathbf{p}, t_2, t_1) - G^<(\mathbf{p}, t_1, t_2)] \tag{4.45}$$

$$= i\left[-[G^<(\mathbf{p}, t_1, t_2)]^* - G^<(\mathbf{p}, t_1, t_2)\right]$$

$$= -2i\,\Re\left(G^<(\mathbf{p}, t_1, t_2)\right).$$

The quantity (4.45) is displayed in Fig. 4.5 as a function of $\Delta t = t_1 - t_2$ and $t = (t_1 + t_2)/2$ for a low lying momentum mode in case of the ϕ^4-theory for strong coupling λ. We observe a damped oscillation in $\Delta t = t_1 - t_2$ in all cases with characteristic time scale $1/\Gamma$ which practically does not depend on the average time $t = (t_1 + t_2)/2$. This pattern is very similar for all momentum modes.

The spectral function in energy-momentum representation is obtained by Fourier transformation with respect to the time difference $\Delta t = (t_1 - t_2)$ for each average time t:

$$\tilde{A}(\mathbf{p}, p_0, t) = \int_{-\infty}^{\infty} d\Delta t\, \exp(i\Delta t\, p_0) A(\mathbf{p}, t_1 = t + \Delta t/2, t_2 = t - \Delta t/2) \quad,$$

$$\tag{4.46}$$

where $\bar{A}$ stands for the full spectral function. Since the spectral function essentially shows a damped oscillation (cf. Fig. 4.5) this implies that the Fourier transform (4.46) is of relativistic Breit–Wigner shape with a width Γ (see below).

We note, that a damping of the function $A(\mathbf{p}, t_1, t_2)$ in relative time Δt corresponds to a finite width Γ of the spectral function in Wigner space. This width in turn can be interpreted as the inverse life time of the interacting scalar particle. We recall, that the spectral function —for all times $T \equiv t$ and for all momenta $\mathbf{p}$—obeys the normalization

$$\int_{-\infty}^{\infty} \frac{dp_0}{2\pi} \, p_0 \, A(\mathbf{p}, p_0, T) = 1 \qquad \forall \, \mathbf{p}, \, T \tag{4.47}$$

which is nothing but a reformulation of the equal-time commutation relation.

In Fig. 4.6[5] we display the time evolution of the spectral function for the initial distributions D1, D2, and DT for two different momentum modes $|\mathbf{p}|/m = 0.0$ and $|\mathbf{p}|/m = 2.0$. Since the spectral functions are antisymmetric in energy for the momentum symmetric configurations considered, i.e. $A(\mathbf{p}, -p_0, T) = -A(\mathbf{p}, p_0, T)$, we only show the positive energy part. For our initial value problem in two-times and space the Fourier transformation is restricted for system times T to an interval $\Delta t \in [-2T, 2T]$. Thus in the very early phase the spectral function assumes a finite width already due to the limited support of the Fourier transform in the interval $\Delta t \in [-2T, 2T]$ and a Wigner representation is not very meaningful. We, therefore, present the spectral functions for various system times $T \equiv t$ starting from $t \cdot m = 15$ up to $t \cdot m = 480$.

For the free thermal initialization DT the evolution of the spectral function is very smooth. In this case the spectral function is already close to the equilibrium shape at small times being initially only slightly broader than for late times. The maximum of the spectral function (for all momenta) is higher than the (bare) on-shell value and nearly keeps its position during the whole time evolution. This results from a positive tadpole mass shift, which is only partly compensated by a downward shift originating from the sunset diagram.

The time evolution for the initial distributions D1, D2, and D3 has a richer structure. For the distribution D1 the spectral function is broad for small system times (see the line for $t \cdot m = 15$) and becomes a little sharper in the course of the time evolution (as presented for the momentum mode $|\mathbf{p}|/m = 0.0$ as well as for $|\mathbf{p}|/m = 2.0$). In line with the decrease in width the height of the spectral function is increasing (as demanded by the normalization property (4.47)). This is indicated by the small arrow close to the peak position. Furthermore, the maximum of the spectral function (which is approximately the on-shell energy) is shifted slightly upwards for the zero mode and downwards for the mode with higher momentum. Although the real part of the (retarded) sunset selfenergy leads (in general) to a lowering of the effective mass, the on-shell energy of the momentum modes is still

---

[5] The figures in this section are taken from Ref. [7].

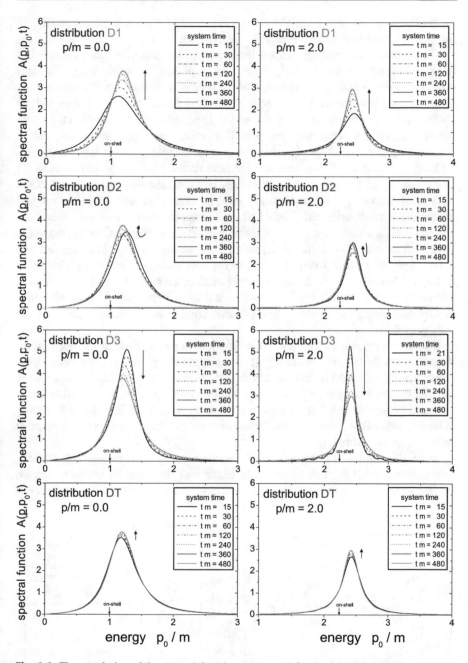

Fig. 4.6 Time evolution of the spectral function $A(\mathbf{p}, p_0, t)$ for the initial distributions D1, D2, D3, and DT (from top to bottom) for the two momenta $|\mathbf{p}|/m = 0.0$ (l.h.s.) and $|\mathbf{p}|/m = 2.0$ (r.h.s.). The spectral function is shown for several times $t \cdot m = 15, 30, 60, 120, 240, 360, 480$ as indicated by the different line types

higher than the one for the initial mass m (indicated by the 'on-shell' arrow) due to the positive mass shift from the tadpole contribution.

For the initial distribution D3 we find the opposite behavior. Here the spectral function is quite narrow for early times and increasing its width during the time evolution. Correspondingly, the height of the spectral function decreases with time. This behavior is observed for the zero momentum mode $|\mathbf{p}|/m = 0.0$ as well as for the finite momentum mode $|\mathbf{p}|/m = 2.0$. Especially in the latter case the width for early times is so small that the spectral function shows oscillations originating from the finite range of the Fourier transformation from relative time to energy. Although we have already increased the system time for the first curve to $t \cdot m = 21$ (for $t \cdot m = 15$ the oscillations are much stronger) the spectral function is not fully resolved, i.e. it is not sufficiently damped in relative time Δt in the interval available for the Fourier transform. For later times the oscillations vanish and the spectral function tends to the common equilibrium shape. The time evolution of the spectral function for the initial distribution D2 is somehow in between the last two cases. Here the spectral function develops (at intermediate times) a slightly higher width than in the beginning before it is approaching the narrower static shape again. The corresponding evolution of the maximum is again indicated by the (bent) arrow. Finally, all spectral functions show the (same) equilibrium form represented by the solid gray line.

One has to emphasize that there is no unique time evolution for the nonequilibrium systems. In fact, the evolution of the system during the equilibration process depends on the initial conditions. On the other hand, the time dependence of the spectral function is only moderate such that one might also work with some time-averaged or even the equilibrium spectral function. In order to investigate this issue in more quantitative detail we concentrate on the maxima and widths of the spectral functions in the following.

Since the solution of the Kadanoff–Baym equation provides the full spectral information for all system times the evolution of the on-shell energies can be studied as well as the spectral widths. In Fig. 4.7 we display the time dependence

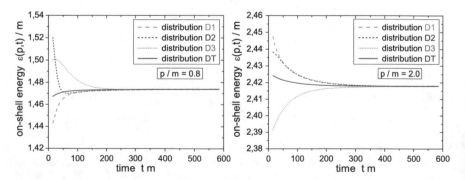

Fig. 4.7 Time evolution of the on-shell energies $\varepsilon(\mathbf{p}, t)$ of the momentum modes $|\mathbf{p}|/m = 0.8$ and $|\mathbf{p}|/m = 2.0$ for the different initializations D1, D2, D3 and DT. The on-shell selfenergies are extracted from the maxima of the time-dependent spectral functions

of the on-shell energies $\varepsilon(\mathbf{p}, t)$ of the momentum modes $|\mathbf{p}|/m = 0.8$ (l.h.s.) and $|\mathbf{p}|/m = 2.0$ (r.h.s.) for the four initial distributions D1, D2, D3, and DT. We see that the on-shell energy for the zero momentum mode increases with time for the initial distribution D1 and to a certain extent for the free thermal distribution DT (as can be also extracted form Fig. 4.6). The on-shell energy of distribution D3 shows a monotonic decrease during the evolution while it passes through a minimum for distribution D2 before joining the line for the initialization D1. For momentum $|\mathbf{p}|/m = 2.0$ a rather opposite behavior is observed. Here the on-shell energy for distribution D1 (and less pronounced for the distribution DT) are reduced in time whereas it is increased in the case of D3. The result for the initialization D2 is monotonous for this mode and matches the one for D1 already for moderate times. Thus we find, that the time evolution of the on-shell energies does not only depend on the initial conditions, but might also be different for various momentum modes. It turns out—for the initial distributions investigated—that the above described characteristics change around $|\mathbf{p}|/m = 1.5$ and are retained for larger momenta (not presented here).

Furthermore, we show in Fig. 4.8 the time evolution of the on-shell width for the usual momentum modes for the different initial distributions. The on-shell width Γ is given by the imaginary part of the retarded sunset selfenergy at the on-shell energy of each respective momentum mode as

$$\Gamma = -2\, Im\, \Sigma^R(\mathbf{p}, \varepsilon(\mathbf{p}, t), t)\, /\, 2\, \varepsilon(\mathbf{p}, t). \tag{4.48}$$

As already discussed in connection with Fig. 4.6 we observe for both momentum modes a strong decrease of the on-shell width for the initial distribution D1 associated with a narrowing of the spectral function. In contrast, the on-shell widths of distribution D3 increase with time such that the corresponding spectral functions broaden towards the common static shape. For the initialization D2 we observe a non-monotonic evolution of the on-shell widths connected with a broadening of the

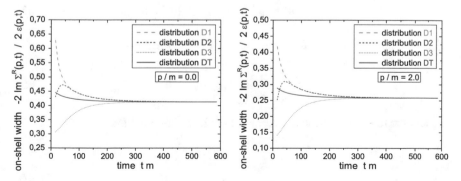

Fig. 4.8 Time evolution of the on-shell widths $-2\, Im\, \Sigma^R(\mathbf{p}, \varepsilon(\mathbf{p}, t), t)/2\, \varepsilon(\mathbf{p}, t)$ of the momentum modes $|\mathbf{p}|/m = 0.8$ and $|\mathbf{p}|/m = 2.0$ for the different initializations D1, D2, D3, and DT

spectral function at intermediate times. Similar to the case of the on-shell energies we find, that the results for the on-shell widths of the distributions D1 and D2 coincide well above a certain system time. As expected from the lower plots of Fig. 4.6 the on-shell width for the free thermal distribution DT exhibits only a weak time dependence with a slight decrease in the initial phase of the time evolution.

In summary, there is no universal time evolution of the spectral functions for the initial distributions considered. Peak positions and widths depend on the initial configuration and evolve differently in time. However, we find only effects in the order of $<10\%$ for the on-shell energies in the initial phase of the system evolution and initial variations of $<50\%$ for the widths of the dominant momentum modes. Thus, depending on the physics problem of interest, one might discard an explicit time dependence of the spectral functions and adopt the equilibrium shape.

4.1.8 Results in First-Order Gradient Expansion

In first order in the gradient expansion the retarded and advanced Green's functions can be written as

$$\tilde{G}^{R/A} = \Re\left(\tilde{G}^R\right) \pm i\,\Im\left(\tilde{G}^R\right) = \Re\left(\tilde{G}^R\right) \mp i\,\tilde{A}/2,$$

$$\tilde{\Sigma}^{R/A} = \Re\left(\tilde{\Sigma}^R\right) \pm i\,\Im\left(\tilde{\Sigma}^R\right) = \Re\left(\tilde{\Sigma}^R\right) \mp i\,\tilde{\Gamma}/2. \tag{4.49}$$

Rewriting the imaginary part of the selfenergy we get

$$\tilde{A}(\mathbf{p}, p_0, t) = \frac{\tilde{\Gamma}}{\left(p_0^2 - \omega_0^2\right)^2 + \tilde{\Gamma}^2/4},$$

$$\tilde{\Gamma} = -2\,\Im\left(\bar{\Sigma}^R\right) = 4p_0\gamma,$$

$$\omega_0^2 = \mathbf{p}^2 + m^2 - \bar{\Sigma}^\delta + \Re\left(\bar{\Sigma}^R\right), \tag{4.50}$$

which is of relativistic Breit–Wigner form. Its normalization is given by

$$\int_{-\infty}^{\infty} \frac{dp_0}{2\pi}\, p_0 \tilde{A}(\mathbf{p}, p_0, t) = 1 \tag{4.51}$$

and reflects the quantization condition for the interacting field ϕ.

4.1.9 The Equilibrium Distribution

Now we introduce the energy- and momentum-dependent distribution function $N(\mathbf{p}, p_0, \tilde{t})$ at any system time $\tilde{t}$ by the definition

$$i\, G^<(\mathbf{p}, p_0, \tilde{t}) = A(\mathbf{p}, p_0, \tilde{t})\ N(\mathbf{p}, p_0, \tilde{t}),$$

$$i\, G^>(\mathbf{p}, p_0, \tilde{t}) = A(\mathbf{p}, p_0, \tilde{t})\ [\,N(\mathbf{p}, p_0, \tilde{t}) + 1\,], \qquad (4.52)$$

since $G^<(\mathbf{p}, p_0, \tilde{t})$ and $G^>(\mathbf{p}, p_0, \tilde{t})$ are known from the integration of the Kadanoff–Baym equations as well as $A(\mathbf{p}, p_0, \tilde{t})$.

In equilibrium (at temperature T) the Green's functions obey the Kubo–Martin–Schwinger (KMS) relation for all momenta $\mathbf{p}$,

$$G^>_{eq}(\mathbf{p}, p_0) = e^{p_0/T}\, G^<_{eq}(\mathbf{p}, p_0) \qquad \forall\ \mathbf{p}. \qquad (4.53)$$

If there exists a conserved quantum number in the theory we have, furthermore, a contribution of the corresponding chemical potential in the exponential function which leads to a shift of arguments: $p_0/T \to (p_0 - \mu)/T$. In the present case, however, there is no conserved quantum number and thus the equilibrium state has the chemical potential $\mu = 0$.

From the KMS condition of the Green's functions (4.53) we obtain the equilibrium form of the distribution function (4.52) at temperature T as

$$N_{eq}(\mathbf{p}, p_0) = N_{eq}(p_0) = \frac{1}{e^{p_0/T} - 1} = N_{bose}(p_0/T), \qquad (4.54)$$

from

$$\frac{G^<}{G^>} = e^{-p_0/T} = \frac{N_{eq}}{N_{eq} + 1},$$

which is the well-known Bose distribution. As is obvious from (4.54) the equilibrium distribution can only be a function of energy p_0 and not of the momentum variable $\mathbf{p}$ in addition.

In Fig. 4.9 (lower part) we present the spectral function $A(\mathbf{p}, p_0)$ for the initial distribution D2 at late times $\tilde{t} \cdot m = 540$ for various momentum modes $|\mathbf{p}|/m = 0.0, 0.8, 1.6, 2.4, 3.2, 4.0$ as a function of the energy p_0. We note, that for all other initial distributions—with equal energy density—the spectral function looks very similar at this time since the systems proceed to the same stationary state. We recognize that the spectral function is quite broad, especially for the low momentum modes, while for the higher momentum modes its width is slightly lower.

The distribution function $N(p_0)$ as extracted from (4.52) is displayed in Fig. 4.9 (upper part) for the same momentum modes as a function of the energy p_0. We find that $N(p_0)$ for all momentum modes can be fitted by a single Bose function

Fig. 4.9 Spectral function $A(p_0)$ for various momentum modes $|\mathbf{p}|/m=$ 0.0, 0.8, 1.6, 2.4, 3.2, 4.0 as a function of energy for late times $t \cdot m = 540$ (lower part). The corresponding distribution function $N(p_0)$ at the same time for the same momentum modes is displayed in the upper part. All momentum modes can be fitted with a single Bose function of temperature $T_{eq}/m = 1.836$ and a chemical potential close to zero

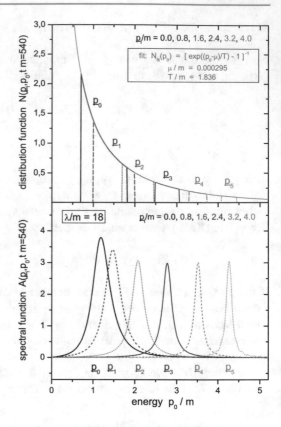

with temperature $T/m = 1.836$. Thus the distribution function emerging from the Kadanoff–Baym time evolution for $t \rightarrow \infty$ approaches a Bose function in the energy that is independent of the momentum as demanded by the equilibrium form (4.54). Figure 4.9 (upper part) demonstrates, furthermore, that the KMS condition is fulfilled not only for on-shell energies, but for all p_0. We, therefore, have obtained the full off-shell equilibrium state by integrating the Kadanoff–Baym equations in time. In addition, the limiting stationary state is the correct equilibrium state for all energies p_0, i.e. also away from the quasiparticle energies.

4.2   Full Versus Approximate Dynamics

The Kadanoff–Baym equations studied in the previous section represent the full quantum-field theoretical equations on the single-particle level. However, its numerical solution is quite involved and it is of strong interest to investigate, in how far approximate schemes deviate from the full calculation. Nowadays, transport models are widely used in the description of quantum systems out-of-equilibrium. Most of these models work in the "quasiparticle" picture, where all particles obey a fixed energy-momentum relation and the energy is no independent degree of freedom anymore; it is determined by the momentum and the (effective) mass of the particle (cf. Chaps. 2 and 3). Accordingly, these particles are treated with their δ-function spectral shape as infinitely long living, i.e. stable objects. This assumption is very questionable e.g. for high-energy heavy-ion reactions, where the particles achieve a large width due to the frequent collisions with other particles in the high-density and/or high-energy regime. Furthermore, this is doubtful for particles that are unstable even in the vacuum. The question, in how far the quasiparticle approximation influences the dynamics in comparison to the full Kadanoff–Baym calculation, is of widespread interest.

4.2.1 Derivation of the Quantum Boltzmann Approximation

In the following we will present a short derivation of the quantum Boltzmann equation starting directly from the Kadanoff–Baym dynamics in the two-time and momentum-space representation. This derivation is briefly reviewed since we want (1) to emphasize the link of the full Kadanoff–Baym equation with its approximated version and (2) to clarify the assumptions that enter the Boltzmann equation .

Since the Boltzmann equation describes the time evolution of distribution functions for quasiparticles we first consider the quasiparticle Green's functions in two-time representation for homogeneous systems

$$
G^{\lessgtr}_{\phi\phi,qp}(\mathbf{p}, t, t') = \frac{-i}{2\omega_{\mathbf{p}}}\{\, N_{qp}(\mp\mathbf{p}) \quad \exp(\pm i\omega_{\mathbf{p}}(t - t')) \tag{4.55}
$$

$$
+[\, N_{qp}(\pm\mathbf{p})+1\,] \quad \exp(\mp i\omega_{\mathbf{p}}(t - t'))\,\}
$$

$$
G^{\lessgtr}_{\phi\pi,qp}(\mathbf{p}, t, t') = \frac{1}{2}\{\, \mp N_{qp}(\mp\mathbf{p}) \quad \exp(\pm i\omega_{\mathbf{p}}(t - t'))
$$

$$
\pm[\, N_{qp}(\pm\mathbf{p})+1\,] \quad \exp(\mp i\omega_{\mathbf{p}}(t - t'))\,\}
$$

$$
G^{\lessgtr}_{\pi\phi,qp}(\mathbf{p}, t, t') = \frac{1}{2}\{\, \pm N_{qp}(\mp\mathbf{p}) \quad \exp(\pm i\omega_{\mathbf{p}}(t - t'))
$$

$$
\mp[\, N_{qp}(\pm\mathbf{p})+1\,] \quad \exp(\mp i\omega_{\mathbf{p}}(t - t'))\,\}
$$

$$
G^{\lessgtr}_{\pi\pi,qp}(\mathbf{p}, t, t') = \frac{-i\,\omega_{\mathbf{p}}}{2}\{\, N_{qp}(\mp\mathbf{p}) \quad \exp(\pm i\omega_{\mathbf{p}}(t - t'))
$$

$$
+[\, N_{qp}(\pm\mathbf{p})+1\,] \quad \exp(\mp i\omega_{\mathbf{p}}(t - t'))\,\}
$$

where for each momentum $\mathbf{p}$ the Green's functions are freely oscillating in relative time $t - t'$ with the on-shell energy $\omega_{\mathbf{p}}$. The time-dependent quasiparticle distribution functions are given with the energy variable fixed to the on-shell energy as $N_{qp}(\mathbf{p}, \tilde{t}) \equiv N(\mathbf{p}, p_0 = \omega_{\mathbf{p}}, \tilde{t})$, where the on-shell energies $\omega_{\mathbf{p}}$ can depend on time as well. Such a time variation e.g. might be due to an effective mass as generated by the time-dependent tadpole selfenergy . In this case the on-shell energy reads

$$\omega_{\mathbf{p}}(\tilde{t}) = \sqrt{\mathbf{p}^2 + m^2 + \bar{\Sigma}^\delta_{ren}(\tilde{t})}. \tag{4.56}$$

Vice versa we can define the quasiparticle distribution function by means of the quasiparticle Green's functions at equal times $\tilde{t}$ as

$$N_{qp}(\mathbf{p}, \tilde{t}) = \left[\frac{\omega_{\mathbf{p}}(\tilde{t})}{2} \, i \, G^<_{\phi\phi,qp}(\mathbf{p}, \tilde{t}, \tilde{t}) \; + \; \frac{1}{2\omega_{\mathbf{p}}(\tilde{t})} \, i \, G^<_{\pi\pi,qp}(\mathbf{p}, \tilde{t}, \tilde{t}) \right] \tag{4.57}$$

$$- \frac{1}{2} \left[\; G^<_{\pi\phi,qp}(\mathbf{p}, \tilde{t}, \tilde{t}) \; - \; G^<_{\phi\pi,qp}(\mathbf{p}, \tilde{t}, \tilde{t}) \; \right].$$

Using the equations of motions for the Green's functions in diagonal time direction (4.39) (exploiting $G^<_{\phi\pi}(\mathbf{p}, \tilde{t}, \tilde{t}) = -[\, G^<_{\pi\phi}(\mathbf{p}, \tilde{t}, \tilde{t})\,]^*$) the time evolution of the distribution function is given by

$$\partial_{\tilde{t}} \, N_{qp}(\mathbf{p}, \tilde{t}) = -\, \Re \left\{ I^<_{1;\,qp}(\mathbf{p}, \tilde{t}, \tilde{t}) \right\} - \frac{1}{\omega_{\mathbf{p}}(\tilde{t})} \, \Im \left\{ I^<_{1,2;\,qp}(\mathbf{p}, \tilde{t}, \tilde{t}) \right\}. \tag{4.58}$$

The time derivatives of the on-shell energies cancel out since the quasiparticle Green's functions obey

$$G^<_{\pi\pi}(\mathbf{p}, \tilde{t}, \tilde{t}) = \omega^2_{\mathbf{p}}(\tilde{t}) \, G^<_{\phi\phi}(\mathbf{p}, \tilde{t}, \tilde{t}) \tag{4.59}$$

as seen from (4.55). Furthermore, it is remarkable that contributions containing the energy $\omega^2_{\mathbf{p}}$—as present in the equation of motion for the Green's functions (4.39)—no longer show up. The time evolution of the distribution function is entirely determined by (equal-time) collision integrals containing (time derivatives of the) Green's functions and selfenergies.

$$I^<_{1;qp}(\mathbf{p}, \tilde{t}, \tilde{t}) = \int_{t_0}^{\tilde{t}} dt' \; \left(\Sigma^<_{qp}(\mathbf{p}, \tilde{t}, t') \, G^>_{\phi\phi,qp}(\mathbf{p}, t', \tilde{t}) \right.$$

$$\left. - \Sigma^>_{qp}(\mathbf{p}, \tilde{t}, t') \, G^<_{\phi\phi,qp}(\mathbf{p}, t', \tilde{t}) \right),$$

$$I^<_{1,2;qp}(\mathbf{p}, \tilde{t}, \tilde{t}) = \int_{t_0}^{\tilde{t}} dt' \; \left(\Sigma^<_{qp}(\mathbf{p}, \tilde{t}, t') \, G^>_{\phi\pi,qp}(\mathbf{p}, t', \tilde{t}) \right. \tag{4.60}$$

$$\left. - \Sigma^>_{qp}(\mathbf{p}, \tilde{t}, t') \, G^<_{\phi\pi,qp}(\mathbf{p}, t', \tilde{t}) \right).$$

Since we are dealing with a system of on-shell quasiparticles within the Boltzmann approximation, the Green's functions in the collision integrals (4.60) are given by the respective quasiparticle quantities of (4.55). Moreover, the collisional selfenergies (4.36) are obtained in accordance with the quasiparticle approximation as

$$\Sigma_{qp}^{\lessgtr}(\mathbf{p}, t, t') = -i\frac{\lambda^2}{6} \int \frac{d^d q}{(2\pi)^d} \int \frac{d^d k}{(2\pi)^d} \int \frac{d^d l}{(2\pi)^d} \ (2\pi)^d \ \delta^{(d)}(\mathbf{p}-\mathbf{q}-\mathbf{k}-\mathbf{l}) \ \frac{1}{2\omega_\mathbf{q} \ 2\omega_\mathbf{k} \ 2\omega_\mathbf{l}}$$

$$\left\{ N_{qp}(\mp\mathbf{q}) N_{qp}(\mp\mathbf{k}) N_{qp}(\mp\mathbf{l}) \ \exp(+i\,[\,t - t'\,]\,[\,\pm\omega_\mathbf{q} \pm \omega_\mathbf{k} \pm \omega_\mathbf{l}\,]) \right.$$

$$+ 3N_{qp}(\mp\mathbf{q}) N_{qp}(\mp\mathbf{k}) [\, N_{qp}(\pm\mathbf{l})+1\,] \ \exp(+i\,[\,t - t'\,]\,[\,\pm\omega_\mathbf{q} \pm \omega_\mathbf{k} \mp \omega_\mathbf{l}\,])$$

$$+ 3N_{qp}(\mp\mathbf{q}) [\, N_{qp}(\pm\mathbf{k})+1\,][\, N_{qp}(\pm\mathbf{l})+1\,] \exp(+i\,[\,t - t'\,]\,[\,\pm\omega_\mathbf{q} \mp \omega_\mathbf{k} \mp \omega_\mathbf{l}\,])$$

$$\left. + [\, N_{qp}(\pm\mathbf{q})+1\,][\, N_{qp}(\pm\mathbf{k})+1\,][\, N_{qp}(\pm\mathbf{l})+1\,] \ \exp(+i\,[\,t - t'\,]\,[\,\mp\omega_\mathbf{q} \mp \omega_\mathbf{k} \mp \omega_\mathbf{l}\,]) \ \right\}.$$

$$(4.61)$$

For a free theory the distribution functions $N_{qp}(\mathbf{p})$ are obviously constant in time which, of course, is no longer valid for an interacting system out-of-equilibrium . Thus one has to specify the above expressions for the quasiparticle Green's functions (4.55) to account for the time dependence of the distribution functions.

The quantum Boltzmann approximation is defined in the limit, that the distribution functions have to be taken always at *the latest time argument* of the two-time Green's function . Accordingly, for the general nonequilibrium case, we introduce the ansatz for the Green's functions in the collision term

$$G_{\phi\phi,qp}^{\lessgtr}(\mathbf{p}, t, t') = \frac{-i}{2\omega_\mathbf{p}} \{ N_{qp}(\mp\mathbf{p}, t_{max}) \ \exp(\pm i\omega_\mathbf{p}(t - t')) \tag{4.62}$$

$$+ [\, N_{qp}(\pm\mathbf{p}, t_{max})+1\,] \ \exp(\mp i\omega_\mathbf{p}(t - t')) \}$$

$$G_{\phi\pi,qp}^{\lessgtr}(\mathbf{p}, t, t') = \frac{1}{2} \{ \mp N_{qp}(\mp\mathbf{p}, t_{max}) \ \exp(\pm i\omega_\mathbf{p}(t - t'))$$

$$\pm [\, N_{qp}(\pm\mathbf{p}, t_{max})+1\,] \ \exp(\mp i\omega_\mathbf{p}(t - t')) \},$$

with the maximum time $t_{max} = max(t, t')$. The same ansatz is employed for the time-dependent on-shell energies which enter the representation of the quasiparticle two-time Green's functions (4.62) with their value at t_{max}, i.e. $\omega_\mathbf{p} = \omega_\mathbf{p}(t_{max} = max(t, t'))$.

The collision term contains a time integration which extends from an initial time t_0 to the current time $\tilde{t}$. All two-time Green's functions and selfenergies depend on the current time $\tilde{t}$ as well as on the integration time $t' \leq \tilde{t}$. Thus only distribution functions at the current time, i.e. the maximum time of all appearing two-time functions, enter the collision integrals and the evolution equation for the distribution function becomes local in time. Since the distribution functions are given at fixed time $\tilde{t}$, they can be taken out of the time integral. When inserting the expressions for the selfenergies and the Green's functions in the collision integrals the evolution equation for the quasiparticle distribution function reads

$$
\partial_{\tilde{t}} N_{qp}(\mathbf{p}, \tilde{t}) = \frac{\lambda^2}{3} \int \frac{d^d q}{(2\pi)^d} \int \frac{d^d k}{(2\pi)^d} \int \frac{d^d l}{(2\pi)^d}
$$

$$
(2\pi)^d \, \delta^{(d)}(\mathbf{p} - \mathbf{q} - \mathbf{k} - \mathbf{l}) \, \frac{1}{2\omega_p \, 2\omega_q \, 2\omega_k \, 2\omega_l} \tag{4.63}
$$

$$
\left\{ [\bar{N}_{\mathbf{p},\tilde{t}} \bar{N}_{-\mathbf{q},\tilde{t}} \bar{N}_{-\mathbf{k},\tilde{t}} \bar{N}_{-\mathbf{l},\tilde{t}} - N_{\mathbf{p},\tilde{t}} N_{-\mathbf{q},\tilde{t}} N_{-\mathbf{k},\tilde{t}} N_{-\mathbf{l},\tilde{t}}] \int_{t_0}^{\tilde{t}} dt' \, \cos([\tilde{t} - t'][\omega_p + \omega_q + \omega_k + \omega_l]) \right.
$$

$$
+ 3 [\bar{N}_{\mathbf{p},\tilde{t}} \bar{N}_{-\mathbf{q},\tilde{t}} \bar{N}_{-\mathbf{k},\tilde{t}} N_{\mathbf{l},\tilde{t}} - N_{\mathbf{p},\tilde{t}} N_{-\mathbf{q},\tilde{t}} N_{-\mathbf{k},\tilde{t}} \bar{N}_{\mathbf{l},\tilde{t}}] \int_{t_0}^{\tilde{t}} dt' \, \cos([\tilde{t} - t'][\omega_p + \omega_q + \omega_k - \omega_l])
$$

$$
+ 3 [\bar{N}_{\mathbf{p},\tilde{t}} \bar{N}_{-\mathbf{q},\tilde{t}} N_{\mathbf{k},\tilde{t}} N_{\mathbf{l},\tilde{t}} - N_{\mathbf{p},\tilde{t}} N_{-\mathbf{q},\tilde{t}} \bar{N}_{\mathbf{k},\tilde{t}} \bar{N}_{\mathbf{l},\tilde{t}}] \int_{t_0}^{\tilde{t}} dt' \, \cos([\tilde{t} - t'][\omega_p + \omega_q - \omega_k - \omega_l])
$$

$$
\left. + [\bar{N}_{\mathbf{p},\tilde{t}} N_{\mathbf{q},\tilde{t}} N_{\mathbf{k},\tilde{t}} N_{\mathbf{l},\tilde{t}} - N_{\mathbf{p},\tilde{t}} \bar{N}_{\mathbf{q},\tilde{t}} \bar{N}_{\mathbf{k},\tilde{t}} \bar{N}_{\mathbf{l},\tilde{t}}] \int_{t_0}^{\tilde{t}} dt' \, \cos([\tilde{t} - t'][\omega_p - \omega_q - \omega_k - \omega_l]) \right\},
$$

where we have introduced the abbreviation $N_{\mathbf{p},\tilde{t}} = N_{qp}(\mathbf{p}, \tilde{t})$ for the distribution function at current time $\tilde{t}$ and $\bar{N}_{\mathbf{p},\tilde{t}} = N_{qp}(\mathbf{p}, \tilde{t}) + 1$ for the according Bose factor. Furthermore, a possible time dependence of the on-shell energies is suppressed in the above notation.

The contributions in the collision term (4.63) for particles of momentum $\mathbf{p}$ are ordered as they describe different types of scattering processes where, however, we always find the typical gain and loss structure. The first line in (4.63) corresponds to the production and annihilation of four on-shell particles ($0 \to 4$, $4 \to 0$), where a particle of momentum $\mathbf{p}$ is produced or destroyed simultaneous with three other particles with momenta $\mathbf{q}, \mathbf{k}, \mathbf{l}$. The second line and the fourth line describe ($1 \to 3$) and ($3 \to 1$) processes where the quasiparticle with momentum $\mathbf{p}$ is the single one or appears with two other particles. The relevant contribution in the Boltzmann limit is the third line which represents ($2 \to 2$) scattering processes; quasiparticles with momentum $\mathbf{p}$ can be scattered out of their momentum cell by collisions with particles of momenta $\mathbf{q}$ (second term) or can be produced within a reaction of on-shell particles with momenta $\mathbf{k}, \mathbf{l}$ (first term).

The time evolution of the quasiparticle distribution is given as an initial value problem for the function $N_{qp}(\mathbf{p})$ prepared at initial time t_0. For large system times $\tilde{t}$ (compared to the initial time) the time integration over the trigonometric function results in an energy conserving δ-function:[6]

$$\lim_{\tilde{t}-t_0 \to \infty} \int_{t_0}^{\tilde{t}} dt' \cos((\tilde{t}-t')\,\hat{\omega}) = \lim_{\tilde{t}-t_0 \to \infty} \frac{1}{\hat{\omega}} \sin((\tilde{t}-t_0)\,\hat{\omega}) = \pi\,\delta(\hat{\omega}) . \quad (4.64)$$

Here $\hat{\omega} = \omega_\mathbf{p} \pm \omega_\mathbf{q} \pm \omega_\mathbf{k} \pm \omega_\mathbf{l}$ represents the energy sum which is conserved in the limit $\tilde{t} - t_0 \to \infty$ where the initial time t_0 is considered as fixed. In this limit the time evolution of the distribution function amounts to

$$\partial_{\tilde{t}} N_{qp}(\mathbf{p}, \tilde{t}) = \frac{\lambda^2}{6} \int \frac{d^d q}{(2\pi)^d} \int \frac{d^d k}{(2\pi)^d} \int \frac{d^d l}{(2\pi)^d} \ (2\pi)^{d+1} \ \frac{1}{2\omega_\mathbf{p}\,2\omega_\mathbf{q}\,2\omega_\mathbf{k}\,2\omega_\mathbf{l}} \quad (4.65)$$

$$\Big\{ \ [\,\bar{N}_{\mathbf{p},\tilde{t}}\,\bar{N}_{\mathbf{q},\tilde{t}}\,\bar{N}_{\mathbf{k},\tilde{t}}\,\bar{N}_{\mathbf{l},\tilde{t}} - N_{\mathbf{p},\tilde{t}}\,N_{\mathbf{q},\tilde{t}}\,N_{\mathbf{k},\tilde{t}}\,N_{\mathbf{l},\tilde{t}}\,]\ \delta^{(d)}(\mathbf{p}+\mathbf{q}+\mathbf{k}+\mathbf{l})\ \delta(\omega_\mathbf{p}+\omega_\mathbf{q}+\omega_\mathbf{k}+\omega_\mathbf{l})$$

$$+3[\,N_{\mathbf{p},\tilde{t}}\,\bar{N}_{\mathbf{q},\tilde{t}}\,\bar{N}_{\mathbf{k},\tilde{t}}\,N_{\mathbf{l},\tilde{t}} - N_{\mathbf{p},\tilde{t}}\,N_{\mathbf{q},\tilde{t}}\,N_{\mathbf{k},\tilde{t}}\,\bar{N}_{\mathbf{l},\tilde{t}}\,]\ \delta^{(d)}(\mathbf{p}+\mathbf{q}+\mathbf{k}-\mathbf{l})\ \delta(\omega_\mathbf{p}+\omega_\mathbf{q}+\omega_\mathbf{k}-\omega_\mathbf{l})$$

$$+3[\,\bar{N}_{\mathbf{p},\tilde{t}}\,\bar{N}_{\mathbf{q},\tilde{t}}\,N_{\mathbf{k},\tilde{t}}\,N_{\mathbf{l},\tilde{t}} - N_{\mathbf{p},\tilde{t}}\,N_{\mathbf{q},\tilde{t}}\,\bar{N}_{\mathbf{k},\tilde{t}}\,\bar{N}_{\mathbf{l},\tilde{t}}\,]\ \delta^{(d)}(\mathbf{p}+\mathbf{q}-\mathbf{k}-\mathbf{l})\ \delta(\omega_\mathbf{p}+\omega_\mathbf{q}-\omega_\mathbf{k}-\omega_\mathbf{l})$$

$$+ [\,\bar{N}_{\mathbf{p},\tilde{t}}\,N_{\mathbf{q},\tilde{t}}\,N_{\mathbf{k},\tilde{t}}\,N_{\mathbf{l},\tilde{t}} - N_{\mathbf{p},\tilde{t}}\,\bar{N}_{\mathbf{q},\tilde{t}}\,\bar{N}_{\mathbf{k},\tilde{t}}\,\bar{N}_{\mathbf{l},\tilde{t}}\,]\ \delta^{(d)}(\mathbf{p}-\mathbf{q}-\mathbf{k}-\mathbf{l})\ \delta(\omega_\mathbf{p}-\omega_\mathbf{q}-\omega_\mathbf{k}-\omega_\mathbf{l}) \Big\} .$$

In the energy conserving long-time limit (4.64) only the $2 \to 2$ scattering processes are nonvanishing, because all other terms do not contribute since the energy δ-functions cannot be fulfilled for massive on-shell quasiparticles. Furthermore, the system evolution is explicitly local in time because it depends only on the current configuration; there are no memory effects from the integration over past times as present in the full Kadanoff–Baym equation.

In the following we will solve the energy conserving Boltzmann equation for on-shell particles:

$$\partial_{\tilde{t}} N_{qp}(\mathbf{p}, \tilde{t}) = \frac{\lambda^2}{2} \int \frac{d^d q}{(2\pi)^d} \int \frac{d^d k}{(2\pi)^d} \int \frac{d^d l}{(2\pi)^d} \ (2\pi)^{d+1} \ \frac{1}{2\omega_\mathbf{p}\,2\omega_\mathbf{q}\,2\omega_\mathbf{k}\,2\omega_\mathbf{l}}$$

$$(4.66)$$

$$[\,\bar{N}_{\mathbf{p},\tilde{t}}\,\bar{N}_{\mathbf{q},\tilde{t}}\,N_{\mathbf{k},\tilde{t}}\,N_{\mathbf{l},\tilde{t}} - N_{\mathbf{p},\tilde{t}}\,N_{\mathbf{q},\tilde{t}}\,\bar{N}_{\mathbf{k},\tilde{t}}\,\bar{N}_{\mathbf{l},\tilde{t}}\,]$$

$$\times \quad \delta^{(d)}(\mathbf{p}+\mathbf{q}-\mathbf{k}-\mathbf{l})\ \delta(\omega_\mathbf{p}+\omega_\mathbf{q}-\omega_\mathbf{k}-\omega_\mathbf{l}) .$$

---

[6] This is equivalent to Eq. (2.115).

The numerical algorithm employed for the solution of (4.66) is basically the same as for the solution of the Kadanoff–Baym equation. We explicitly calculate the time integral in (4.63). Energy conservation can be assured by a precalculation including a shift of the lower boundary t_0 to earlier times. We note that in contrast to the Kadanoff–Baym equation no correlation energy is generated in the Boltzmann limit!

In addition to the procedure presented above we calculate the actual momentum-dependent on-shell energy for every momentum mode by a solution of the dispersion relation including contributions from the tadpole and the real part of the (retarded) sunset selfenergy . In this way one can guarantee that at every time t the particles are treated as quasiparticles with the correct energy-momentum relation.

Before presenting the actual numerical results we comment on the derivation of the Boltzmann equation within the conventional scheme that is different from the one presented above. Here, at first the Kadanoff–Baym equation (in coordinate space) is transformed to the Wigner representation by Fourier transformation with respect to the relative coordinates in space and time. The problem then is formulated in terms of energy and momentum variables together with a single system time. For non-homogeneous systems a mean spatial coordinate is necessary as well. As a next step the "semiclassical approximation" is introduced, which consists of a gradient expansion of the convolution integrals in coordinate space within the Wigner transformation. For the time evolution only contributions up to first order in the gradients are kept. Finally, the quasiparticle assumption is introduced as follows: The Green's functions appearing in the transport equation—explicitly or implicitly via the selfenergies—are written in Wigner representation as a product of a distribution function N and the spectral function A. The quasiparticle assumption is then realized by employing a δ-like form for the spectral function which connects the energy variable to the three-momentum. By integrating the first-order transport equation over all (positive) energies, furthermore, the quantum Boltzmann equation for the time evolution of the on-shell distribution function (4.66) is obtained.

In spite of the fact, that the quantum Boltzmann equation (4.66) can be obtained in different subsequent approximation schemes, it is of basic interest, how its actual solutions compare to those from the full Kadanoff–Baym dynamics.

4.2.2   Boltzmann vs. Kadanoff–Baym dynamics

In the following we compare the solutions of the quantum Boltzmann equation with the solution of the Kadanoff–Baym theory in two spatial dimensions. We start with a presentation of the nonequilibrium time evolution of two colliding particle accumulations (tsunamis) within the full Kadanoff–Baym calculation (see Fig. 4.10).

During the time evolution the bumps at finite momenta (in p_x direction) slowly disappear, while the one close to zero momentum—which initially stems from the vacuum contribution to the Green's function—is decreased as seen for different snapshots at times $t \cdot m = 0, 15, 30, 45, 75, 150$ in Fig. 4.10. The system with

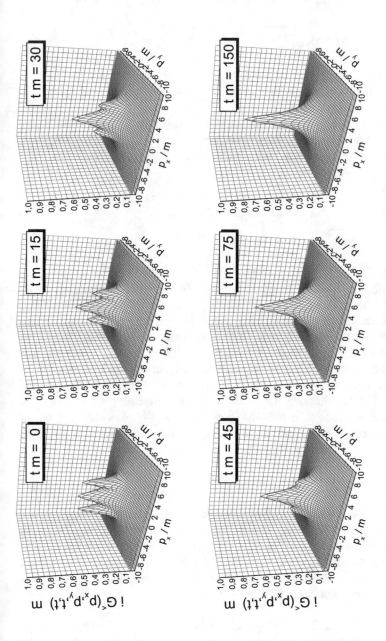

Fig. 4.10 Time evolution of the Green's function $iG^<(p_x, p_y, t, t)$ in momentum space for the initial distribution d2 for λ/m=18. The equal-time Green's function is displayed for various times $t \cdot m = 0, 15, 30, 45, 75, 150$. Starting from an initially non-isotropic shape it develops towards a rotational symmetric distribution in momentum space

initially apparent collision axis slowly merges—as expected—into an isotropic final distribution in momentum space.

The relation to the occupation density in momentum space is given by

$$n(\mathbf{p}, \tilde{t}) = \sqrt{G^<_{\phi\phi}(\mathbf{p}, \tilde{t}, \tilde{t}) \, G^<_{\pi\pi}(\mathbf{p}, \tilde{t}, \tilde{t})} - \frac{1}{2}. \tag{4.67}$$

The actual results for $n(p_x, p_y; t)$ are displayed in Fig. 4.11 and demonstrate the equilibration of the momentum distribution from two separated Fermi spheres to the final equilibrium distribution.

For the comparison between the full Kadanoff–Baym dynamics and the Boltzmann approximation we concentrate on equilibration times. To this aim we define a "quadrupole" moment for a given momentum distribution $n(\mathbf{p}, \tilde{t})$ at time $\tilde{t}$ as

$$Q(\tilde{t}) = \frac{\displaystyle\int \frac{d^d p}{(2\pi)^d} \, (p_x^2 - p_y^2) \, N(\mathbf{p}, \tilde{t})}{\displaystyle\int \frac{d^d p}{(2\pi)^d} \, N(\mathbf{p}, \tilde{t})}, \tag{4.68}$$

which vanishes for the equilibrium state. For the Kadanoff–Baym case we employ the actual distribution function by the relation (4.67). Note that when constructing the distribution function by means of equal-time Green's functions the energy variable has been effectively integrated out. This has the advantage that the distribution function is given independently of the actual on-shell energies.

The relaxation of the quadrupole moment (4.68) has been studied for two different initial distributions: The evolution of distribution d2—which is practically identical to the distribution D2 in Fig. 4.4—is displayed in Fig. 4.10 while for distribution d1 the position and the width of the two-particle bumps have been slightly modified compared to the distribution D1 in Fig. 4.4. The calculated quadrupole moment (4.68) shows a nearly exponential decrease with time (cf. Fig. 4.12) and one can extract a relaxation rate Γ_Q via the relation

$$Q(\tilde{t}) \sim \exp\left(-\Gamma_Q \tilde{t}\right). \tag{4.69}$$

Figure 4.12 shows for both initializations that the relaxation in the full quantum (KB) calculation occurs faster for large coupling constants than in the Boltzmann approximation, whereas for small couplings the equilibration times of the full and the approximate evolutions are comparable. We find that the scaled relaxation rate Γ_Q/λ^2 is nearly constant in the Boltzmann case (cf. Fig. 4.13), but increases with the coupling strength in the Kadanoff–Baym calculation (especially for initial distribution d2).

These findings are readily explained: Since the free Green's function—as used in the Boltzmann calculation—has only support on the mass shell, only $(2 \leftrightarrow 2)$ scattering processes are described in the Boltzmann limit. All other processes with a different number of incoming and outgoing particles vanish (as noted before).

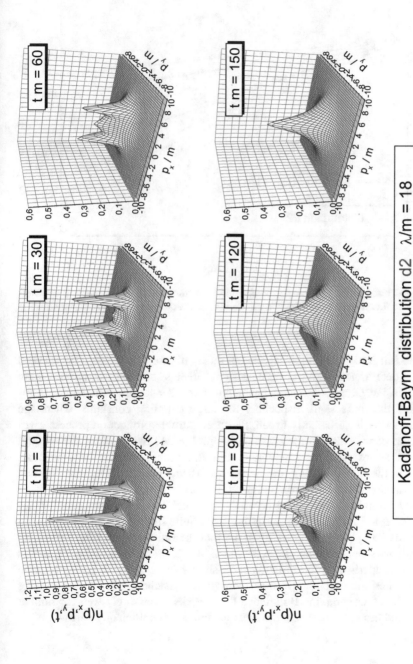

Fig. 4.11 Time evolution of the occupation density $n(p_x, p_y; t)$ in momentum space. The occupation density is displayed for the same times as the Green's function in Fig. 4.10

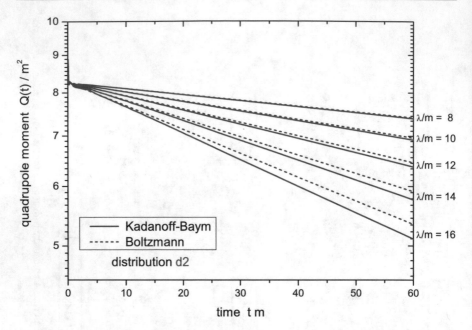

Fig. 4.12 Decrease of the quadrupole moment (4.68) in time for different coupling constants $\lambda/m = 8 - 16$ for the full Kadanoff–Baym calculation and the quantum Boltzmann approximation

Within the full Kadanoff–Baym calculation this is different, since here the spectral function – determined from the selfconsistent Green's function—acquires a finite width. Thus the Green's function has support at all energies although it drops fast far off the mass shell. Especially for large coupling constants, where the spectral function is sufficiently broad, the three particle production process gives a significant contribution to the collision integral. Since the width of the spectral function increases with the interaction strength, such processes become more important in the high coupling regime. As a consequence the difference between both approaches is larger for stronger interactions as observed in Fig. 4.12. For small couplings λ/m in both approaches basically the usual $2 \leftrightarrow 2$ scattering contributes and the results for the thermalization rate Γ_Q are quite similar.

In summarizing this section we point out that the full solution of the Kadanoff–Baym equations does include $1 \leftrightarrow 3$ and $2 \leftrightarrow 2$ off-shell collision processes which—in comparison to the Boltzmann on-shell $2 \leftrightarrow 2$ collision limit—become important when the spectral width of the particles reaches $\sim 1/3$ of the particle mass. On the other hand, the Boltzmann limit works surprisingly well for smaller couplings and those cases, where the spectral function is sufficiently narrow.

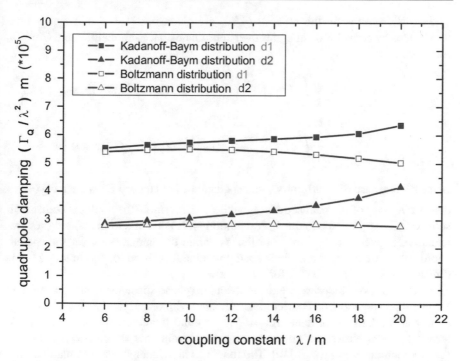

Fig. 4.13 Relaxation rate (divided by the coupling λ squared) for Kadanoff–Baym and Boltzmann calculations as a function of the interaction strength λ/m. For the two different initial configurations d_1 (upper lines) and d_2 (lower lines) the full Kadanoff–Baym evolution leads to a faster equilibration

4.3 Derivation of Off-Shell Relativistic Transport Theory

Formal derivations of covariant off-shell transport equations have been presented more than 50 years ago by Kadanoff and Baym [8] but actual solutions have been addressed only in a limited number of cases. This section is devoted to a transparent derivation of generalized transport equations in first-order gradient expansion in phase space including a generalized testparticle ansatz for the solution of the off-shell transport equations. For a nonrelativistic formulation of related off-shell transport equations we refer the reader to Ref. [9].

The derivation of generalized transport equations starts by rewriting the Kadanoff–Baym equation for the Wightman functions in coordinate space $(x_1 = (t_1, \mathbf{x}_1), x_2 = (t_2, \mathbf{x}_2))$ (4.35) as

$$[\partial_{x_1}^{\mu} \partial_{\mu}^{x_1} + m^2 + \Sigma^{\delta}(x_1)] \, iG^{\lessgtr}(x_1, x_2) = i \, I_1^{\lessgtr}(x_1, x_2). \qquad (4.70)$$

The collision terms on the r.h.s. of (4.70) are given in $D = d + 1$ space-time dimensions by convolution integrals over coordinate-space selfenergies and Green's functions :

$$I_1^{\lessgtr}(x_1, x_2) = -\int_{t_0}^{t_1} d^D z \ \left[\Sigma^>(x_1, z) - \Sigma^<(x_1, z) \right] \ G^{\lessgtr}(z, x_2) \quad (4.71)$$

$$+\int_{t_0}^{t_2} d^D z \ \Sigma^{\lessgtr}(x_1, z) \ \left[G^>(z, x_2) - G^<(z, x_2) \right].$$

In the general case of an arbitrary (scalar) quantum-field theory Σ^δ is the local (non-dissipative) part of the path selfenergy while $\Sigma^{\lessgtr}$ resemble the nonlocal collisional selfenergy contributions. In the representation (4.71) the integration boundaries are exclusively given for the time coordinates, while the integration over the spatial coordinates extends over the whole spatial volume from $-\infty$ to $+\infty$ in $d = D - 1$ dimensions, i.e. $d^D z = dt \, d^d z$ for $z = (t, \mathbf{z})$.

Since transport theories are formulated in phase-space one changes to the Wigner representation via Fourier transformation with respect to the rapidly varying ('intrinsic') relative coordinate $\Delta x = x_1 - x_2$ and treats the system evolution in terms of the ('macroscopic') mean space-time coordinate $x = (x_1 + x_2)/2$ and the four-momentum $p = (p_0, \mathbf{p})$ [10]. The functions in Wigner space are obtained as

$$\bar{F}(p, x) = \int_{-\infty}^{\infty} d^D \Delta x \ e^{+i \, \Delta x_\mu \, p^\mu} \ F(x_1 = x + \Delta x/2, \ x_2 = x - \Delta x/2). \quad (4.72)$$

For the formulation of transport theory in the Wigner representation we have to focus not only on the transformation properties of ordinary two-point functions as given in (4.72), but also of convolution integrals as appearing in Eq. (4.71). A convolution integral in D dimensions (for arbitrary functions F, G),

$$H(x_1, x_2) = \int_{-\infty}^{\infty} d^D z \ F(x_1, z) \ G(z, x_2) \quad (4.73)$$

transforms as

$$\bar{H}(p, x) = \int_{-\infty}^{\infty} d^D \Delta x \ e^{+i \, \Delta x_\mu \, p^\mu} \ H(x_1, x_2) \quad (4.74)$$

$$= \int_{-\infty}^{\infty} d^D \Delta x \ e^{+i \, \Delta x_\mu \, p^\mu} \int_{-\infty}^{\infty} d^D z \ F(x_1, z) \ G(z, x_2)$$

$$= e^{+i \frac{1}{2} (\partial_p^\mu \cdot \partial_\mu^{x'} - \partial_x^\mu \cdot \partial_\mu^{p'})} \ \left[\bar{F}(p, x) \ \bar{G}(p', x') \right] \Big|_{x'=x, \ p'=p}.$$

In accordance with the standard assumption of transport theory we assume that all functions only smoothly evolve in the mean space-time coordinates and thus restrict to first-order derivatives. All terms proportional to second or higher order derivatives in the mean space-time coordinates (also mixed ones) will be dropped. Thus the Wigner transformed convolution integrals (4.73) are given in first-order gradient approximation by,

$$\bar{H}(p,x) = \bar{F}(p,x)\,\bar{G}(p,x) + i\,\frac{1}{2}\,\{\bar{F}(p,x),\,\bar{G}(p,x)\}_P + \mathcal{O}(\partial_x^2), \quad (4.75)$$

using the relativistic generalization of the Poisson bracket

$$\{\bar{F}(p,x),\,\bar{G}(p,x)\}_P := \partial_\mu^p \bar{F}(p,x) \cdot \partial_x^\mu \bar{G}(p,x)$$
$$-\partial_x^\mu \bar{F}(p,x) \cdot \partial_\mu^p \bar{G}(p,x)\,. \quad (4.76)$$

In order to obtain the dynamics for the spectral functions within the approximate scheme we start with the Dyson–Schwinger equations for the retarded and advanced Green's functions in coordinate space (4.21). Note that the convolution integrals in (4.21) extend over the whole space and time range in contrast to the equations of motion for the Wightman functions given in (4.22) and (4.23)!—The further procedure consists in the following steps: First we

1. transform the above equations into the Wigner representation and apply the first-order gradient approximation. In this limit the convolution integrals yield the product terms and the general Poisson bracket of the selfenergies and the Green's functions $\{\Sigma^{R/A}, G^{R/A}\}_P$. We, further on, represent both equations in terms of real quantities by the decomposition of the retarded and advanced Green's functions and selfenergies as

$$\bar{G}^{R/A} = \Re\,\bar{G}^R \pm i\,\Im\,\bar{G}^R = \Re\,\bar{G}^R \mp i\,\bar{A}/2\,, \qquad \bar{A} = \mp 2\,\Im\,\bar{G}^{R/A}\,,$$
$$(4.77)$$
$$\bar{\Sigma}^{R/A} = \Re\,\bar{\Sigma}^R \pm i\,\Im\,\bar{\Sigma}^R = \Re\,\bar{\Sigma}^R \mp i\,\bar{\Gamma}/2\,, \qquad \bar{\Gamma} = \mp 2\,\Im\,\bar{\Sigma}^{R/A}\,.$$

We find that in Wigner space the real parts of the retarded and advanced Green's functions and selfenergies are equal, while the imaginary parts have opposite sign and are proportional to the spectral function $\bar{A}$ and the width $\bar{\Gamma}$, respectively. The next step consists in

2. the separation of the real part and the imaginary part of the two equations for the retarded and advanced Green's functions, that have to be fulfilled independently. Thus we obtain four real-valued equations for the selfconsistent retarded and advanced Green's functions. In the last step

3. we get simple relations by linear combination of these equations, i.e. by adding/subtracting the relevant equations.

This finally leads to two algebraic relations for the spectral function $\bar{A}$ and the real part of the retarded Green's function $Re\,\bar{G}^R$ in terms of the width $\bar{\Gamma}$ and the real part of the retarded selfenergy $\Re\,\bar{\Sigma}^R$ as:

$$[\, p_0^2 - \mathbf{p}^2 - m^2 - \bar{\Sigma}^\delta + \Re\,\bar{\Sigma}^R\,]\,\Re\,\bar{G}^R = 1 + \frac{1}{4}\,\bar{\Gamma}\,\bar{A}, \tag{4.78}$$

$$[\, p_0^2 - \mathbf{p}^2 - m^2 - \bar{\Sigma}^\delta + \Re\,\bar{\Sigma}^R\,]\,\bar{A} = \bar{\Gamma}\,\Re\,\bar{G}^R. \tag{4.79}$$

Note that all terms with first-order gradients have disappeared in (4.78) and (4.79). A first consequence of (4.79) is a direct relation between the real and the imaginary parts of the retarded/advanced Green's function, which reads (for $\bar{\Gamma} \neq 0$):

$$\Re\,\bar{G}^R = \frac{p_0^2 - \mathbf{p}^2 - m^2 - \bar{\Sigma}^\delta - \Re\,\bar{\Sigma}^R}{\bar{\Gamma}}\,\bar{A}. \tag{4.80}$$

Inserting (4.80) in (4.78) we end up with the following analytical result for the spectral function and the real part of the retarded Green's function

$$\bar{A} = \frac{\bar{\Gamma}}{[\, p_0^2 - \mathbf{p}^2 - m^2 - \bar{\Sigma}^\delta - \Re\,\bar{\Sigma}^R\,]^2 + \bar{\Gamma}^2/4} = \frac{\bar{\Gamma}}{\bar{M}^2 + \bar{\Gamma}^2/4}, \tag{4.81}$$

$$\Re\,\bar{G}^R = \frac{[\, p_0^2 - \mathbf{p}^2 - m^2 - \bar{\Sigma}^\delta - \Re\,\bar{\Sigma}^R\,]}{[\, p_0^2 - \mathbf{p}^2 - m^2 - \bar{\Sigma}^\delta - \Re\,\bar{\Sigma}^R\,]^2 + \bar{\Gamma}^2/4} = \frac{\bar{M}}{\bar{M}^2 + \bar{\Gamma}^2/4}, \tag{4.82}$$

where we have introduced the mass-function $\bar{M}(p, x)$ in Wigner space:

$$\bar{M}(p, x) = p_0^2 - \mathbf{p}^2 - m^2 - \bar{\Sigma}^\delta(x) - \Re\,\bar{\Sigma}^R(p, x). \tag{4.83}$$

The spectral function (4.81) shows a typical Breit–Wigner shape with energy- and momentum-dependent selfenergy terms. Although the above equations are purely algebraic solutions and contain no derivative terms, they are valid up to the first order in the gradients!

Exercise 4.2: Show that the negative imaginary part of the propagator

$$G_F(p) = 1/(p_0^2 - \mathbf{p}^2 - M^2 + i2\gamma p_0)$$

(continued)

(for $\gamma > 0$) can be written in relativistic Breit–Wigner form as

$$\frac{2\gamma p_0}{(p_0^2 - \mathbf{p}^2 - M^2)^2 + 4\gamma^2 p_0^2}$$

and is normalized to unity for all momenta $\mathbf{p}$, i.e.

$$-\int_{-\infty}^{\infty} \frac{dp_0}{2\pi} \, 2p_0 \, \Im G_F(p_0, \mathbf{p}) = 1.$$

In addition, subtraction of the real parts and adding up the imaginary parts lead to the time-evolution equations

$$p^\mu \, \partial_\mu^x \, \bar{A} = \frac{1}{2} \{ \bar{\Sigma}^\delta + \Re \bar{\Sigma}^R , \, \bar{A} \}_P + \frac{1}{2} \{ \bar{\Gamma} , \, \Re \bar{G}^R \}_P , \qquad (4.84)$$

$$p^\mu \, \partial_\mu^x \, \Re \bar{G}^R = \frac{1}{2} \{ \bar{\Sigma}^\delta + \Re \bar{\Sigma}^R , \, \Re \bar{G}^R \}_P - \frac{1}{8} \{ \bar{\Gamma} , \, \bar{A} \}_P . \qquad (4.85)$$

The Poisson bracket containing the mass-function $\bar{M}$ leads to the well-known drift operator $p^\mu \, \partial_\mu^x \, \bar{F}$ (for an arbitrary function $\bar{F}$), i.e.

$$\{ \bar{M} , \, \bar{F} \}_P = \{ p_0^2 - \mathbf{p}^2 - m^2 - \bar{\Sigma}^\delta - \Re \bar{\Sigma}^R , \, \bar{F} \}_P \qquad (4.86)$$

$$= 2 \, p^\mu \, \partial_\mu^x \, \bar{F} - \{ \bar{\Sigma}^\delta + \Re \bar{\Sigma}^R , \, \bar{F} \}_P , \qquad (4.87)$$

such that the first-order equations (4.84) and (4.85) can be written in a more comprehensive form as

$$\{ \bar{M} , \, \bar{A} \}_P = \{ \bar{\Gamma} , \, \Re \bar{G}^R \}_P , \qquad (4.88)$$

$$\{ \bar{M} , \, \Re \bar{G}^R \}_P = -\frac{1}{4} \{ \bar{\Gamma} , \, \bar{A} \}_P . \qquad (4.89)$$

When inserting (4.81) and (4.82) we find that these first-order time-evolution equations are *solved* by the algebraic expressions. In this case the following relations hold:

$$\{ \bar{M} , \, \bar{A} \}_P = \{ \bar{\Gamma} , \, \Re \bar{G}^R \}_P = \{ \bar{M} , \, \bar{\Gamma} \}_P \, \frac{\bar{M}^2 - \bar{\Gamma}^2/4}{[\, \bar{M}^2 + \bar{\Gamma}^2/4 \,]^2} , \qquad (4.90)$$

$$\{ \bar{M} , \, \Re \bar{G}^R \}_P = -\frac{1}{4} \{ \bar{\Gamma} , \, \bar{A} \}_P = \{ \bar{M} , \, \bar{\Gamma} \}_P \, \frac{\bar{M} \, \bar{\Gamma}/2}{[\, \bar{M}^2 + \bar{\Gamma}^2/4 \,]^2} . \qquad (4.91)$$

Thus we have derived the proper structure of the spectral function (4.81) within the first-order gradient (or semiclassical) approximation. Together with the explicit form for the real part of the retarded Green's function (4.82) we now have fixed the dynamics of the spectral properties, which is consistent up to first order in the gradients.

4.3.1 Kadanoff–Baym Transport

As a next step we rewrite the memory terms in the collision integrals such that the time integrations extend from $-\infty$ to $+\infty$. In this respect we consider the initial time $t_0 = -\infty$ whereas the upper time boundaries t_1, t_2 are taken into account by Θ-functions, i.e.

$$I_1^{\lessgtr}(x_1, x_2) = -\int_{-\infty}^{\infty} d^D x' \; \Theta(t_1 - t') \left[\Sigma^{>}(x_1, x') - \Sigma^{<}(x_1, x') \right] \; G^{\lessgtr}(x', x_2)$$

$$+ \int_{-\infty}^{\infty} d^D x' \; \Sigma^{\lessgtr}(x_1, x') \; \Theta(t_2 - t') \left[G^{>}(x', x_2) - G^{<}(x', x_2) \right]$$

$$= -\int_{-\infty}^{\infty} d^D x' \; \left\{ \Sigma^{R}(x_1, x') \, G^{\lessgtr}(x', x_2) \right.$$

$$\left. + \Sigma^{\lessgtr}(x_1, x') \, G^{A}(x', x_2) \right\} . \tag{4.92}$$

We now perform the analogous steps as invoked before for the retarded and advanced Dyson–Schwinger equations. We start with a first-order gradient expansion of the Wigner transformed Kadanoff–Baym equation using (4.92) for the memory integrals. Again we separate the real and the imaginary parts in the resulting equation, which have to be satisfied independently. At the end of this procedure we obtain a generalized transport equation:

$$\underbrace{2 p^{\mu} \, \partial_{\mu}^{x} \, i\bar{G}^{\lessgtr} - \{ \bar{\Sigma}^{\delta} + \Re \bar{\Sigma}^{R}, \, i\bar{G}^{\lessgtr} \}_{P}}_{\{\bar{M}, \, i\bar{G}^{\lessgtr}\}_{P}} - \{ i\bar{\Sigma}^{\lessgtr}, \, \Re \bar{G}^{R} \}_{P} = i\bar{\Sigma}^{<} \, i\bar{G}^{>} - i\bar{\Sigma}^{>} \, i\bar{G}^{<}$$

$$\qquad\qquad\qquad\qquad\qquad - \{ i\bar{\Sigma}^{\lessgtr}, \, \Re \bar{G}^{R} \}_{P} = i\bar{\Sigma}^{<} \, i\bar{G}^{>} - i\bar{\Sigma}^{>} \, i\bar{G}^{<} \tag{4.93}$$

as well as a generalized mass-shell equation

$$\underbrace{[\,p^2 - m^2 - \bar{\Sigma}^\delta - \Re\,\bar{\Sigma}^R\,]}_{\bar{M}}\; i\bar{G}^{\lessgtr} = i\bar{\Sigma}^{\lessgtr}\,\Re\,\bar{G}^R$$

$$+\frac{1}{4}\{i\bar{\Sigma}^>, \, i\bar{G}^<\}_P - \frac{1}{4}\{i\bar{\Sigma}^<, \, i\bar{G}^>\}_P \tag{4.94}$$

with the mass-function $\bar{M}$ specified in (4.83). Since the Green's function $G^{\lessgtr}(x_1, x_2)$ consists of an antisymmetric real part and a symmetric imaginary part with respect to the relative coordinate $x_1 - x_2$, the Wigner transform of this function is purely imaginary. It is thus convenient to represent the Wightman functions in Wigner space by the real-valued quantities $i\bar{G}^{\lessgtr}(p, x)$. Since the collisional selfenergies obey the same symmetry relations in coordinate space and in phase-space, they will be kept also as $i\bar{\Sigma}^{\lessgtr}(p, x)$ further on.

In the transport equation (4.93) one recognizes on the l.h.s. the drift term $p^\mu\,\partial^x_\mu\,i\bar{G}^{\lessgtr}$, as well as the Vlasov term with the local selfenergy $\bar{\Sigma}^\delta$ and the real part of the retarded selfenergy $\Re\,\bar{\Sigma}^R$. On the other hand the r.h.s. represents the collision term with its typical "gain and loss" structure. The loss term $i\bar{\Sigma}^>\,i\bar{G}^<$ (proportional to the Green's function itself) describes the scattering out of a respective phase-space cell whereas the gain term $i\bar{\Sigma}^<\,i\bar{G}^>$ takes into account scatterings into the actual cell. The last term on the l.h.s. $\{i\bar{\Sigma}^{\lessgtr}, \Re\,\bar{G}^R\}_P$ is very *peculiar* since it does not contain directly the distribution function $i\bar{G}^<$. This second Poisson bracket vanishes in the quasiparticle approximation and thus does not appear in the on-shell Boltzmann limit. The second Poisson bracket $\{i\bar{\Sigma}^{\lessgtr}, \Re\,\bar{G}^R\}_P$ governs the evolution of the off-shell dynamics for nonequilibrium systems.

> **Exercise 4.3:** Compute the propagator for static massive fields, i.e. the solution of
>
> $$(-\Delta + M^2)G_f(\mathbf{x}) = \delta^3(\mathbf{x}).$$

Although the generalized transport equation (4.93) and the generalized mass-shell equation (4.94) have been derived from the same Kadanoff–Baym equation in a first-order gradient expansion, both equations are not exactly equivalent. Instead, they deviate from each other by contributions of second gradient order, which are hidden in the term $\{i\bar{\Sigma}^{\lessgtr}, \Re\,\bar{G}^R\}_P$. This raises the question: which one of these two equations has to be considered of higher priority? The question is answered in practical applications by the prescription of solving the generalized transport equation (4.93) for $i\bar{G}^<$ in order to study the dynamics of the nonequilibrium system

in phase-space. Since the dynamical evolution of the spectral properties is taken
into account by the equations derived in first-order gradient expansion from the
retarded and advanced Dyson–Schwinger equations, one can neglect the generalized
mass-shell equation (4.94). Thus for actual numerical studies one should use the
generalized transport equation (4.93) supported by the algebraic relations (4.81)
and (4.82).

Exercise 4.4: Derive the propagator for a massive scalar field of finite lifetime
γ^{-1} in its rest frame, i.e. the solution of

$$\left(\frac{\partial^2}{\partial t^2} - \Delta + M^2 + 2\gamma \frac{\partial}{\partial t} \right) G_{ret}(x - x') = \delta^4(x - x')$$

for $\gamma > 0$.

4.3.2 Transport in the Botermans–Malfliet Scheme

Furthermore, one recognizes by subtraction of the $i\bar{G}^>$ and $i\bar{G}^<$ mass-shell and
transport equations, that the dynamics of the spectral function $\bar{A} = i\bar{G}^> - i\bar{G}^<$ is determined in the same way as derived from the retarded and advanced
Dyson–Schwinger equations (4.81) and (4.88). The inconsistency between the two
Eqs. (4.93) and (4.94) vanishes since the differences are contained in the collisional
contributions on the r.h.s. of (4.93).

In order to evaluate the $\{ i\bar{\Sigma}^<, Re\, \bar{G}^R \}_P$-term on the l.h.s. of (4.93) and to
explore the differences between the KB- and Botermans–Malfliet (BM)-form of the
transport equations (see below) it is useful to introduce distribution functions for the
Green's functions and selfenergies as

$$i\bar{G}^<(p, x) = \bar{N}(p, x)\, \bar{A}(p, x)\,, \quad i\bar{G}^>(p, x) = [1 + \bar{N}(p, x)]\, \bar{A}(p, x)\,, \qquad (4.95)$$

$$i\bar{\Sigma}^<(p, x) = \bar{N}^\Sigma(p, x)\, \bar{\Gamma}(p, x)\,, \quad i\bar{\Sigma}^>(p, x) = [1 + \bar{N}^\Sigma(p, x)]\, \bar{\Gamma}(p, x)\,. \qquad (4.96)$$

In equilibrium the distribution function with respect to the Green's functions $\bar{N}$
and the selfenergies $\bar{N}^\Sigma$ are given as Bose functions in the energy p_0 at given
temperature; they thus are equal in equilibrium but in general might differ out-
of-equilibrium. Following the argumentation of Botermans and Malfliet [11] the
distribution functions $\bar{N}$ and $\bar{N}^\Sigma$ in (4.95) should be identical within the second term
of the l.h.s. of (4.93) in order to obtain a consistent first-order gradient expansion
(without hidden higher order gradient terms). In order to demonstrate their argument
we write

$$i\bar{\Sigma}^< = \bar{\Gamma}\, \bar{N}^\Sigma = \bar{\Gamma}\, \bar{N} + \bar{K}\,. \qquad (4.97)$$

The "correction" term

$$\bar{K} = \bar{\Gamma} (\bar{N}^{\Sigma} - \bar{N}) = (i\bar{\Sigma}^{<} i\bar{G}^{>} - i\bar{\Sigma}^{>} i\bar{G}^{<}) \bar{A}^{-1} \qquad (4.98)$$

is proportional to the collision term of the generalized transport equation (4.93), which itself is already of first order in the gradients. Thus, whenever a distribution function $\bar{N}^{\Sigma}$ appears within a Poisson bracket, the difference term $(\bar{N}^{\Sigma} - \bar{N})$ becomes of second order in the gradients and should be omitted for consistency. As a consequence $\bar{N}^{\Sigma}$ can be replaced by $\bar{N}$ and thus the selfenergy $\bar{\Sigma}^{<}$ by $\bar{G}^{<} \cdot \bar{\Gamma}/\bar{A}$ in the Poisson bracket term $\{\bar{\Sigma}^{<}, Re\, \bar{G}^{R}\}_P$. The generalized transport equation (4.93) then can be written in shorthand notation as

$$\frac{1}{2} \bar{A}\, \bar{\Gamma} \left[\{\bar{M},\, i\bar{G}^{<}\}_P - \frac{1}{\bar{\Gamma}} \{\bar{\Gamma},\, \bar{M} \cdot i\bar{G}^{<}\}_P \right] = i\bar{\Sigma}^{<} i\bar{G}^{>} - i\bar{\Sigma}^{>} i\bar{G}^{<} \quad (4.99)$$

with the mass-function $\bar{M}$ (4.83). The transport equation (4.99) within the Botermans–Malfliet (BM) form resolves the discrepancy between the generalized mass-shell equation (4.94) and the generalized transport equation in its original Kadanoff–Baym (KB) form (4.93).

In summarizing this section we have derived a covariant transport equation (4.99) that incorporates the off-shell propagation of the degrees of freedom as well as off-shell scattering and transitions. Since all quantities in Eq. (4.99) depend on (p, x) its general solution on an eight-dimensional phase-space grid will be quite involved.

4.3.3   Testparticle Representation

The generalized transport equation (4.99) allows to extend the traditional on-shell transport approaches (cf. Chaps. 2 and 3) for which efficient numerical recipes have been set up. In order to obtain a practical solution to the transport equation (4.99) we use an extended testparticle Ansatz for the Green's function $G^{<}$, more specifically for the real and positive semi-definite quantity

$$F(x, p) = i\, G^{<}(x, p)$$

$$\sim \sum_{i=1}^{N} \delta^{(3)}(\mathbf{x} - \mathbf{X}_i(t)) \delta^{(3)}(\mathbf{p} - \mathbf{P}_i(t))\ \delta(p_0 - \epsilon_i(t)) . \quad (4.100)$$

In the most general case (where the selfenergies depend on four-momentum P, time t and the spatial coordinates $\mathbf{X}$) the equations of motion for the testparticles read

$$\frac{d\mathbf{X}_i}{dt} = \frac{1}{1 - C_{(i)}} \frac{1}{2\epsilon_i}$$

$$\times \left[2\mathbf{P}_i + \nabla_{P_i} \Re\Sigma_{(i)}^{R} + \frac{\epsilon_i^2 - \mathbf{P}_i^2 - M_0^2 - \Re\Sigma_{(i)}^{R}}{\Gamma_{(i)}} \nabla_{P_i} \Gamma_{(i)} \right], \quad (4.101)$$

$$\frac{d\mathbf{P}_i}{dt} = -\frac{1}{1 - C_{(i)}} \frac{1}{2\epsilon_i} \left[\nabla_{X_i} \Re\Sigma_i^R + \frac{\epsilon_i^2 - \mathbf{P}_i^2 - M_0^2 - \Re\Sigma_{(i)}^R}{\Gamma_{(i)}} \nabla_{X_i} \Gamma_{(i)} \right], \quad (4.102)$$

$$\frac{d\epsilon_i}{dt} = \frac{1}{1 - C_{(i)}} \frac{1}{2\epsilon_i} \left[\frac{\partial \Re\Sigma_{(i)}^R}{\partial t} + \frac{\epsilon_i^2 - \mathbf{P}_i^2 - M_0^2 - \Re\Sigma_{(i)}^R}{\Gamma_{(i)}} \frac{\partial \Gamma_{(i)}}{\partial t} \right], \quad (4.103)$$

where the notation $F_{(i)}$ implies that the function is taken at the coordinates of the testparticle, i.e. $F_{(i)} \equiv F(t, \mathbf{X}_i(t), \mathbf{P}_i(t), \epsilon_i(t))$.

In (4.101)–(4.103) a common multiplication factor $(1 - C_{(i)})^{-1}$ appears, which contains the energy derivatives of the retarded selfenergy

$$C_{(i)} = \frac{1}{2\epsilon_i} \left[\frac{\partial}{\partial \epsilon_i} \Re\Sigma_{(i)}^R + \frac{\epsilon_i^2 - \mathbf{P}_i^2 - M_0^2 - \Re\Sigma_{(i)}^R}{\Gamma_{(i)}} \frac{\partial}{\partial \epsilon_i} \Gamma_{(i)} \right]. \quad (4.104)$$

It yields a shift of the system time t to the "eigentime" of particle i defined by $\tilde{t}_i = t/(1 - C_{(i)})$. As the reader immediately verifies, the derivatives with respect to the "eigentime," i.e. $d\mathbf{X}_i/d\tilde{t}_i$, $d\mathbf{P}_i/d\tilde{t}_i$ and $d\epsilon_i/d\tilde{t}_i$ then emerge without this renormalization factor for each testparticle i when neglecting higher order time derivatives in line with the semiclassical approximation scheme.

Some limiting cases should be mentioned explicitly: In case of a momentum-independent "width" $\Gamma(x)$ we take $M^2 = p^2 - \Re\Sigma^R$ as an independent variable instead of p_0, which then fixes the energy (for given $\mathbf{p}$ and M^2) to

$$p_0^2 = \mathbf{p}^2 + M^2 + \Re\Sigma^R(x, \mathbf{p}, M^2). \quad (4.105)$$

Equation (4.103) then turns to ($\Delta M_i^2 = M_i^2 - M_0^2$)

$$\frac{d\Delta M_i^2}{dt} = \frac{\Delta M_i^2}{\Gamma_{(i)}} \frac{d\Gamma_{(i)}}{dt} \quad \leftrightarrow \quad \frac{d}{dt} \ln\left(\frac{\Delta M_i^2}{\Gamma_{(i)}}\right) = 0 \quad (4.106)$$

for the time evolution of the testparticle i in the invariant mass squared. In case of $\Gamma = const.$ the familiar equations of motion for testparticles in on-shell transport approaches are regained.

Accordingly, we have reduced the solution of the l.h.s. of Eq. (4.99) to a coupled set of first-order differential equations in time for the off-shell testparticles.

4.3.4 Model Studies

For our present purpose to demonstrate the physical implications of Eqs. (4.101)–(4.103) we consider the propagation of particles in a complex potential of Woods-Saxon form, i.e.

$$Re\Sigma_X^{ret} - \frac{i}{2}\Gamma_X = 2P_0 \left\{ \frac{V_0}{1 + \exp\{(|\mathbf{r}| - R)/a_0\}} \right.$$
$$\left. -i\left(\frac{W_0}{1 + \exp\{(|\mathbf{r}| - R)/a_0\}} + \frac{\Gamma_V}{2}\right)\right\} \qquad (4.107)$$

where we have used $R = 5$ fm, $a_0 = 0.6$ fm throughout the model studies. Equations (4.101)–(4.103) allow to represent the distribution function in terms of the testparticle distribution (4.100) where $\mathbf{r}_i(t)$, $\mathbf{P}_i(t)$ and $M_i^2(t)$ are the corresponding solutions of Eqs. (4.101)–(4.103). We initialize all testparticles i with a fixed energy P_0 at some distance ($|\mathbf{r}(t = 0)| \approx -15$ fm) on the z-axis with a three-momentum vector in positive z-direction. The mass parameters $M_i(t = 0)$ are selected according to the Breit–Wigner distribution

$$F(M) = \frac{1}{2\pi} \frac{\Gamma_V}{(M - M_0)^2 + \Gamma_V^2/4}, \qquad (4.108)$$

where Γ_V denotes the vacuum width which might be arbitrarily small but finite (see below). The particles are then propagated in time according to Eqs. (4.101)–(4.103) and all one-body quantities can be evaluated from (4.100).

In Fig. 4.14 (l.h.s., upper part)[7] the results for $P_{i0}(z(t))$, $M_i(z(t))$ and $P_{iz}(z(t))$ are displayed as a function of $z(t)$ instead of the time t. We show the evolution of 21 testparticles with mass parameters that are initially separated by $\Delta M = 0.05 \cdot \Gamma_V$ in the case of a nonvanishing imaginary part of the potential ($W_0 = 70$ MeV, $\Gamma_V = 0.8$ MeV) but vanishing real part of the potential ($V_0 = 0$ MeV) (see Fig. 4.14 (l.h.s., lower part)). One recognizes that the differences between the mass parameters increase when reaching the potential region, which corresponds directly to a broadening of the spectral function. The same spreading behavior is observed for the three-momentum of the testparticles, such that the energy P_0 is conserved throughout the whole calculation (upper line). When leaving the potential region the splitting decreases and the correct asymptotic solution is restored.

In Fig. 4.14 (r.h.s., upper part) we show a calculation for a nonvanishing real part of the potential (i.e. $V_0 = -20$ MeV, (r.h.s., lower part)). While the spreading of the mass parameter is not affected by this change, we find a shift of the testparticle momenta where the real part of the potential deviates from zero since here the particles are accelerated.

---

[7] The figures in this Subsection are taken from [12].

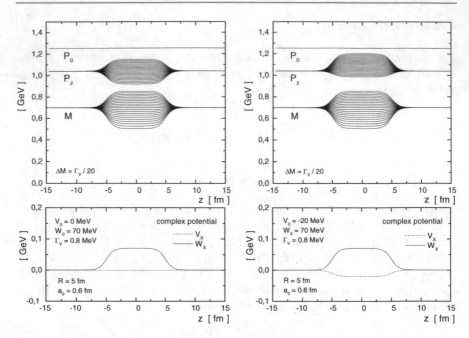

Fig. 4.14 (left) upper part: P_{i0}, M_i, and P_{iz} as a function of $z(t)$ for a purely imaginary potential $W_0 = 70$ MeV (lower part). The vacuum width is $\Gamma_V = 0.8$ MeV and the initial separation in mass of the testparticles is given by $\Delta M = 0.05 \cdot \Gamma_V$. (right) upper part: P_{i0}, M_i, and P_{iz} as a function of $z(t)$ for a complex potential with $V_0 = -20$ MeV, $W_0 = 70$ MeV (lower part)

In the next example of this model study we show in Fig. 4.15 (upper part) the case of a broad vacuum spectral function entering a (time-independent) nonrelativistic potential with $V_0 = -20$ MeV and $W_0 = 100$ MeV. The vacuum width is chosen as $\Gamma_V = 160$ MeV, while the testparticle trajectories are shown with an initial separation of the masses $\Delta M = 0.05 \cdot \Gamma_V$. One observes that the spectral function is further broadened in the complex potential zone and reaches its initial dispersion in mass again after passing the diffractive and absorptive area.

The question remains if the testparticle distribution (4.100) reproduces the local splitting in mass as expected due to quantum mechanics, i.e. in our case a Breit–Wigner distribution (4.108) with a local width $\Gamma_X = 2\ W_0(z) + \Gamma_V$. This is demonstrated in Fig. 4.16 where we show the spectral function as a function of mass M from the testparticle evolution at fixed coordinate z in comparison to the quantum Breit–Wigner distribution with local width Γ_X (full lines) for a pure imaginary potential with parameters $W_0 = 50$ MeV and vacuum width $\Gamma_V = 2$ MeV. The differences from the exact results in Fig. 4.16 are practically not visible for all values of z from -8 to 8 fm. The width of the distribution increases from 1 MeV in the vacuum ($z = \pm 8$ fm) to 102 MeV ($= 2\ W_0 + \Gamma_V$) in the center of the absorptive potential ($z = 0$). Thus our off-shell quasiparticle propagation is fully in line with the quantum mechanical result at least for quasi-stationary quantum states.

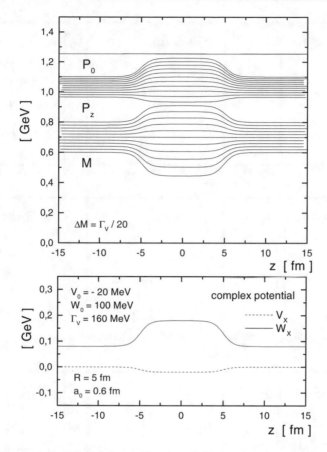

Fig. 4.15 upper part: P_{i0}, M_i, and P_{iz} as a function of $z(t)$ for a broad vacuum spectral function in a time-independent potential $V_0 = -20$ MeV, $W_0 = 100$ MeV (lower part). We have chosen a vacuum width $\Gamma_V = 160$ MeV and an initial mass separation of $\Delta M = 0.05 \cdot \Gamma_V$ for the testparticle trajectories displayed

To summarize our model results for the simple complex potential of Woods-Saxon-type, we find that energy conservation is guaranteed during the propagation and that the correct asymptotic solutions for the spectral functions are restored. Furthermore, in the potential region we observe a broadening of the width of the spectral function due to the space-time dependent imaginary part of the potential in line with quantum mechanics.

4.4 Kadanoff–Baym Dynamics for Fermions

The relativistic description of fermion fields in general follows that for boson fields in Sect. 4.1, however, with a different quantum statistics due to the anti-commutator algebra for fermion fields. Furthermore, the two-point functions are 4×4 matrices in

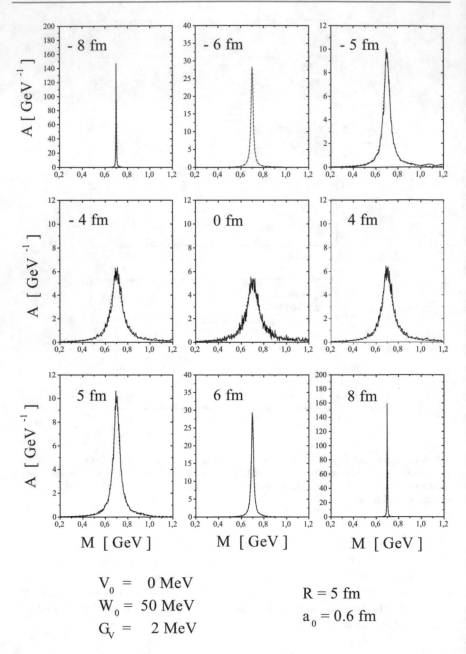

Fig. 4.16 The spectral distribution at different coordinates z from the testparticle distribution in comparison to the analytical result (solid lines) for $V_0 = 0$, $W_0 = 50$ MeV and $\Gamma_V = 2$ MeV. The analytical result is practically identical to the histograms from the testparticle distribution and thus hardly visible

the Dirac indices since the spin $1/2(\hbar)$ Dirac fields $\psi_\alpha(x)$ and Pauli-adjoint spinors $\bar\psi_\alpha$ ($\alpha = 1, .., 4$) are four-component spinors describing simultaneously two spin projections of particles as well as antiparticles. Basic elements of the Dirac-Clifford algebra (and conventions) are recalled in Appendix G.

4.4.1 Two-Point Functions on the CTP

Again the Green's functions on the contour may have time arguments on the same branch of the contour or on opposite branches giving four possibilities for the Green's functions[8] defined by

$$i S^c_{\alpha\beta}(x, y) \quad = i S^{++}_{\alpha\beta}(x, y) = \langle\, \tilde T^c(\psi_\alpha(x)\bar\psi_\beta(y))\,\rangle \tag{4.109}$$

$$i S^<_{\alpha\beta}(x, y) \quad = i S^{+-}_{\alpha\beta}(x, y) = -\langle\bar\psi_\beta(y)\psi_\alpha(x)\rangle \tag{4.110}$$

$$i S^>_{\alpha\beta}(x, y) \quad = i S^{-+}_{\alpha\beta}(x, y) = \langle\psi_\alpha(x)\bar\psi_\beta(y)\rangle \tag{4.111}$$

$$i S^a_{\alpha\beta}(x, y) = i S^{--}_{\alpha\beta}(x, y) = \langle\, \tilde T^a(\psi_\alpha(x)\bar\psi_\beta(y))\,\rangle \quad, \tag{4.112}$$

which are not independent![9] Time-ordering has to be fulfilled if both time arguments are on the same axis. The causal time-ordering operator $\tilde T^c$ places fields at later times to the left while the anticausal operator $\tilde T^a$ places fields at later times to the right (with a (-) sign for each exchange). One may again write the Green's function on the Keldysh contour in terms of a 2×2 matrix

$$S_{\alpha\beta}(x, y) = \begin{matrix} + \\ - \end{matrix}\begin{pmatrix} S^c_{\alpha\beta}(x, y) & S^<_{\alpha\beta}(x, y) \\ S^>_{\alpha\beta}(x, y) & S^a_{\alpha\beta}(x, y) \end{pmatrix} \tag{4.113}$$

in analogy to (4.9), however, now the elements are 4×4 matrices in the Dirac indices each. Similar representations hold for the "free" Green's function $S_0(x, y)$ as well as the selfenergy $\hat\Sigma$,

$$\hat\Sigma_{\alpha\beta}(x, y) = \begin{matrix} + \\ - \end{matrix}\begin{pmatrix} \hat\Sigma^c_{\alpha\beta}(x, y) & \hat\Sigma^<_{\alpha\beta}(x, y) \\ \hat\Sigma^>_{\alpha\beta}(x, y) & \hat\Sigma^a_{\alpha\beta}(x, y) \end{pmatrix} \tag{4.114}$$

---

[8] In order to distinguish the Green's functions for fermions from those for bosons we use the notation S instead of G.

[9] Here α, $\beta = 1,...,4$ denote the Dirac indices.

which has to be evaluated in some truncation scheme (of higher order). In lowest order for fermions it is given by the Dirac–Hartree–Fock selfenergy , however, a nonperturbative Dirac–Brueckner scheme—in analogy to the nonrelativistic case in Sect. 2.4—should be mandatory.

The further derivation again starts with a Dyson equation (cf. (C.7)) (suppressing now the Dirac indices)

$$S(x, y) = S_0(x, y) + S_0(x, y') \odot \hat{\Sigma}(y', z') \odot S(z', y) \quad , \tag{4.115}$$

where $\odot$ stands for the integration over repeated space-time arguments including the contour matrix $\eta = \mathrm{diag}(1, -1)$ that is acting on the contour index, i.e. $\eta = 1$ for the upper branch and $\eta = -1$ for the lower branch, respectively. Note that Eq. (4.115) is of one-body type and higher order correlations are included in the selfenergy $\hat{\Sigma}$ (which should be evaluated in a nonperturbative scheme as pointed out above). The inverse of the free Green's function in coordinate space is given by

$$S_{0,x}^{-1} = i\gamma^\mu \partial_\mu^x - M \tag{4.116}$$

and fulfills the equation of motion

$$S_{0,x}^{-1} S_{0,x}(x, y) = \eta \, \delta^4(x - y). \tag{4.117}$$

The Dyson equation (4.115) thus can also be written as

$$S_{0,x}^{-1} S(x, y) = \eta \, \delta^4(x - y) + \hat{\Sigma}(x, y') \odot S(y', y). \tag{4.118}$$

For the further considerations it is again useful to introduce retarded and advanced quantities by

$$S^A(x, y) = S^c(x, y) - S^>(x, y) = S^<(x, y) - S^a(x, y), \tag{4.119}$$

$$S^R(x, y) = S^c(x, y) - S^<(x, y) = S^>(x, y) - S^a(x, y),$$

$$\hat{\Sigma}^A(x, y) = \hat{\Sigma}^c(x, y) - \hat{\Sigma}^>(x, y) = \hat{\Sigma}^<(x, y) - \hat{\Sigma}^a(x, y), \tag{4.120}$$

$$\hat{\Sigma}^R(x, y) = \hat{\Sigma}^c(x, y) - \hat{\Sigma}^<(x, y) = \hat{\Sigma}^>(x, y) - \hat{\Sigma}^a(x, y).$$

From Eq. (4.118) we then get

$$S_{0,x}^{-1} S^{R/A}(x, y) = \delta^4(x - y) + \hat{\Sigma}^{R/A}(x, z) \odot S^{R/A}(z, y) \tag{4.121}$$

containing only information from retarded/advanced quantities. The non-diagonal elements of (4.115) give

$$S_{0,x}^{-1} S^<(x, y) = \hat{\Sigma}^R(x, z) \odot S^<(z, y) + \hat{\Sigma}^<(x, z) \odot S^A(z, y) \tag{4.122}$$

and provide the space-time evolution of the system as well as the time evolution of the spectral properties. This will become evident after Wigner transformation of (4.122) and a gradient expansion in phase space (see below), whereas (4.121) will provide the spectral properties of the fields (particles).

In the relativistic case the scalar density $\rho_s(x)$ and the vector four-current $j^\mu(x)$ are of central interest and defined by

$$\rho_s(x) = \langle \bar{\psi}(x)\psi(x) \rangle = -iTr(S^<(x,x)) = -i\sum_\alpha S^<(x,x)_{\alpha\alpha}, \qquad (4.123)$$

$$j^\mu(x) = \langle \bar{\psi}(x)\gamma^\mu\psi(x) \rangle = -iTr(\gamma^\mu S^<(x,x)) = -i\sum_{\alpha\beta} (\gamma^\mu)_{\alpha\beta} S^<(x,x)_{\beta\alpha}.$$

4.4.2 Definition of Selfenergies

For the solution of the Kadanoff–Baym equations the computation/fixing of the selfenergies $\hat{\Sigma}$ is mandatory. Here a Φ-derivable scheme—as used in Sect. 4.1 for bosons—is of substantial advantage (or necessary) since it leads to equations of motions that conserve energy-momentum, angular momentum and charge-like quantum numbers and is simultaneously thermodynamically consistent, i.e. the bulk properties of the system fulfill the Maxwell relations between the corresponding thermodynamic potentials in equilibrium. We recall that in case of an "effective" relativistic approach like QHD (Quantum-Hadro-Dynamics) [13, 14] this thermodynamic consistency had to be shown explicitly.[10] The Φ-functional itself is a function of the full propagators and mean fields and the selfenergy is given by the variational derivative [15]

$$\hat{\Sigma}^c(x,y) = -i\frac{\delta\Phi[S^c]}{\delta S^c(y,x)}, \qquad (4.124)$$

where the functional $\Phi[S]$ is defined diagrammatically as the sum of all closed Feynman diagrams, where the internal limes represent the full propagator S, and which cannot be separated into disconnected parts by cutting two propagator lines (cf. Sect. 4.1.4 in case of the scalar ϕ^4-theory). This scheme is denoted by a two-particle irreducible (2PI) approximation. The variational derivative then leads to a skeleton expansion for the selfenergy in terms of full propagators, i.e. as a sum over all selfenergy diagrams with propagator lines that do not contain selfenergy insertions. This avoids double-counting of interactions. In this context a partial resummation of ring-diagrams leads to Landau-Fermi-liquid theory while a partial resummation of ladder diagrams leads to Dirac–Brueckner theory. Furthermore, the definition of the selfenergy (4.124) differs from the case of bosons by a factor (-2)

---

[10] cf. Appendix H.

since the Klein–Gordon equation is of second order in the space-time derivatives whereas the Dirac equation is of first order. Note however, that each component ψ_α of the free Dirac spinor is also a solution of the Klein–Gordon equation expressing the mass-shell constraint $E^2 = \mathbf{p}^2 + M^2$. On the other hand any partial resummation of diagrams violates the Ward–Takahashi identities, which are a counterpart to the trace relations discussed in Appendix B in the nonrelativistic case. Also the latter are violated in the different truncation schemes (cf. Sect. 2.2).

The transport equation for the Green's function—or Kadanoff–Baym equation— is obtained from Eq. (4.122) and its adjoint equation using the identities

$$\hat{\Sigma}^R(x, y) = \Re\hat{\Sigma}^R(x, y) + \frac{1}{2}(\hat{\Sigma}^>(x, y) - \hat{\Sigma}^<(x, y)), \tag{4.125}$$

$$\hat{\Sigma}^A(x, y) = \Re\hat{\Sigma}^R(x, y) - \frac{1}{2}(\hat{\Sigma}^>(x, y) - \hat{\Sigma}^<(x, y)),$$

i.e.

$$i(\gamma^\mu \partial_\mu^x + \gamma^\mu \partial_\mu^y)S^<(x, y) - [\Re\hat{\Sigma}^R \odot S^<](x, y) + [S^< \odot \Re\hat{\Sigma}^R](x, y) \tag{4.126}$$

$$- [\hat{\Sigma}^< \odot \Re S^R](x, y) + [\Re S^R \odot \hat{\Sigma}^<](x, y)$$

$$= \frac{1}{2}\left([\hat{\Sigma}^> \odot S^<](x, y) + [S^< \odot \hat{\Sigma}^>](x, y)\right.$$

$$\left. - [\hat{\Sigma}^< \odot S^>](x, y) - [S^> \odot \hat{\Sigma}^<](x, y)\right).$$

Since the coupled equations (4.125) and (4.126) are difficult to solve in space-time representation one proceeds with a Wigner transformation of these equations as in case of the Bose system in Sect. 4.3 and restricts to first-order gradients in phase space thus assuming that the Wigner functions only smoothly depend on the average space-time coordinate $X = (x + y)/2$ and four-momentum P.

In case one is interested only in spin averaged quantities—as for relativistic heavy-ion reactions—one takes additionally the spinor trace of the Kadanoff–Baym equation (4.126) in first-order gradient expansion. Thus defining the vector-current density by

$$\mathcal{V}^\mu(X, P) = -iTr[S^<(X, P)\gamma^\mu] \tag{4.127}$$

and using the cyclic invariance of the trace one ends up with the transport equation for spin $1/2\hbar$ Dirac particles:

$$\partial_\mu \mathcal{V}^\mu(X, P) - Tr\{\Re\hat{\Sigma}^R(X, P), -iS^<(X, P)\}_P + Tr\{\Re S^R(X, P), \tag{4.128}$$

$$- i\hat{\Sigma}^<(X, P)\}_P = I_{coll}(X, P)$$

$$= Tr[\hat{\Sigma}^<(X, P)S^>(X, P) - \hat{\Sigma}^>(X, P)S^<(X, P)],$$

with the collision term I_{coll} describing the dissipative part with separate gain and loss terms. In Eq. (4.128) we have used again the relativistic generalization of the Poisson bracket (3.27),

$$\{F(X, P), G(X, P)\}_P := \partial^P_\mu F(X, P) \cdot \partial^\mu_X G(X, P) - \partial^\mu_X F(X, P) \cdot \partial^P_\mu G(X, P).$$
(4.129)

The l.h.s. of Eq. (4.128) contains the free streaming term $\partial_\mu \mathcal{V}^\mu(X, P)$, the regular mean-field term with the derivatives of $\Re\hat{\Sigma}^R(X, P)$ and $S^<(X, P)$ as well as the backflow term with the derivatives of $\Re S^R(X, P)$ and $\hat{\Sigma}^<(X, P)$.

In principle the 4×4 matrices $\Re\hat{\Sigma}^R(X, P)$, $S^<(X, P)$, $\Re S^R(X, P)$, and $\hat{\Sigma}^<(X, P)$ may have nonvanishing off-diagonal terms that even change in X and P in a complicated fashion. Since the many-body truncation scheme is incomplete and only a partial resummation of interaction terms can be carried out a full and accurate solution might be practically impossible. Accordingly, we will assume in the following that these matrices are approximately diagonal such that the trace simply gives the sum of Dirac particles and antiparticles with two spin projections each. This allows to write down four transport equations in the form of (4.128) for each component separately, however, with separate selfenergies for particles and antiparticles. Adopting the simplified Lorentz decomposition

$$\hat{\Sigma}^R_u(X, P) \approx \Sigma^s(X, P) - \gamma_\mu \Sigma^\mu(X, P)$$
(4.130)

for the upper diagonal elements (denoted by the index u) we get

$$\hat{\Sigma}^R_d(X, P) \approx \Sigma^s(X, P) + \gamma_\mu \Sigma^\mu(X, P)$$
(4.131)

for the lower diagonal elements (denoted by the index d) due to the time-reflection properties. This, however, will severely violate dispersion relations[11] such that different scalar and vector selfenergies should be employed for particles and antiparticles in practice in line with coupled-channel Dirac–Brueckner calculations.

4.4.3 Spectral Functions

The spectral properties of the fermion fields can be specified via (4.119) which gives

$$\Im S^R(X, P) = -\frac{i}{2}(S^>(X, P) - S^<(X, P)),$$
(4.132)

$$\Im S^A(X, P) = \frac{i}{2}(S^>(X, P) - S^<(X, P)),$$

---

[11] The Kramers–Kronig relations are recalled in Appendix F.

leading to the spectral function in terms of a 4×4 matrix (in phase-space representation)

$$\hat{A}(X, P)_{\alpha\beta} = -\frac{1}{\pi} \Im S^R(X, P)_{\alpha\beta} = \frac{i}{2\pi} \left(S^>(X, P)_{\alpha\beta} - S^<(X, P)_{\alpha\beta}\right) \qquad (4.133)$$

$$= \frac{i}{2\pi} \left(S^R(X, P)_{\alpha\beta} - S^A(X, P)_{\alpha\beta}\right).$$

An integration over P_0 leads to the equal-time commutation relations for Dirac spinors, i.e.

$$\int_{-\infty}^{\infty} dP_0 \, \hat{A}(X, P)_{\alpha\beta} = (\gamma^0)_{\alpha\beta}. \qquad (4.134)$$

Furthermore, the retarded and advanced Green's functions follow the dispersion relations

$$S^{R/A}(X, P)_{\alpha\beta} = \int_{-\infty}^{\infty} dP_0' \, \frac{\hat{A}(X, P_0', \mathbf{P})_{\alpha\beta}}{P_0 - P_0' \pm i\eta}. \qquad (4.135)$$

By taking the trace over Dirac indices one can define a spin averaged spectral function

$$A(X, P) := \frac{1}{2} Tr\left(\hat{A}(X, P)\gamma^0\right) = -\frac{1}{2\pi} Tr\left(\Im(S^R(X, P))\gamma^0\right). \qquad (4.136)$$

The quantity (4.136) transform as the temporal component of a Lorentz four-vector and has the interpretation of an energy distribution due to the normalization (4.134) (or quantization).

A spin averaged Wigner function $F(X, P)$, furthermore, is defined by

$$F(X, P) = -2f(X, P) \, Tr\left(\Im(S^R(X, P))\gamma^0\right) \qquad (4.137)$$

and also transforms as the temporal component of a Lorentz four-vector. It is the generalization of the phase-space density in Wigner representation for the relativistic case of Dirac fermions. Furthermore, using (4.136) this function can be rewritten as

$$F(X, P) = 2\pi \, D f(X, P) \, A(X, P), \qquad (4.138)$$

(with $D = 2$) such that the spectral information is separated out in $F(X, P)$ (4.137) in analogy to the bosonic case in Sect. 4.3. In thermal equilibrium (of a homogenous system in space) the function $f(P)$, however, becomes a Fermi-distribution instead of the Bose distribution. Accordingly, $f(X, P)$ has the interpretation of "local occupation numbers." Furthermore, the spectral function $A(X, P)$ has the dimension [1/energy] in case of fermions whereas in case of bosons the dimension is [1/energy^2]. On the other hand the selfenergies have the dimension [energy] in case of fermions whereas the dimension is [energy^2] for bosons. This has to be taken care of in actual transport calculations that simultaneously propagate fermions and bosons.

4.5 Spectral Evolution of Vector Mesons in Heavy-Ion Collisions at 2 A GeV

The theory of quantum-chromo-dynamics (QCD) describes hadrons as many-body bound or resonant states of partonic constituents, i.e. quarks, antiquarks, and gluons. While the properties of hadrons are rather well known in free space (embedded in a nonperturbative QCD vacuum) the mass and lifetimes of hadrons in a baryonic and/or mesonic environment are subject of research in order to achieve a better understanding of the strong interaction and the nature of confinement. In this context the modification of hadron properties in nuclear matter is of fundamental interest since QCD sum rules as well as QCD inspired effective Lagrangian models [16] have predicted significant changes e.g. of the vector mesons (ρ, ω and ϕ) with the nuclear density ρ_N and/or temperature T. Due to the electromagnetic decay of vector mesons to e^+e^- or $\mu^+\mu^-$ pairs and the weak interaction of leptons with hadrons in the dense medium the spectroscopy of lepton pairs provides a closer view on the spectral properties of vector mesons in dense and hot nuclear matter. Indeed, some direct evidence for the modification of vector meson spectral functions has been obtained experimentally from the enhanced production of lepton pairs above known sources in nucleus-nucleus collisions [17]. The observed enhancement in the invariant mass range $0.3 \leq M \leq 0.7$ GeV might be due to a downward shift of the ρ-meson mass or simply due to a substantial broadening of the vector meson spectral functions due to frequent interactions with nucleons and mesons or their excited states. A microscopic description of such medium modifications requires an explicit dynamical evolution of their spectral functions, i.e. off-shell transport as derived in the previous sections.

In order to demonstrate the importance of off-shell transport dynamics in particular for vector mesons we present in Fig. 4.17 the time evolution of the mass distribution of ρ (upper part) and ω (lower part) mesons for central C+C collisions (at impact parameter b=1 fm) at 2 A GeV for the case of a dropping mass + collisional broadening scenario (as an example). The transport approach is conceptually equivalent to the RBUU approach in Sect. 3.1, where the baryons are treated on-shell, however, the vector mesons are treated off-shell in line with the generalized equations of motion (4.101)–(4.103). In this scenario the pole masses of the vector mesons drop linearly with local baryon density while the width of the vector mesons is broadened in the medium due to the local interaction rate with

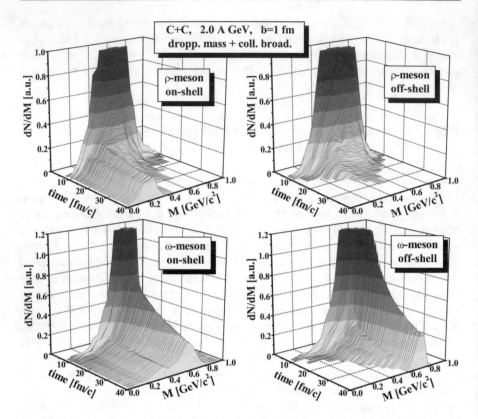

Fig. 4.17 Time evolution of the mass distribution of ρ (upper part) and ω (lower part) mesons for central $C + C$ collisions (b=1 fm) at 2 A GeV for the dropping mass + collisional broadening scenario. The l.h.s. of Fig. 4.17 correspond to the calculations with on-shell dynamics whereas the r.h.s. show the off-shell results

the medium.[12] The l.h.s. of Fig. 4.17 corresponds to the calculations with on-shell propagation whereas the r.h.s. show the results for the off-shell dynamics for the vector mesons.

As seen from Fig. 4.17 the initial ρ- and ω-mass distributions are quite broad even for a small system such as $C + C$ where, however, the baryon density at 2 A GeV may reach (in some local cells) up to twice nuclear matter density ρ_0. The number of vector mesons initially is enhanced—as compared to the free case—due to the downward shift of the ρ and ω spectral functions and decreases with time due to their decays (to 2 or 3 pions) and the absorption by baryons ($\rho N \rightarrow \pi N$, $\rho N \rightarrow \pi \pi N$, $\omega N \rightarrow \pi N$, etc.). Most of the ρ mesons decay/disappear already inside the 'fireball' for density $\rho_N > 0$. Due to the "fireball" expansion the baryon density drops quite fast, so some amount of ρ mesons reach the very low density zone or

---

[12] The actual details are not of interest here but can be taken from Ref. [18].

even the "vacuum." Since for the off-shell case (r.h.s. of Fig. 4.17) the ρ spectral function changes dynamically by propagation in the dense medium according to Eqs. (4.101)–(4.103), it regains the vacuum shape for vanishing density. This does not happen for the on-shell treatment (l.h.s. of Fig. 4.17); the ρ spectral function does not change its shape by propagation but only by explicit collisions with other particles. Indeed, there is a number of ρ's which survive the decay or absorption and leave the "fireball" with masses below $2m_\pi$. Accordingly, the on-shell treatment leads to the appearance of ρ mesons in the vacuum with $M \leq 2m_\pi$, which can not decay to two pions; thus they live practically "forever" since the probability to decay to other channels is very small. Such ρ's will continuously shine low mass dilepton (e^+e^-) pairs which leads to an unphysical "enhancement/divergence" of the dilepton yield at low masses.

The same statements are valid for the ω mesons (cf. lower part of Fig. 4.17): since the ω-meson is a longer living resonance, a larger amount of ω's survives with an in-medium like spectral function in the vacuum (in case of on-shell dynamics). Such ω's with $M < 3m_\pi$ can decay only to $\pi\gamma$ or electromagnetically (if $M < m_\pi$). Since such unphysical phenomena appear in on-shell transport descriptions (including an explicit vector meson propagation) an off-shell treatment is mandatory.

In summarizing this section we note that the proper dynamical evolution of short-lived particles or strongly-interacting degrees of freedom requires off-shell transport equations and the inclusion of off-shell transitions in particular for those processes that are kinematically forbidden in an on-shell treatment. Further applications of off-shell transport—in case of heavy-ion collisions—address the dynamics of antikaons at high baryon density and temperature [19, 20] as well as partonic degrees of freedom with broad spectral functions [21–23] in the quark-gluon plasma phase.

4.6   Electromagnetic Field Evolution in Ultra-Relativistic Collisions

Heavy-ion collisions at ultra-relativistic energies represent the only laboratory on Earth for the investigation of the deconfined phase of strongly-interacting matter, which is denoted as Quark-Gluon-Plasma (QGP). Its formation and evolution shows some of the most extreme properties of matter ever observed, i.e. very high temperatures T even several times the critical temperature $T_c \approx 155$ MeV for the transition between hadronic and partonic matter, which is about five orders of magnitude higher than the temperature in the center of the sun; a very low ratio of the shear viscosity over entropy density η/s close to T_c (more than twenty times smaller than that of water at the corresponding T_c); strong magnetic fields up to $eB \sim 50m_\pi^2 \sim 10^{19}$G (some order of magnitude larger than that expected on the surface of magnetars) and an intense vorticity up to $\omega \sim 0.1$ c/fm $\sim 10^{22}$ s^{-1} which is about 14 orders of magnitude higher than that of any other fluid observed.[13]

---

[13] cf. the review [24].

In ultra-relativistic collisions of heavy ions at Relativistic Heavy-Ion Collider (RHIC) or Large Hadron Collider (LHC) energies one observes a rather clear-cut separation of "participants" and "spectators" where the latter move practically "free" with their initial velocity in the nucleon-nucleon center-of-mass system. Since the spectators are electrically charged in line with the proton content the latter generate an electromagnetic field which is contracted in beam direction by the Lorentz γ-factor of the spectators[14] and can achieve a high field strength in the region of the participants. Let us assume that the impact parameter of the collision is in x-direction while the beam is in z-direction. In this case the electric field $\mathbf{E}(\mathbf{r}; t)$ approximately cancels in the center of the participants ($\mathbf{r} \approx 0$) in case of symmetric systems, however, the magnetic field $\mathbf{B}(\mathbf{r}; t)$ adds up the contribution from charged projectile and target spectators and is directed preferentially in y-direction. On the other hand, in asymmetric collision systems the electric field may also be high at $\mathbf{r} \approx 0$ due to the different number of positive charges in the projectile and target spectators. It is thus of basic interest to know the strength and time dependence of electromagnetic fields in (ultra-) relativistic nucleus-nucleus collisions in order to learn about the electromagnetic forces and their impact on the charged degrees of freedom in the participant zone. Since the electromagnetic interaction and coupling strength is well known the extension of transport theory to electromagnetic interactions is rather straight forward and does not involve any new parameter.

To describe an ultra-relativistic collision of heavy ions we start from the relativistic Boltzmann equation for an on-shell phase-space distribution function $f \equiv f(x, p)$ for nucleons (and mesons) discarding selfenergies of the particles which are by far subleading as compared to the huge kinetic energy of the particles,

$$p^\mu \frac{\partial}{\partial x^\mu} f = p^0 \hat{C}[f] \tag{4.139}$$

where $\hat{C}[f]$ is the collision integral and x, p are the 4-coordinate and 4-momentum of a particle with p^0 taken on-shell. A background electromagnetic field may be taken into account by including an electromagnetic tensor $F_{\mu\nu}$ into Eq.(4.139) as

$$p^\mu \left(\frac{\partial}{\partial x^\mu} - F_{\mu\nu} \frac{\partial}{\partial p^\nu} \right) f = p^0 \hat{C}[f], \tag{4.140}$$

where

$$e F_{\mu\nu} = \partial_\mu A_\nu - \partial_\nu A_\mu \tag{4.141}$$

with the electromagnetic four-vector potential $A_\mu = (\Phi, \mathbf{A})$. Note that the left-hand side of Eq. (4.140) is gauge invariant since it involves only the electromagnetic field

---

[14] This γ-factor is in the order of 100 at top RHIC energies and > 1000 at LHC energies.

strength tensor $F_{\mu\nu}(x)$. Equation (4.140) can be rewritten in terms of components of A_μ (after dividing by p^0):

$$\left\{ \frac{\partial}{\partial t} + \frac{\mathbf{p}}{p_0} \cdot \nabla_{\mathbf{r}} - \left(e\nabla\Phi + e\frac{\partial \mathbf{A}}{\partial t} - e\mathbf{v} \times (\nabla \times \mathbf{A}) \right) \nabla_{\mathbf{p}} \right\} f = \hat{C}[f] \qquad (4.142)$$

with $\mathbf{v} = \mathbf{p}/p^0$. The strength of the magnetic $\mathbf{B}$ and electric $\mathbf{E}$ fields is, respectively,

$$\mathbf{B} = \nabla \times \mathbf{A}, \quad \mathbf{E} = -\nabla\Phi - \frac{\partial \mathbf{A}}{\partial t}. \qquad (4.143)$$

We note that the electromagnetic field generated by moving nuclei may be considered as an external field: the value of the electromagnetic field at a given point is determined by the global charge current of the colliding nuclei and thus, in good approximation, independent of the local strong interaction dynamics. However, the presence of the electromagnetic field can affect the interactions among particles, which simultaneously carry electric and (possibly) color charges. Using the relations (4.143) the system (4.142) is reduced to a more familiar form:

$$\left\{ \frac{\partial}{\partial t} + \frac{\mathbf{p}}{p_0} \cdot \nabla_{\mathbf{r}} + (e\mathbf{E} + e\mathbf{v} \times \mathbf{B}) \nabla_{\mathbf{p}} \right\} f = \hat{C}[f] \qquad (4.144)$$

for particles of charge e containing explicitly the well-known electromagnetic Lorentz force.[15]

> Exercise 4.5: Derive the retarded propagator for the wave equation
>
> $$\partial_\mu \partial^\mu G_{ret}(x, x') = \delta^4(x - x') \equiv \left(\frac{1}{c^2} \frac{\partial^2}{\partial t^2} - \Delta_{\mathbf{r}} \right) G_{ret}(\mathbf{r} - \mathbf{r}'; t - t')$$
>
> $$= \delta(t - t')\delta^3(\mathbf{r} - \mathbf{r}').$$

The transport equations for strongly-interacting particles with electric charge e (4.142) have to be supplemented by the electromagnetic field equations

$$\nabla \times \mathbf{E} = -\frac{1}{c} \frac{\partial \mathbf{B}}{\partial t}, \quad \nabla \cdot \mathbf{B} = 0. \qquad (4.145)$$

---

[15] In case of quarks and antiquarks the charge may also be $\pm e/3$ or $\pm 2e/3$.

The general solution of the wave equations (4.145) for the charge distribution $\rho(\mathbf{r}, t)$ and electric current $\mathbf{j}(\mathbf{r}, t)$ is

$$\Phi(\mathbf{r}, t) = \frac{1}{4\pi} \int \frac{\rho(\mathbf{r}', t')\, \delta(t - t' - |\mathbf{r} - \mathbf{r}'|/c)}{|\mathbf{r} - \mathbf{r}'|}\, d^3r'dt' \tag{4.146}$$

for the electromagnetic scalar potential $\Phi(\mathbf{r}, t)$ and

$$\mathbf{A}(\mathbf{r}, t) = \frac{1}{4\pi} \int \frac{\mathbf{j}(\mathbf{r}', t')\, \delta(t - t' - |\mathbf{r} - \mathbf{r}'|/c)}{|\mathbf{r} - \mathbf{r}'|}\, d^3r'dt' \tag{4.147}$$

for the vector potential. For a single moving point-like particle with charge e one gets

$$\rho(\mathbf{r}, t) = e\, \delta(\mathbf{r} - \mathbf{r}(t)), \quad \mathbf{j}(\mathbf{r}, t) = e\, \mathbf{v}(t)\, \delta(\mathbf{r} - \mathbf{r}(t)) \tag{4.148}$$

and, after integration of Eq. (4.147) using

$$\int_{-\infty}^{\infty} g(x)\, \delta(f(x))\, dx = \sum_i \frac{g(x_i)}{|f'(x_i)|}, \tag{4.149}$$

we obtain (after summing over all charged particles i):

$$\Phi(\mathbf{r}, t) = \frac{e}{4\pi} \sum_i \frac{1}{R(t_i')\kappa(t_i')} \tag{4.150}$$

with the definitions

$$\mathbf{R}(t_i') = \mathbf{r} - \mathbf{r}(t_i'), \qquad \kappa(t_i') = 1 - \frac{\mathbf{R}(t_i') \cdot \mathbf{v}(t_i')}{cR(t_i')} = \left| \left(\frac{df}{dt'} \right)_{t'=t_i'} \right|. \tag{4.151}$$

In (4.151) the times t_i' are solutions of the retardation equation $(R(t') = |\mathbf{R}(t')|)$

$$f(t') = t' - t + R(t')/c = 0, \tag{4.152}$$

which has an easy solution for straight line trajectories. By analogy we get for the vector potential:

$$\mathbf{A}(\mathbf{r}, t) = \frac{e}{4\pi} \sum_i \frac{\mathbf{v}(t_i')}{R(t_i')\kappa(t_i')}. \tag{4.153}$$

Thus, Eqs. (4.146) and (4.147) lead to the retarded Liénard–Wiechert potentials (4.150) and (4.153) acting at the point $\mathbf{R} = \mathbf{r} - \mathbf{r}'$ at the time t. The electromagnetic potentials $\Phi(\mathbf{r}, t)$ and $\mathbf{A}(\mathbf{r}, t)$ are generated by every moving charged particle and describe elastic Coulomb scattering as well as inelastic bremsstrahlung processes. Here we have used $\epsilon_0\mu_0 = 1/c^2 \equiv 1$ by setting $\epsilon_0 = \mu_0 = 1$ such that $eA \sim e^2/(4\pi) = \alpha_{em} \approx 1/137$.

With the help of the solutions (4.150) and (4.153) the retarded electric and magnetic fields can be derived from (4.143) by evaluating the proper space-time derivatives:

$$\mathbf{E}(\mathbf{r}, t) = \frac{e}{4\pi}\left(\frac{\mathbf{n}}{\kappa R^2} + \frac{-\mathbf{b}/c - [(\mathbf{n} \cdot \mathbf{v})\mathbf{n} - \mathbf{v}]/R}{\kappa^2 cR}\right)_{ret}$$
$$- \frac{e}{4\pi}\left(\frac{(-\mathbf{v}(t')/c + \mathbf{n}(t'))(v^2/c - \mathbf{n} \cdot \mathbf{v} - R/c(\mathbf{n} \cdot \mathbf{b}))}{\kappa^3 cR^2}\right)_{ret} ,(4.154)$$

and

$$\mathbf{B}(\mathbf{r}, t) = \frac{e}{4\pi}\left(\frac{\mathbf{v} \times \mathbf{n}}{\kappa R^2} + (\frac{\mathbf{b}(t') \times \mathbf{n}(t') + \mathbf{v}(t') \times [(\mathbf{n} \cdot \mathbf{v})\mathbf{n} - \mathbf{v}]/R}{\kappa^2 cR})\right)_{ret}$$
$$- \frac{e}{4\pi}\left(\frac{(\mathbf{v}(t') \times \mathbf{n}(t'))(v^2/c - \mathbf{n} \cdot \mathbf{v} - R/c(\mathbf{n} \cdot \mathbf{b}))}{\kappa^3 cR^2}\right)_{ret} \qquad(4.155)$$

with the acceleration $\mathbf{b} = d/dt' \mathbf{v}$ and the unit vector $\mathbf{n} = \mathbf{R}/R$. When neglecting the acceleration $\mathbf{b}$—being practically zero for spectator nucleons—we arrive at

$$e\,\mathbf{E} = \frac{sign(e)\,\alpha\,\mathbf{R}(t)\,(1 - v^2/c^2)}{\left[(\mathbf{R}(t) \cdot \mathbf{v}/c)^2 + R^2(t)(1 - v^2/c^2)\right]^{3/2}} , \qquad(4.156)$$

$$e\,\mathbf{B} = \frac{sign(e)\,\alpha\,[\mathbf{v} \times \mathbf{R}(t)]\,(1 - v^2/c^2)}{c\left[(\mathbf{R}(t) \cdot \mathbf{v}/c)^2 + R^2(t)(1 - v^2/c^2)\right]^{3/2}}, \qquad(4.157)$$

where in the left-hand side an additional charge e is introduced to get the electromagnetic fine-structure constant $\alpha = e^2/4\pi \approx 1/137$ in the right-hand side of these equations. The last expression is reduced to the familiar form of the retarded Liénard–Wiechert equation for the magnetic field of a moving charge,

$$\mathbf{B}(\mathbf{r}, t) = \frac{e}{4\pi}\frac{[\mathbf{v} \times \mathbf{R}]}{cR^3}\frac{(1 - v^2/c^2)}{[1 - (v/c)^2 \sin^2 \phi_{Rv}]^{3/2}}, \qquad(4.158)$$

with $\mathbf{R} = \mathbf{R}(t)$, while ϕ_{Rv} is the angle between $\mathbf{R}(t)$ and $\mathbf{v}$.

The set of transport equations (4.142) in the following is solved in the quasiparticle approximation (as before) while the electromagnetic fields are computed on a suitable space-time grid. The quasiparticle propagation in the electromagnetic fields follows from Eq.(4.144) as

$$\frac{d\mathbf{p}}{dt} = e\mathbf{E} + \frac{e}{c}\mathbf{v} \times \mathbf{B} \qquad(4.159)$$

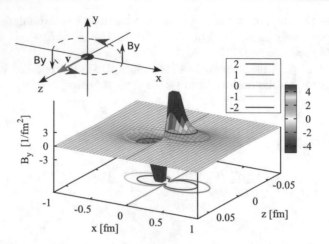

Fig. 4.18 Snapshot of the B_y distribution for the magnetic field and its projection on the $(z - x)$ plane for a single charge e moving along the z axis

and may be supplemented by strong forces from scalar and vector selfenergies as in Sect. 3.1. The change of the electromagnetic energy is $(e/c)(\mathbf{v} \cdot \mathbf{E})$, i.e. the magnetic field does not change the quasiparticle energy. In order to avoid singularities and self-interaction effects arising from point-like particles, all particles within a given Lorentz-contracted cell are excluded from the field calculations.

4.6.1 Electromagnetic Fields in Au+Au Collisions at $\sqrt{s}$= 200 GeV

The scheme described above for the computation of the electromagnetic field is applied here to ultra-relativistic heavy-ion collisions at the invariant energy per nucleon pair $\sqrt{s}$= 200 GeV,[16] which is the top energy achieved at the Relativistic Heavy-Ion Collider (RHIC) at Brookhaven. For transparency, however, we will start with the magnetic field created by a single freely moving charge. As seen in Fig. 4.18, the charge creates a cylindrically symmetric field with the symmetry axis along the direction of motion. If one follows the magnetic field direction, it appears to be torqued around the direction of motion. Therefore, the magnetic field on the left and right sides with respect to the moving charge has opposite signs, resulting in some maximum and minimum of the magnetic field at a given instant of time. The opposite field signs directly follow from Eq. (4.158) if one takes into account that the vector $\mathbf{R}$ (Eq. (4.151)) in this situation has opposite signs.

---

[16] The figures in this Subsection are taken from Ref. [25].

Fig. 4.19 The transverse
plane of a noncentral
heavy-ion collision. The
impact parameter of the
collision is denoted by b. The
magnetic field is displayed by
the dashed lines

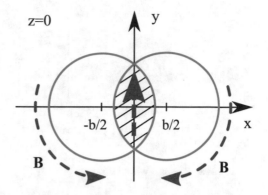

In a nucleus-nucleus collision the magnetic field will be a superposition of solenoidal fields from different moving charges. The collision geometry for a peripheral collision is shown in Fig. 4.19 in the transverse plane. The overlapping strongly-interacting region (participants) has an "almond"-like shape. The nuclear region outside this almond (shaded in Fig. 4.19) corresponds to spectator matter which is the dominant source of the electromagnetic field at the very beginning of the nuclear collision. Note that in the transport calculation the particles are subdivided into target and projectile spectators and participants not geometrically but dynamically: spectators are nucleons which did not yet suffer a collision.

The time evolution of the magnetic field $eB_y(x, y = 0, z)$ for Au+Au collisions at the colliding energy $\sqrt{s_{NN}} = 200$ GeV and the impact parameter $b = 10$ fm is shown in Fig. 4.20. If the impact parameter direction is taken as the x axis (as in the present calculations), then the magnetic field will be directed along the y-axis perpendicularly to the reaction plane $(z - x)$. The geometry of the colliding system at the moment considered is demonstrated by points in the $(z - x)$ plane where every point corresponds to a spectator nucleon. It is seen that the largest values of $eB_y \sim 5m_\pi^2$ are reached in the beginning of the collision for a very short time corresponding to the maximal overlap of the colliding ions. This is an extremely high magnetic field since $m_\pi^2 \approx 10^{18} G$. The first panel in Fig. 4.20 is taken at a very early compression stage with $t = 0.01$ fm/c. The time $t = 0.05$ fm/c is close to the maximal overlap and the magnetic field here is maximal, too. Then the system expands (note the different z-scales in different panels of Fig. 4.20) and the magnetic field decreases. For $b = 0$ the overlap time is maximal and roughly given by $2R/\gamma_c$— with γ_c denoting the Lorentz γ factor—which for our case is about 0.15 fm/c. For peripheral collisions this time is even shorter.

Globally, the spatial distribution of the magnetic field is inhomogeneous and Lorentz-contracted along the z-axis. At the compression stage there is a single maximum which in the expansion stage is splitted into two parts associated with the spectators. In the transverse direction the bulk magnetic field is limited by two minima coming from the torqued structure of the single-charge field (see Fig. 4.18).

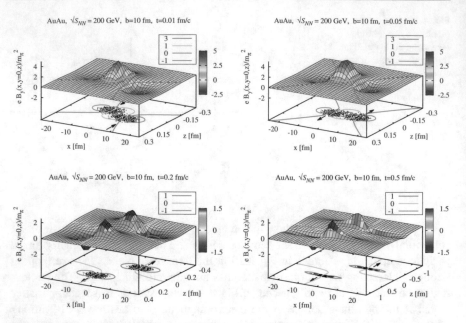

Fig. 4.20 Time dependence of the spatial distribution of the magnetic field B_y at times t created in Au+Au ($\sqrt{s}$=200 GeV) collisions with the impact parameter b =10 fm. The location of spectator protons is shown by dots in the $(x - z)$-plane. The level $B_y = 0$ and the projection of its location on the $(x - z)$ plane are shown by the solid lines

4.6.2   Electromagnetic Fields in Cu+Au Collisions at $\sqrt{s}$= 200 GeV

Before coming to the actual results for asymmetric systems in comparison to those for symmetric Au+Au reactions some comments with respect to the participant zone are in order. The initial conversion of energy happens during roughly 0.15 fm/c at the top RHIC energy for Cu+Au and Au+Au collisions when the nuclei pass through each other. At this time the energy density in between the leading baryons is very high due to the fact that the spatial volume is very small and $\sim$ 0.3 fm in longitudinal extension; the transverse contribution to this volume is given by the overlap area. Accordingly, the energy density in the participant zone is above critical ($\approx$ 0.5 GeV/fm^3), i.e. the degrees of freedom should change from hadronic to partonic ones according to lattice QCD calculations. Due to the Heisenberg uncertainty relation this energy density cannot be specified as being due to "particles" since the latter may form only much later on a timescale of their inverse transverse mass (in their rest frame). More specifically, only jets at midrapidity with transverse momenta above $p_T = 2$ GeV/c are expected to appear at $t \approx 0.1$ fm/c while a soft parton with transverse momentum $p_T = 0.5$ GeV/c should be formed after $t \geq 0.5$ fm/c. Although it is not clear what the actual nature of the degrees of freedom is in this very initial state, there will naturally be a small amount of electric charges due to charge conservation. On the other hand, if a large amount of electric charges

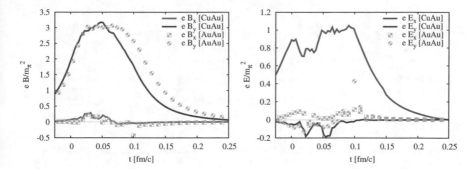

Fig. 4.21 Time evolution of event-averaged components of the magnetic (l.h.s.) and electric (r.h.s.) fields in the center of the overlap region of colliding Cu+Au (solid lines) and Au+Au (dotted lines) systems at $\sqrt{s_{NN}} = 200$ GeV and $b = 7$ fm. The distributions are averaged over 70 events

(e. g. from the conversion of energy to quarks and antiquarks) are present in the very beginning of the reaction, then there should be observable signals from this early electric accelerator. A promising observable here is the directed flow $v_1(y)$ of charged particles with opposite charge, e.g. π^+ and π^- or protons and antiprotons. This offers the specific property of the early electric field to check experimentally if electric charges are already present at this very early instant.

The time evolution of transverse electromagnetic field components is compared between asymmetric Cu+Au (solid lines) and symmetric Au+Au systems (dotted lines) in Fig. 4.21 where the l.h.s. displays the magnetic field components and the r.h.s. the electric field components.[17] The maximal values of the magnetic field components $\langle eB_y \rangle$ are on the level of a few m_π^2 being comparable for both systems. However, in case of the Cu+Au reaction the $\langle eE_x \rangle$ component of the electric field is by a factor of ~5 larger than that for symmetric Au+Au collisions at the same energy. This strong electric field eE_x is only present for about 0.25 fm/c during the overlap phase of the heavy ions and will act as an electric accelerator on charges that are present during this time. Note that when charges appear only later together with the formation of soft partons ($t \geq 0.5$ fm/c) there will be no corresponding charge separation effect on the directed flow! In the case of symmetric collisions $\langle E_x \rangle \approx \langle B_y \rangle$, however, this approximate equality is broken for asymmetric Cu+Au collisions where $\langle eB_y \rangle > \langle eE_x \rangle$.

Figure 4.22, furthermore, shows the distribution in the strength and direction of electric field components for off-central Cu+Au and Au+Au collisions. This snapshot is made for the time when both nuclear centers are in the same transverse plane. This condition corresponds to different times for the two systems considered which is confirmed by a shift of the component $\langle eB_y \rangle$ in time (cf. l.h.s. of Fig. 4.21) where the maximum is reached earlier in Cu+Au collisions. Here we take t ~0.05 fm/c in view of Fig. 4.21. Note that in Cu+Au collisions a significant electric field

---

[17] The figures in this Subsection are taken from Ref. [26].

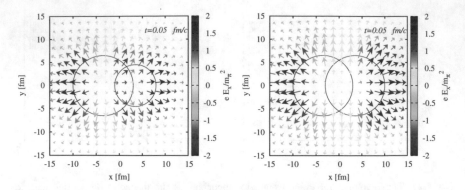

Fig. 4.22 Event-averaged electric field in the transverse plane for a Cu+Au (left panel) and Au+Au (right panel) collision at $\sqrt{s_{NN}} = 200$ GeV at time $t = 0.05$ fm/c for the impact parameter $b = 7$ fm. Each vector represents the direction and magnitude of the electric field at that point

eE_x is generated in the overlap region of the two nuclei in x-direction, i.e. directed towards the lighter copper nucleus. The situation is different in collisions of nuclei of the same size as illustrated in Fig. 4.22 (r.h.s.). In symmetric collisions like Au+Au or Cu+Cu, the event-averaged electric field does not show a preferential direction and the magnitude of the electric fields generated in each event is lower, too.

The strong electric field eE_x towards the Cu nucleus at the early stage induces an electric current in the medium (if electric charges are present). As a result, the charge distribution is modified and a charge dipole is formed. In central Cu+Au collisions the Cu nucleus is completely embedded within the Au-nucleus and due to the absence of Cu spectators no sizeable electric current is formed. We note in passing that the electric field sharply drops after $t \gtrsim 0.25$ fm/c in free space, while in conducting matter the time dependence of the field strength is flattening out and reaches some plateau even up to $t \sim 10$ fm/c. The level of this plateau is proportional to the electric conductivity σ_0 and therefore the conductivity effect could be sizeable in the case of a weakly interacting QGP. However, the electric conductivity—as evaluated within lattice QCD calculations—is much lower for temperatures from 170 MeV to 250 MeV, which are of relevance for the actual cases. Accordingly, the strong electric field eE_x provides an ultra-short electric pulse on the participant matter and allows to explore the early time electromagnetic response.

In closing this section we note that the production of real photons and virtual photons (e^+e^- or $\mu^+\mu^-$ pairs) can be calculated within the transport approach by including explicitly the bremsstrahlung from elastic collisions of charged particles as well as the electromagnetic decays of the hadrons produced as well as those from annihilation channels which are of hadronic or partonic nature.[18]

---

[18] cf. the review [27].

Solution of Exercises

Exercise 4.1: Derive Eqs. (4.21)–(4.23) starting from (4.17).

The Dyson–Schwinger equation on the closed-time-path (CTP) reads in matrix form (4.17):

$$\begin{pmatrix} G^c(x, y) & G^<(x, y) \\ G^>(x, y) & G^a(x, y) \end{pmatrix} = \begin{pmatrix} G_0^c(x, y) & G_0^<(x, y) \\ G_0^>(x, y) & G_0^a(x, y) \end{pmatrix} \tag{4.160}$$

$$+ \begin{pmatrix} G_0^c(x, x') & G_0^<(x, x') \\ G_0^>(x, x') & G_0^a(x, x') \end{pmatrix} \odot \begin{pmatrix} \Sigma^c(x', y') & -\Sigma^<(x', y') \\ -\Sigma^>(x', y') & \Sigma^a(x', y') \end{pmatrix}$$

$$\odot \begin{pmatrix} G^c(y', y) & G^<(y', y) \\ G^>(y', y) & G^a(y', y) \end{pmatrix},$$

where the symbol $\odot$ stands for an intermediate integration over space-time on the CTP. Multiplying from the left with the negative inverse Klein–Gordon operator in space-time representation

$$\hat{G}_{0x}^{-1} = -(\partial_\mu^x \partial_x^\mu + m^2), \tag{4.161}$$

and using (4.20) we obtain

$$-(\partial_\mu^x \partial_x^\mu + m^2) \begin{pmatrix} G^c(x, y) & G^<(x, y) \\ G^>(x, y) & G^a(x, y) \end{pmatrix} = \delta(\mathbf{x} - \mathbf{y}) \begin{pmatrix} \delta(x_0 - y_0) & 0 \\ 0 & -\delta(x_0 - y_0) \end{pmatrix}$$

$$+ \begin{pmatrix} \Sigma^c(x, y') & -\Sigma^<(x, y') \\ \Sigma^>(x, y') & -\Sigma^a(x, y') \end{pmatrix} \odot \begin{pmatrix} G^c(y', y) & G^<(y', y) \\ G^>(y', y) & G^a(y', y) \end{pmatrix}. \tag{4.162}$$

Multiplying out the matrices we obtain for the independent matrix elements:

$$- (\partial_\mu^x \partial_x^\mu + m^2) G^c(x, y)$$
$$= \delta(x - y) + \Sigma^c(x, y') \odot G^c(y', y) - \Sigma^<(x, y') \odot G^>(y', y), \tag{4.163}$$

$$- (\partial_\mu^x \partial_x^\mu + m^2) G^<(x, y)$$
$$= \Sigma^c(x, y') \odot G^<(y', y) - \Sigma^<(x, y') \odot G^a(y', y), \tag{4.164}$$

$$- (\partial_\mu^x \partial_x^\mu + m^2) G^>(x, y)$$
$$= \Sigma^>(x, y') \odot G^c(y', y) - \Sigma^a(x, y') \odot G^>(y', y), \tag{4.165}$$

$$- (\partial_\mu^x \partial_x^\mu + m^2) G^a(x, y)$$
$$= -\delta(x - y) + \Sigma^>(x, y') \odot G^<(y', x) - \Sigma^a(x, y') \odot G^a(y', y). \qquad (4.166)$$

These equations are identical to Eqs. (4.21)–(4.23) when employing the identities (4.14)

$$F^R(x, y) = F^c(x, y) - F^<(x, y) = F^>(x, y) - F^a(x, y) \quad , \qquad (4.167)$$
$$F^A(x, y) = F^c(x, y) - F^>(x, y) = F^<(x, y) - F^a(x, y) \quad ,$$

which will be shown in the following.

We obtain for the retarded propagator

$$- (\partial_\mu^x \partial_x^\mu + m^2) G^R(x, y) = -(\partial_\mu^x \partial_x^\mu + m^2)(G^c(x, y) - G^<(x, y)) \qquad (4.168)$$
$$= \delta(x - y) + \Sigma^c(x, y') \odot G^c(y', y) - \Sigma^<(x, y') \odot G^>(y', y)$$
$$- \Sigma^c(x, y') \odot G^<(y', y) + \Sigma^<(x, y') \odot G^a(y', y)$$
$$= \delta(x - y) + \Sigma^c(x, y') \odot (G^c(y', y) - G^<(y', y))$$
$$- \Sigma^<(x, y') \odot (G^>(y', y) - G^a(y', y))$$
$$= \delta(x - y) + (\Sigma^c(x, y') - \Sigma^<(x, y')) \odot G^R(y', y)$$
$$= \delta(x - y) + \Sigma^R(x, y') \odot G^R(y', y).$$

For the advanced propagator we obtain

$$- (\partial_\mu^x \partial_x^\mu + m^2) G^A(x, y) = -(\partial_\mu^x \partial_x^\mu + m^2)(G^c(x, y) - G^>(x, y)) \qquad (4.169)$$
$$= \delta(x - y) + \Sigma^c(x, y') \odot G^c(y', y) - \Sigma^<(x, y') \odot G^>(y', y)$$
$$- \Sigma^>(x, y') \odot G^c(y', y) + \Sigma^a(x, y') \odot G^>(y', y)$$
$$= \delta(x - y) + (\Sigma^c(x, y') - \Sigma^>(x, y')) \odot G^c(y', y) - (\Sigma^<(x, y')$$
$$- \Sigma^a(x, y')) \odot G^>(y', y)\delta(x - y)$$
$$+ \Sigma^A(x, y') \odot G^c(y', y) - \Sigma^A(x, y') \odot G^>(y', y)$$
$$= \delta(x - y) + \Sigma^A(x, y') \odot G^A(y', y).$$

For the off-diagonal elements we obtain:

$$- (\partial_\mu^x \partial_x^\mu + m^2) G^<(x, y) = \Sigma^c(x, y') \odot G^<(y', y) \qquad (4.170)$$
$$- \Sigma^<(x, y') \odot G^a(y', y)$$
$$- \Sigma^<(x, y') \odot G^<(y', y) + \Sigma^<(x, y') \odot G^<(y', y)$$

$$= (\Sigma^c(x, y') - \Sigma^<(x, y')) \odot G^<(y', y)$$
$$+ \Sigma^<(x, y') \odot (G^<(y', y) - G^a(y', y))$$
$$= \Sigma^R(x, y') \odot G^<(y', y) + \Sigma^<(x, y') \odot G^A(y', y)$$

and

$$- (\partial_\mu^x \partial_x^\mu + m^2) G^>(x, y) = \Sigma^>(x, y') \odot G^c(y', y) - \Sigma^a(x, y') \odot G^>(y', y) \tag{4.171}$$

$$+ \Sigma^>(x, y') \odot G^>(y', y) - \Sigma^>(x, y') \odot G^>(y', y)$$
$$= (\Sigma^>(x, y') - \Sigma^a(x, y')) \odot G^>(y', y)$$
$$+ \Sigma^>(x, y') \odot (G^c(y', y) - G^>(y', y))$$
$$= \Sigma^R(x, y') \odot G^>(y', y) + \Sigma^>(x, y') \odot G^A(y', y),$$

which completes the proof.

Exercise 4.2: Show that the negative imaginary part of the propagator $G_F(p) = 1/(p_0^2 - \mathbf{p}^2 - M^2 + i2\gamma p_0)$ (for $\gamma > 0$) can be written in relativistic Breit–Wigner form as

$$\frac{2\gamma p_0}{(p_0^2 - \mathbf{p}^2 - M^2)^2 + 4\gamma^2 p_0^2}$$

and is normalized to unity for all momenta $\mathbf{p}$, i.e.

$$- \int_{-\infty}^{\infty} \frac{dp_0}{2\pi} 2p_0 \, \Im G_F(p_0, \mathbf{p}) = 1.$$

The propagator in momentum space

$$G_F(p) = \frac{1}{p_0^2 - \mathbf{p}^2 - M^2 + i2\gamma p_0} = \frac{p_0^2 - \mathbf{p}^2 - M^2 - i2\gamma p_0}{(p_0^2 - \mathbf{p}^2 - M^2)^2 + 4\gamma^2 p_0^2} \tag{4.172}$$

has the negative imaginary part

$$- \Im G_F(p) = \frac{2\gamma p_0}{(p_0^2 - \mathbf{p}^2 - M^2)^2 + 4\gamma^2 p_0^2}. \tag{4.173}$$

To prove the normalization it is useful to rewrite (4.173) as

$$-\Im G_F(p) = \frac{2\gamma p_0}{(p_0^2 - \mathbf{p}^2 - M^2)^2 + 4\gamma^2 p_0^2} \tag{4.174}$$

$$= \frac{\gamma}{2E} \left(\frac{1}{(p_0 - E)^2 + \gamma^2} - \frac{1}{(p_0 + E)^2 + \gamma^2} \right)$$

$$= \frac{\gamma}{2E} \left(\frac{1}{(p_0 - E - i\gamma)(p_0 - E + i\gamma)} \right.$$

$$\left. - \frac{1}{(p_0 + E - i\gamma)(p_0 + E + i\gamma)} \right)$$

with $E = \sqrt{\mathbf{p}^2 + M^2 - \gamma^2}$ which allows to identify the poles in the complex p_0 plane at $\pm E \pm i\gamma$. In (4.174) we have used the identity

$$(p_0^2 - \mathbf{p}^2 - M^2)^2 + 4\gamma^2 p_0^2 = (p_0^2 + 2p_0 E + E^2 + \gamma^2)(p_0^2 - 2p_0 E + E^2 + \gamma^2) \tag{4.175}$$

which can be shown in a straight forward way.

We note in passing that the real part of the propagator can be written as:

$$\Re G_F(p) = \frac{p_0^2 - \mathbf{p}^2 - M^2}{(p_0^2 - \mathbf{p}^2 - M^2)^2 + 4\gamma^2 p_0^2} \tag{4.176}$$

$$= \frac{p_0^2 - \mathbf{p}^2 - M^2}{4 p_0 E} \left(\frac{1}{(p_0 - E)^2 + \gamma^2} - \frac{1}{(p_0 + E)^2 + \gamma^2} \right)$$

$$= \frac{p_0^2 - \mathbf{p}^2 - M^2}{4 p_0 E} \left(\frac{1}{(p_0 - E - i\gamma)(p_0 - E + i\gamma)} \right.$$

$$\left. - \frac{1}{(p_0 + E - i\gamma)(p_0 + E + i\gamma)} \right).$$

The normalization of the imaginary part is obtained by closing the contour in the upper half plane and inserting the residues:

$$-\int_{-\infty}^{\infty} \frac{dp_0}{2\pi} 2p_0 \, \Im G_F(p_0, \mathbf{p}) \tag{4.177}$$

$$= \oint \frac{dp_0}{2\pi} 2p_0 \frac{\gamma}{2E} \left(\frac{1}{(p_0 - E - i\gamma)(p_0 - E + i\gamma)} \right.$$

$$-\frac{1}{(p_0 + E - i\gamma)(p_0 + E + i\gamma)}\Bigg)$$

$$= \frac{2\pi i}{2\pi}\frac{-i}{2E}\left((E + i\gamma) + (E - i\gamma)\right) = \frac{2E}{2E} = 1$$

for all **p** which guarantees a quasiparticle interpretation.

Exercise 4.3: Compute the propagator for static massive fields, i.e. the solution of

$$(-\Delta + M^2)G_f(\mathbf{x}) = \delta^3(\mathbf{x}).$$

The equation for G_f in case of static massive fields,

$$(-\Delta + M^2)G_f(\mathbf{x}) = \delta^3(\mathbf{x}) , \tag{4.178}$$

reads in momentum space

$$(\mathbf{p}^2 + M^2)G_f(\mathbf{p}) = 1 \tag{4.179}$$

and leads to

$$G_f(\mathbf{p}) = \frac{1}{p^2 + M^2}. \tag{4.180}$$

The propagator of the system in coordinate space (for $R = |\mathbf{x}|$, $p = |\mathbf{p}|$) then is given by a Fourier transformation:

$$G_f(\mathbf{x}) = \int \frac{d^3p}{(2\pi)^3}\frac{\exp(i\mathbf{p}\cdot\mathbf{x})}{p^2 + M^2} = \int_0^\infty p^2\frac{dp}{(2\pi)^2}\int_{-1}^1 d\xi\,\frac{\exp(ipR\xi)}{p^2 + M^2} \tag{4.181}$$

$$= \int_0^\infty p^2\frac{dp}{(2\pi)^2}\frac{\exp(ipR) - \exp(-ipR)}{ipR}\frac{1}{p^2 + M^2}$$

$$= \int_{-\infty}^\infty p\frac{dp}{(2\pi)^2}\frac{\exp(ipR) - \exp(-ipR)}{2iR}\frac{1}{(p - iM)(p + iM)}$$

$$= 2\pi i\frac{1}{(2\pi)^2}\left(iM\frac{\exp(-MR)}{2iM2iR} + (iM)\frac{\exp(-MR)}{2iR2iM}\right)$$

$$= \frac{\exp(-MR)}{4\pi R} = G_f(R)$$

by integrating over the poles at $p = iM$ in the upper half plane and $p = -iM$ in the lower half plane. Furthermore, we have used that the integrand is even in p.

Exercise 4.4: Derive the propagator for a massive scalar field of finite lifetime γ^{-1} in its rest frame, i.e. the solution of

$$\left(\frac{\partial^2}{\partial t^2} - \Delta + M^2 + 2\gamma \frac{\partial}{\partial t} \right) G_{ret}(x - x') = \delta^4(x - x')$$

for $\gamma > 0$.

For the solution of

$$\left(\frac{\partial^2}{\partial t^2} - \Delta + M^2 + 2\gamma \frac{\partial}{\partial t} \right) G_{ret}(x - x') = \delta^4(x - x') \tag{4.182}$$

we perform a Fourier transformation to obtain the propagator in momentum space:

$$G_{ret}(p_0, \mathbf{p}) = -\frac{1}{p_0^2 - \mathbf{p}^2 - M^2 + 2i\gamma p_0}. \tag{4.183}$$

The retarded propagator then reads (for $y = (t, \mathbf{y}) = x - x'$, $R = |\mathbf{y}|$)

$$G_{ret}(y) = -\int_{-\infty}^{\infty} \frac{dp_0}{2\pi} \int \frac{d^3 p}{(2\pi)^3} \frac{\exp(-ip_0 t + i\mathbf{p} \cdot \mathbf{y})}{p_0^2 - \mathbf{p}^2 - M^2 + i2\gamma p_0} \tag{4.184}$$

$$= -\int_{-\infty}^{\infty} \frac{dp_0}{2\pi} \int \frac{d^3 p}{(2\pi)^3} \frac{\exp(-ip_0 t + i\mathbf{p} \cdot \mathbf{y})}{(p_0 - E_1(p^2))(p_0 - E_2(p^2))}$$

with

$$E_1(p^2) = \sqrt{\mathbf{p}^2 + M^2 - \gamma^2} - i\gamma = E_0(p^2) - i\gamma; \quad E_2(p^2) = -E_0(p^2) - i\gamma. \tag{4.185}$$

The integration over p_0 gives (for $t \geq 0$—two poles in the lower plane—since $\Im p_0 < 0$)

$$G_{ret}(y) = i \int \frac{d^3 p}{(2\pi)^3} \left(\frac{\exp(-iE_1(p^2)t)}{E_1(p^2) - E_2(p^2)} + \frac{\exp(-iE_2(p^2)t)}{E_2(p^2) - E_1(p^2)} \right) \exp(+i\mathbf{p} \cdot \mathbf{y}) \tag{4.186}$$

$$= i \int \frac{d^3 p}{(2\pi)^3} \left(\frac{\exp(-iE_0(p^2)t)}{2E_0(p^2)} - \frac{\exp(iE_0(p^2)t)}{2E_0(p^2)} \right) \exp(-\gamma t + i\mathbf{p} \cdot \mathbf{y}) \, \Theta(t)$$

$$= \int_0^\infty \frac{dp}{(2\pi)^2} p^2 \int_{-1}^1 d\xi \, \frac{\sin(E_0(p^2)t)}{E_0(p^2)} \exp(-\gamma t + ipR\,\xi) \, \Theta(t)$$

$$= \int_0^\infty \frac{dp}{(2\pi)^2} p^2 \, \frac{\exp(ipR) - \exp(-ipR)}{ipR} \, \frac{\sin(E_0(p^2)t)}{E_0(p^2)} \exp(-\gamma t) \, \Theta(t)$$

$$= 2 \int_0^\infty \frac{dp}{(2\pi)^2} \, p \, \frac{\sin(pR)}{R} \, \frac{\sin(E_0(p^2)t)}{E_0(p^2)} \exp(-\gamma t) \, \Theta(t)$$

$$= \frac{\exp(-\gamma t)}{R} \, \Theta(t) \int_{-\infty}^\infty \frac{dp}{(2\pi)^2} \, \frac{p}{E_0(p^2)} \, \sin(pR) \, \sin(E_0(p^2)t)$$

for $pR = |\mathbf{p}||\mathbf{y}|$.

The remaining integral over dp has a singular contribution on the light cone and a regular part on and within the light cone. The singular part is easily seen in the limit $E_0(p^2) = p$, i.e. for $M = \gamma = 0$:

$$\frac{1}{R} \int_{-\infty}^\infty \frac{dp}{(2\pi)^2} \, \sin(pR) \, \sin(pt) \tag{4.187}$$

$$= \frac{-1}{4R} \int_{-\infty}^\infty \frac{dp}{(2\pi)^2} \, (\exp(ipR) - \exp(-ipR)) \, (\exp(ipt) - \exp(-ipt))$$

$$= \frac{-1}{4R} \int_{-\infty}^\infty \frac{dp}{(2\pi)^2} \, (\exp(ip(R+t)) - \exp(ip(R-t))$$

$$- \exp(-ip(R-t) + \exp(-ip(R+t))$$

$$= \frac{-1}{4\pi R} \, (\delta(t+R) - \delta(t-R)) = \frac{1}{2\pi} \, \delta(t^2 - R^2).$$

The regular part is obtained by subtracting the δ-term (4.187) from the propagator

$$I_{reg} = \int_{-\infty}^\infty \frac{dp}{(2\pi)^2} \, \frac{\sin(pR)}{R} \left(\frac{p}{E_0(p^2)} \sin(E_0(p^2)t) - \sin(pt) \right) \tag{4.188}$$

and one obtains

$$I_{reg} = -\Theta(t^2 - R^2) \frac{M^*}{4\pi \sqrt{t^2 - R^2}} \, J_1(M^* \sqrt{t^2 - R^2}), \tag{4.189}$$

where $J_1(x)$ is the Bessel function and $M^* = \sqrt{M^2 - \gamma^2}$. Note that the integral (4.189) vanishes outside the light cone. By adding (4.187) and (4.189) we obtain $G_{ret}(t^2 - R^2, R, M)$ with the additional factor $\exp(-\gamma t)\Theta(t)$, using

$$\delta(t^2 - R^2)\Theta(t) = \delta((t - R)(t + R))\Theta(t) = \frac{\delta(t - R)}{2R}, \tag{4.190}$$

i.e.

$$G_{ret}(y) = \left(\frac{\delta(t-R)}{4\pi R} - \Theta(t^2 - R^2) \frac{M^*}{4\pi \sqrt{t^2 - R^2}} \, J_1(M^* \sqrt{t^2 - R^2}) \right)$$

$$\times \exp(-\gamma t)\Theta(t).$$
(4.191)

We may also obtain the result for the regular part of the propagator by searching for the solution of the homogenous equation

$$\left(\frac{\partial^2}{\partial t^2} - \Delta + M^2 \right) G_h(x) = 0$$
(4.192)

with $x = (t, \mathbf{r})$. To this aim we introduce the variable

$$\xi = \sqrt{t^2 - R^2}$$
(4.193)

with $R = |\mathbf{r}|$ and search for a solution depending only on ξ, i.e.

$$G_h(x) = \frac{f(\xi)}{\xi}.$$
(4.194)

In this case (4.192) reduces to

$$\left(\frac{\partial^2}{\partial t^2} - \frac{\partial^2}{\partial R^2} - \frac{2}{R} \frac{\partial}{\partial R} + M^2 \right) \frac{f(\xi)}{\xi}$$

$$= f(\xi) \left(\frac{\partial^2}{\partial t^2} - \frac{\partial^2}{\partial R^2} \right) \frac{1}{\xi} + \frac{1}{\xi} \left(\frac{\partial^2}{\partial t^2} - \frac{\partial^2}{\partial R^2} \right) f(\xi)$$

$$+ 2 \left(\frac{\partial}{\partial t} \frac{1}{\xi} \right) \frac{\partial}{\partial t} f(\xi) - 2 \left(\frac{\partial}{\partial R} \frac{1}{\xi} \right) \frac{\partial}{\partial R} f(\xi)$$
(4.195)

$$- \frac{2}{R} f(\xi) \frac{\partial}{\partial R} \frac{1}{\xi} - \frac{2}{R\xi} \frac{\partial}{\partial R} f(\xi) + M^2 \frac{f(\xi)}{\xi} = 0.$$

We need

$$\frac{\partial}{\partial t} f(\xi) = \frac{t}{\xi} \frac{\partial}{\partial \xi} f(\xi); \qquad\qquad \frac{\partial}{\partial R} f(\xi) = -\frac{R}{\xi} \frac{\partial}{\partial \xi} f(\xi);$$
(4.196)

$$\frac{\partial^2}{\partial t^2} f(\xi) = \frac{t^2}{\xi^2} \frac{\partial^2}{\partial \xi^2} f(\xi) - \frac{t^2}{\xi^3} \frac{\partial}{\partial \xi} f(\xi) + \frac{1}{\xi} \frac{\partial}{\partial \xi} f(\xi);$$

$$\frac{\partial^2}{\partial R^2} f(\xi) = \frac{R^2}{\xi^2} \frac{\partial^2}{\partial \xi^2} f(\xi) - \frac{R^2}{\xi^3} \frac{\partial}{\partial \xi} f(\xi) - \frac{1}{\xi} \frac{\partial}{\partial \xi} f(\xi),$$

giving

$$\left(\frac{\partial^2}{\partial t^2} - \frac{\partial^2}{\partial R^2}\right) f(\xi) = \frac{\partial^2}{\partial \xi^2} f(\xi) - \frac{1}{\xi} \frac{\partial}{\partial \xi} f(\xi) + \frac{2}{\xi} \frac{\partial}{\partial \xi} f(\xi)$$

$$= \frac{\partial^2}{\partial \xi^2} f(\xi) + \frac{1}{\xi} \frac{\partial}{\partial \xi} f(\xi). \qquad (4.197)$$

Furthermore,

$$\frac{\partial}{\partial t} \frac{1}{\xi} = -\frac{t}{\xi^3}; \qquad\qquad \frac{\partial}{\partial R} \frac{1}{\xi} = \frac{R}{\xi^3}; \qquad (4.198)$$

$$\frac{\partial^2}{\partial t^2} \frac{1}{\xi} = -\frac{1}{\xi^3} + \frac{3t^2}{\xi^5}, \quad \frac{\partial^2}{\partial R^2} \frac{1}{\xi} = \frac{1}{\xi^3} + \frac{3R^2}{\xi^5}, \quad \left(\frac{\partial^2}{\partial t^2} - \frac{\partial^2}{\partial R^2}\right) \frac{1}{\xi} = -\frac{2}{\xi^3} + \frac{3}{\xi^3} = \frac{1}{\xi^3}.$$

$$\left(\frac{\partial}{\partial t} \frac{1}{\xi}\right) \frac{\partial}{\partial t} f(\xi) = -\frac{t}{\xi^3} \frac{t}{\xi} \frac{\partial}{\partial \xi} f(\xi) = -\frac{t^2}{\xi^4} \frac{\partial}{\partial \xi} f(\xi),$$

$$\left(\frac{\partial}{\partial R} \frac{1}{\xi}\right) \frac{\partial}{\partial R} f(\xi) = -\frac{R}{\xi^3} \frac{R}{\xi} \frac{\partial}{\partial \xi} f(\xi) = -\frac{R^2}{\xi^4} \frac{\partial}{\partial \xi} f(\xi),$$

$$2 \left(\frac{\partial}{\partial t} \frac{1}{\xi}\right) \frac{\partial}{\partial t} f(\xi) - 2 \left(\frac{\partial}{\partial R} \frac{1}{\xi}\right) \frac{\partial}{\partial R} f(\xi) = -\frac{2}{\xi^2} \frac{\partial}{\partial \xi} f(\xi). \qquad (4.199)$$

We finally use

$$-\frac{2}{R} f(\xi) \frac{\partial}{\partial R} \frac{1}{\xi} - \frac{2}{R\xi} \frac{\partial}{\partial R} f(\xi) = -\frac{2}{R} f(\xi) \frac{R}{\xi^3} + \frac{2}{\xi^2} \frac{\partial}{\partial \xi} f(\xi)$$

$$= -\frac{2}{\xi^3} f(\xi) + \frac{2}{\xi^2} \frac{\partial}{\partial \xi} f(\xi). \qquad (4.200)$$

Adding all terms we end up with

$$\frac{1}{\xi} \frac{\partial^2}{\partial \xi^2} f(\xi) + \frac{1}{\xi^2} \frac{\partial}{\partial \xi} f(\xi) + \frac{1}{\xi^3} f(\xi) - \frac{2}{\xi^2} \frac{\partial}{\partial \xi} f(\xi) - \frac{2}{\xi^3} f(\xi)$$

$$+ \frac{2}{\xi^2} \frac{\partial}{\partial \xi} f(\xi) + M^2 \frac{f(\xi)}{\xi} = 0,$$

i.e. for $\xi \neq 0$:

$$\left(\frac{1}{\xi} \frac{\partial^2}{\partial \xi^2} + \frac{1}{\xi^2} \frac{\partial}{\partial \xi} - \frac{1}{\xi^3} + \frac{M^2}{\xi}\right) f(\xi) = 0. \qquad (4.201)$$

Multiplying (4.201) by ξ^3 we obtain for $\xi \neq 0$:

$$\left(\xi^2 \frac{\partial^2}{\partial \xi^2} + \xi \frac{\partial}{\partial \xi} - 1 + \xi^2 M^2\right) f(\xi) = 0, \tag{4.202}$$

which suggests to introduce the variable $z = M\xi$ giving

$$\left(z^2 \frac{\partial^2}{\partial z^2} + z \frac{\partial}{\partial z} - 1 + z^2\right) f(z) = 0. \tag{4.203}$$

The solution of Eq. (4.203) is given by the Bessel function $J_1(z)$ which has the expansion (and representation)

$$J_1(z) = \sum_{n=0}^{\infty} \frac{(-1)^n}{n!(n+1)!} \left(\frac{z}{2}\right)^{2n+1} = \frac{1}{\pi} \int_0^\pi d\tau \, \cos(\tau - z \sin(\tau)). \tag{4.204}$$

One thus obtains for the regular part

$$G_h(x) = -\Theta(\xi^2) \frac{M}{4\pi \xi} J_1(M\xi) = -\Theta(t^2 - R^2) \frac{M}{4\pi \sqrt{t^2 - R^2}} J_1(M\sqrt{t^2 - R^2}) \tag{4.205}$$

using (for $t > R$)

$$\delta(\xi^2) = \delta(t^2 - R^2) = \delta((t - R)(t + R)) = \frac{\delta(t - R)}{2R}.$$

Replacing M by $M^* = \sqrt{M^2 - \gamma^2}$ in case of $\gamma > 0$ we regain the result in Eq. (4.189). Note that the regular part (inside the light cone) vanishes for critical damping ($M^* = 0$), however, the solution stays valid also in case of overcritical damping $\gamma^2 > M^2$.

Exercise 4.5: Derive the retarded propagator for the wave equation

$$\partial_\mu \partial^\mu G_{ret}(x, x') = \delta^4(x - x') \equiv \left(\frac{1}{c^2} \frac{\partial^2}{\partial t^2} - \Delta_{\mathbf{r}}\right) G_{ret}(\mathbf{r} - \mathbf{r}'; t - t') = \delta(t - t')\delta^3(\mathbf{r} - \mathbf{r}').$$

This problem corresponds to the previous exercise for the case $M = \gamma = 0$ and leads directly to ($x = (t, \mathbf{r})$)

$$G_{ret}(\tau, R) = \frac{\delta(\tau - R/c)}{4\pi R} \Theta(\tau) \tag{4.206}$$

with $\tau = t - t'$ and $R = |\mathbf{r} - \mathbf{r}'|$. However, it is instructive to work out the solution alternatively. Since the wave equation

$$\partial_\mu \partial^\mu G_{ret}(x, x') = \delta^4(x - x') \tag{4.207}$$

is invariant with respect to space-time shifts the propagator $G_{ret}(x, x')$ is a function of $\mathbf{r} - \mathbf{r}'$ and $\tau = (t - t')$. Due to causality, furthermore, it must vanish for $t < t'$.

Since the inhomogeneity of the wave equation is a δ-function in space-time its solution is a spherical wave. Accordingly we start with the Ansatz:

$$G_{ret}(R, \tau) = \frac{g(\tau - R/c)}{R} = \frac{g(t - t' - |\mathbf{r} - \mathbf{r}'|/c)}{|\mathbf{r} - \mathbf{r}'|} \tag{4.208}$$

with $R = |\mathbf{r} - \mathbf{r}'|$. In order to determine the function g we consider:

$$\Delta G_{ret} = g \Delta \left(\frac{1}{R}\right) + \frac{1}{R} \Delta g + 2\nabla\left(\frac{1}{R}\right) \cdot \nabla g$$

$$= -4\pi g\, \delta(\mathbf{R}) + \frac{1}{R} \frac{\partial^2}{\partial R^2} g + \frac{2}{R^2} \frac{\partial}{\partial R} g - \frac{2}{R^2} \frac{\partial}{\partial R} g \tag{4.209}$$

using

$$\Delta = \frac{\partial^2}{\partial R^2} + \frac{2}{R} \frac{\partial}{\partial R} + \frac{1}{R^2 \sin\theta} \frac{\partial}{\partial\theta}\left(\sin\theta \frac{\partial}{\partial\theta}\right) + \frac{1}{R^2 \sin^2\theta} \frac{\partial^2}{\partial\phi^2} \tag{4.210}$$

$$= \frac{\partial^2}{\partial R^2} + \frac{2}{R} \frac{\partial}{\partial R} + \quad \text{angular part}$$

and

$$\frac{\partial g}{\partial x} = \frac{\partial g}{\partial R} \cdot \frac{x}{R}, \quad \frac{\partial g}{\partial y} = \frac{\partial g}{\partial R} \cdot \frac{y}{R}, \quad \frac{\partial g}{\partial z} = \frac{\partial g}{\partial R} \cdot \frac{z}{R}. \tag{4.211}$$

This leads to

$$\partial_\mu \partial^\mu G = 4\pi g(\tau - R/c)\, \delta^3(\mathbf{R}), \tag{4.212}$$

since

$$\frac{1}{R}\left(\frac{\partial^2}{\partial R^2} - \frac{1}{c^2} \frac{\partial^2}{\partial \tau^2}\right) g(\tau - R/c) = 0 \tag{4.213}$$

for arbitrary functions $g(\tau - R/c)$. By comparison with (4.207) we obtain:

$$4\pi g(\tau - R/c) = \delta(\tau - R/c) \tag{4.214}$$

or:

$$G_{ret}(\mathbf{r} - \mathbf{r}'; t - t') = \frac{\delta(t - t' - |\mathbf{r} - \mathbf{r}'|/c)}{4\pi |\mathbf{r} - \mathbf{r}'|} \qquad \text{for } t > t' \qquad (4.215)$$

$$G(\mathbf{r} - \mathbf{r}'; t - t') = 0 \qquad \text{for } t < t',$$

which is identical to (4.206).

References

1. J. Schwinger, Phys. Rev. **83**, 664 (1951)
2. J. Schwinger, J. Math. Phys. **2**, 407 (1961)
3. P.M. Bakshi, K.T. Mahanthappa, J. Math. Phys. **4**, 12 (1963)
4. L.V. Keldysh, Zh. Eks. Teor. Fiz. **47**, 1515 (1964)
5. J.M. Luttinger, J.C. Ward, Phys. Rev. C **118**, 1417 (1960)
6. Y.B. Ivanov, J. Knoll, D.N. Voskresensky, Nucl. Phys. A **657**, 413 (1999)
7. S. Juchem, W. Cassing, C. Greiner, Phys. Rev. D **69**, 025006 (2004)
8. L.P. Kadanoff, G. Baym, *Quantum Statistical Mechanics* (Benjamin, New York, 1962)
9. S. Leupold, Nucl. Phys. A **672**, 475 (2000)
10. S. Juchem, W. Cassing, C. Greiner, Nucl. Phys. A **743**, 92 (2004)
11. W. Botermans, R. Malfliet, Phys. Rep. **198**, 115 (1990)
12. W. Cassing, S. Juchem, Nucl. Phys. A **665**, 377 (2000)
13. C. Fuchs, H. Lenske, H.H. Wolter, Phys. Rev. C **52**, 3043 (1995)
14. F. Hofmann, C.M. Keil, H. Lenske, Phys. Rev. C **64**, 034314 (2001)
15. O. Buss, T. Gaitanos, K. Gallmeister, H. van Hees, M. Kaskulov, O. Lalakulich, A.B. Larionov, T. Leitner, J. Weil, U. Mosel, Phys. Rep. **512**, 1 (2012)
16. G.E. Brown, M. Rho, Phys. Rev. Lett. **66**, 2720 (1991)
17. G. Agakichiev, et al., Phys. Rev. Lett. **75**, 1272 (1995)
18. E.L. Bratkovskaya, W. Cassing, Nucl. Phys. A **807**, 214 (2008)
19. W. Cassing, L. Tolós, E.L. Bratkovskaya, A. Ramos, Nucl. Phys. A **727**, 59 (2003)
20. T. Song, L. Tolós, J. Wirth, J. Aichelin, E. Bratkovskaya, Phys. Rev. C **103**, 044901 (2021)
21. W. Cassing, E.L. Bratkovskaya, Nucl. Phys. A **831**, 215 (2009)
22. E.L. Bratkovskaya, W. Cassing, V.P. Konchakovski, O. Linnyk, Nucl. Phys. A **856**, 162 (2011)
23. T. Song, W. Cassing, P. Moreau, E.L. Bratkovskaya, Phys. Rev. C **97**, 064907 (2018)
24. L. Oliva, Eur. Phys. J. A **56**, 10 (2020)
25. V. Voronyuk, V.D. Toneev, W. Cassing, E.L. Bratkovskaya, V.P. Konchakovski, S.A. Voloshin, Phys. Rev. C **83**, 054911 (2011)
26. V. Voronyuk, V.D. Toneev, S.A. Voloshin, W. Cassing, Phys. Rev. C **C90**, 064903 (2014)
27. O. Linnyk, E.L. Bratkovskaya, W. Cassing, Prog. Part. Nucl. Phys. **87**, 50 (2016)

The Time Evolution Operator

In order to describe the time evolution of a quantum system different "pictures" may be employed which in principle all are equivalent but in practice are employed in different expansion schemes. We will start with the

A.1   Schrödinger Picture

Here the starting point is the time-dependent Schrödinger equation

$$i\frac{\partial}{\partial t}|\Psi(t)\rangle = H|\Psi(t)\rangle, \tag{A.1}$$

which describes the dynamical evolution of the system by the (N-body) Hamiltonian H. If the vector $|\Psi(t_0)\rangle$, which describes the state of the system at time t_0, is known then Eq. (A.1) uniquely defines $|\Psi(t)\rangle$ for any other time $t \neq t_0$. Accordingly, there must be a unique transformation

$$|\Psi(t)\rangle = U(t, t_0)|\Psi(t_0)\rangle \tag{A.2}$$

with a linear operator $U(t, t_0)$ due to the linearity of Eq. (A.1). Since the Hamiltonian H must be self-adjoint we get:

$$\frac{d}{dt}\langle \Psi(t)|\Psi(t)\rangle = 0, \tag{A.3}$$

or

$$\langle\Psi(t)|\Psi(t)\rangle = \langle\Psi(t_0)|\Psi(t_0)\rangle = \langle\Psi(t_0)|U^\dagger(t, t_0)U(t, t_0)|\Psi(t_0)\rangle. \tag{A.4}$$

© The Author(s), under exclusive license to Springer Nature Switzerland AG 2021
W. Cassing, *Transport Theories for Strongly-Interacting Systems*, Lecture Notes
in Physics 989, https://doi.org/10.1007/978-3-030-80295-0

Accordingly, $U(t, t_0)$ is a unitary operator, i.e.

$$U^\dagger(t, t_0)U(t, t_0) = 1_\mathcal{H},\tag{A.5}$$

where $1_\mathcal{H}$ denotes the identity in the Hilbert space $\mathcal{H}$. $U(t, t_0)$ is denoted as **time evolution operator** and as in case of rotations or spatial translations defines a group with

$$U^{-1}(t, t_0) = U^\dagger(t, t_0).\tag{A.6}$$

As in case of spatial translations this group is abelian contrary to rotations.

Inserting (A.2) in (A.1) we get:

$$i\frac{\partial}{\partial t}U(t, t_0)|\Psi(t_0)\rangle = H\ U(t, t_0)|\Psi(t_0)\rangle.\tag{A.7}$$

Since the time t_0 is arbitrary one obtains the operator equation

$$i\frac{\partial}{\partial t}\ U(t, t_0) = H\ U(t, t_0).\tag{A.8}$$

It is useful to consider two cases separately:

1. If H does not depend on time t the solution is simply given by

$$U(t, t_0) = \exp(-iH(t - t_0)),\tag{A.9}$$

   as one finds out by differentiation of (A.9) with respect to t. In case of an eigenvector $|\Psi_E(t_0)\rangle$ of H with energy E we have

$$U(t, t_0)|\Psi_E(t_0)\rangle = \exp(-iE(t - t_0))|\Psi_E(t_0)\rangle,\tag{A.10}$$

   i.e. the time evolution is described by a phase shift.
2. In case H depends on t, i.e. $H = H(t)$, it is useful to consider the integral equation,

$$U(t, t_0) = 1_\mathcal{H} - i\int_{t_0}^t dt'\ H(t')\ U(t', t_0),\tag{A.11}$$

   where the boundary condition $U(t_0, t_0) = 1_\mathcal{H}$ is employed explicitly. Equation (A.11) practically can be solved by iteration. A formal solution is given by

$$U(t, t_0) = T\left(\exp\left[-i\int_{t_0}^t dt'\ H(t')\right]\right) = \sum_{n=0}^\infty \frac{1}{n!}T\left[-i\int_{t_0}^t dt'\ H(t')\right]^n,\tag{A.12}$$

where T denotes the time-ordering operator which places all operators at lower times to the right.

Instead of time-dependent Hilbert vectors in the Schrödinger picture the

A.2   Heisenberg Picture

employs time-independent Hilbert vectors, however, explicitly time-dependent operators. The equations of motion for the operators then replace equation (A.1). To obtain the latter equations of motion one considers the expectation value of an operator A in the state $|\Psi(t)\rangle$,

$$\langle \Psi(t)|A(t)|\Psi(t)\rangle = \langle \Psi(t_0)|U^\dagger(t, t_0)A(t)U(t, t_0)|\Psi(t_0)\rangle. \tag{A.13}$$

In this case the operator $A(t)$ may also explicitly depend on time t, which is useful for unclosed systems, i.c. for systems in contact with some time-dependent environment. In the **Heisenberg picture** one uses the Hilbert vector

$$|\Psi_h\rangle =: |\Psi(t_0)\rangle, \tag{A.14}$$

which fixes the system at some time t_0, and represents the observable by the time-dependent operator

$$A_h(t) =: U^\dagger(t, t_0)A(t)U(t, t_0). \tag{A.15}$$

The expectation value of A in time then reads

$$\langle A \rangle_t = \langle \Psi_h|A_h(t)|\Psi_h\rangle. \tag{A.16}$$

The time evolution of $A_h(t)$ is obtained from (A.15) and (A.8):

$$i\frac{\partial}{\partial t}A_h(t) = iU^\dagger(t, t_0)\left(\frac{\partial}{\partial t}A(t)\right)U(t, t_0) + U^\dagger(t, t_0)[A(t)H - HA(t)]U(t, t_0) \tag{A.17}$$

$$= [A_h(t), H_h] + i\left(\frac{\partial}{\partial t}A(t)\right)_h,$$

with

$$H_h =: U^\dagger(t, t_0)H\,U(t, t_0) \tag{A.18}$$

and

$$\left(\frac{\partial}{\partial t} A(t)\right)_h =: U^\dagger(t, t_0) \left(\frac{\partial}{\partial t} A(t)\right) U(t, t_0). \tag{A.19}$$

If the observable C does not explicitly depend on time t, i.e. $C \neq C(t)$, it represents a conserved quantity if

$$[C_h, H_h] = 0, \tag{A.20}$$

which is equivalent to

$$[C, H] = 0 \tag{A.21}$$

in the Schrödinger picture.

A.3   Interaction or Dirac Picture

For practical calculations it is useful to change to the **Dirac picture**. To this aim one rewrites the Hamiltonian as

$$H = H_0 + H'(t), \tag{A.22}$$

with a time independent H_0 and known system of eigenvectors. In this case the state

$$|\Psi_D(t)\rangle = \exp(i H_0 t)|\Psi(t)\rangle \tag{A.23}$$

allows to "subtract" the time evolution of the state $|\Psi(t)\rangle$ determined by H_0 such that the time evolution of $|\Psi(t)\rangle$ is essentially determined by $H'(t)$. Inserting (A.23) in (A.1) we get:

$$i\frac{\partial}{\partial t}|\Psi_D(t)\rangle = -H_0|\Psi_D(t)\rangle + i \exp(i H_0 t) \frac{\partial}{\partial t}|\Psi(t)\rangle \tag{A.24}$$

$$= -H_0|\Psi_D(t)\rangle + \exp(i H_0 t) [H_0 + H']|\Psi(t)\rangle$$

$$= \exp(i H_0 t) H'(t) \exp(-i H_0 t)|\Psi_D(t)\rangle$$

or

$$i\frac{\partial}{\partial t}|\Psi_D(t)\rangle = H'_D(t)|\Psi_D(t)\rangle \tag{A.25}$$

with

$$H'_D(t) = \exp(i H_0 t) H'(t) \exp(-i H_0 t). \tag{A.26}$$

Equation (A.25) differs from (A.1) in the respect that only the interaction $H'_D(t)$ appears and not the full Hamiltonian $H(t)$ as in (A.1); the exponential terms in (A.26) are known phase factors in the basis of eigenvectors of H_0.

In order to obtain an iterative solution of Eq. (A.25) we introduce the time evolution operator in the Dirac picture by

$$|\Psi_D(t)\rangle = U_D(t, t_0)|\Psi_D(t_0)\rangle \tag{A.27}$$

and obtain from (A.25)

$$i\frac{\partial}{\partial t} U_D(t, t_0) = H'_D(t)U_D(t, t_0) \tag{A.28}$$

with the boundary condition $U_D(t_0, t_0) = 1_{\mathcal{H}}$. The corresponding integral equation for $U_D(t, t_0)$ reads:

$$U_D(t, t_0) = 1_{\mathcal{H}} - i \int_{t_0}^{t} dt'\, H'_D(t')U_D(t', t_0) \tag{A.29}$$

and contrary to (A.11) only contains the interaction operator $H'_D(t)$. Accordingly, an iterative solution of (A.29) should converge fast if H_0 already provides some reasonable approximation to the system. The formal result of such an iteration (with $U_D(t, t_0) = 1_{\mathcal{H}}$ in 0'th order) is the **Dyson series**

$$U_D(t, t_0) = 1_{\mathcal{H}} - i \int_{t_0}^{t} dt'\, H'_D(t') - \int_{t_0}^{t} dt_2 \int_{t_0}^{t_2} dt_1\, H'_D(t_2)H'_D(t_1) \cdots \tag{A.30}$$

$$= 1_{\mathcal{H}} + \sum_{n=1}^{\infty} (-i)^n \int_{t_0}^{t} dt_n \int_{t_0}^{t_n} dt_{n-1} \int_{t_0}^{t_{n-1}} dt_{n-2} \cdots$$

$$\times \int_{t_0}^{t_2} dt_1\, H'_D(t_n)H'_D(t_{n-1}) \cdots H'_D(t_1)$$

with

$$t \geq t_n \geq \ldots\ldots \geq t_1 \geq t_0. \tag{A.31}$$

One has to take care about the sequence of the operators $H'_D(t_n)\, H'_D(t_{n-1}) \cdots H'_D(t_1)$, etc., since for $t_1 \neq t_2$

$$[H'_D(t_1), H'_D(t_2)] \neq 0, \tag{A.32}$$

except for the trivial case $[H_0, H'] = 0$ and $H' \neq H'(t)$.

Trace Relations for n-Body Correlations

B

In order to explore (and compute) the different trace relations in Sect. 2.2 we expand the one-body density matrix ρ and the correlations c_n within a complete orthonormal single-particle basis $\varphi_\alpha(\mathbf{r}) \equiv \langle \mathbf{r} | \alpha \rangle$ as

$$\rho(11'; t) = \sum_{\lambda \lambda'} \rho_{\lambda \lambda'}(t) \varphi_\lambda(\mathbf{r}) \varphi_{\lambda'}^*(\mathbf{r}'), \tag{B.1}$$

$$c_2(12, 1'2'; t) = \sum_{\lambda \gamma \lambda' \gamma'} C_{\lambda \gamma \lambda' \gamma'}(t) \varphi_\lambda(\mathbf{r}_1) \varphi_\gamma(\mathbf{r}_2) \varphi_{\lambda'}^*(\mathbf{r}_1') \varphi_{\gamma'}^*(\mathbf{r}_2'). \tag{B.2}$$

The cluster decomposition then reads explicitly for the two-body density matrix

$$\rho^2_{\alpha \beta \alpha' \beta'} = \rho_{\alpha \alpha'} \rho_{\beta \beta'} - \rho_{\alpha \beta'} \rho_{\beta \alpha'} + C_{\alpha \beta \alpha' \beta'} \tag{B.3}$$

and for the three-body density matrix

$$
\begin{aligned}
\rho^3_{\alpha \beta \gamma \alpha' \beta' \gamma'} = {} & \rho_{\alpha \alpha'} \rho_{\beta \beta'} \rho_{\gamma \gamma'} - \rho_{\alpha \beta'} \rho_{\beta \alpha'} \rho_{\gamma \gamma'} - \rho_{\alpha \gamma'} \rho_{\beta \beta'} \rho_{\gamma \alpha'} \\
& - \rho_{\alpha \alpha'} \rho_{\beta \gamma'} \rho_{\gamma \beta'} + \rho_{\alpha \gamma'} \rho_{\beta \alpha'} \rho_{\gamma \beta'} + \rho_{\alpha \beta'} \rho_{\beta \gamma'} \rho_{\gamma \alpha'} \\
& + \rho_{\alpha \alpha'} C_{\beta \gamma \beta' \gamma'} - \rho_{\alpha \beta'} C_{\beta \gamma \alpha' \gamma'} - \rho_{\alpha \gamma'} C_{\beta \gamma \beta' \alpha'} + \rho_{\beta \beta'} C_{\alpha \gamma \alpha' \gamma'} \\
& - \rho_{\beta \alpha'} C_{\alpha \gamma \beta' \gamma'} - \rho_{\beta \gamma'} C_{\alpha \gamma \alpha' \beta'} \\
& + \rho_{\gamma \gamma'} C_{\alpha \beta \alpha' \beta'} - \rho_{\gamma \alpha'} C_{\alpha \beta \gamma' \beta'} - \rho_{\gamma \beta'} C_{\alpha \beta \alpha' \gamma'} + C^3_{\alpha \beta \gamma \alpha' \beta' \gamma'}.
\end{aligned}
\tag{B.4}
$$

The trace relation between the one-body and two-body density matrices give

$$\rho_{\alpha \alpha'} = \frac{1}{N-1} \sum_\beta \rho^2_{\alpha \beta \alpha' \beta} = \frac{1}{N-1} \sum_\beta \left(\rho_{\alpha \alpha'} \rho_{\beta \beta} - \rho_{\alpha \beta} \rho_{\beta \alpha'} + C_{\alpha \beta \alpha' \beta} \right), \tag{B.5}$$

© The Author(s), under exclusive license to Springer Nature Switzerland AG 2021
W. Cassing, *Transport Theories for Strongly-Interacting Systems*, Lecture Notes
in Physics 989, https://doi.org/10.1007/978-3-030-80295-0

which implies that

$$\sum_{\beta} C_{\alpha\beta\alpha'\beta} = -\rho_{\alpha\alpha'} + \sum_{\beta} \rho_{\alpha\beta}\rho_{\beta\alpha'}. \tag{B.6}$$

In case of a basis $|\alpha\rangle$ that diagonalizes the one-body density matrix

$$\rho_{\alpha\alpha'} = \delta_{\alpha\alpha'} n_\alpha, \tag{B.7}$$

(with the occupation number n_α) this reduces to

$$\sum_{\beta} C_{\alpha\beta\alpha'\beta} = -\delta_{\alpha\alpha'} n_\alpha + \delta_{\alpha\alpha'} n_\alpha^2 = -\delta_{\alpha\alpha'} n_\alpha (1 - n_\alpha), \tag{B.8}$$

i.e. $\sum_\beta C_{\alpha\beta\alpha\beta}$ gives the (negative) fluctuation in the particle number n_α while the sum vanishes for $\alpha \neq \alpha'$. Only in case of $n_\alpha = 0, 1$ these fluctuations vanish which implies that the many-body state is described by a single Slater determinant. The total trace thus gives

$$\sum_{\alpha\beta} C_{\alpha\beta\alpha\beta} = -\sum_{\alpha} n_\alpha (1 - n_\alpha). \tag{B.9}$$

The trace relation

$$\rho_{\alpha\beta\alpha'\beta'}^2 = \frac{1}{N-2} \sum_{\gamma} \rho_{\alpha\beta\gamma\alpha'\beta'\gamma}^3 \tag{B.10}$$

leads to the condition

$$\sum_{\gamma} C_{\alpha\beta\gamma\alpha'\beta'\gamma}^3 = -\sum_{\gamma=\gamma'} \left(\frac{2}{N} - P_{\alpha\gamma} - P_{\beta\gamma} - P_{\alpha'\gamma'} - P_{\beta'\gamma'} \right) \rho_{\gamma\gamma'} C_{\alpha\beta\alpha'\beta'} \tag{B.11}$$

$$= -2C_{\alpha\beta\alpha'\beta'} + \sum_{\gamma=\gamma'} (P_{\alpha\gamma} + P_{\beta\gamma} + P_{\alpha'\gamma'} + P_{\beta'\gamma'}) \rho_{\gamma\gamma'} C_{\alpha\beta\alpha'\beta'}$$

with the exchange operator $P_{\alpha\beta}$ exchanging the indices to the right. In case of (B.7) this gives

$$\sum_{\gamma} C_{\alpha\beta\gamma\alpha'\beta'\gamma}^3 = (n_\alpha + n_\beta + n_{\alpha'} + n_{\beta'} - 2) C_{\alpha\beta\alpha'\beta'}, \tag{B.12}$$

which in general is nonvanishing. When taking the total trace of C^3 we arrive at

$$\sum_{\alpha\beta\gamma} C^3_{\alpha\beta\gamma\alpha\beta\gamma} = 2\sum_{\alpha}(n_\alpha(1-n_\alpha)^2 - n_\alpha^2(1-n_\alpha)). \tag{B.13}$$

For completeness we mention that the trace relation

$$\rho^3_{\alpha\beta\gamma\alpha'\beta'\gamma'} = \frac{1}{N-3}\sum_\lambda \rho^4_{\alpha\beta\gamma\lambda\alpha'\beta'\gamma'\lambda} \tag{B.14}$$

leads to the condition

$$\begin{aligned}
Tr_{(4=4')}C_4(1234, 1'2'3'4') = &-Tr_{(4=4')}(3/A - P_{14} - P_{24} - P_{34} - P_{1'4'} - P_{2'4'} \\
&- P_{3'4'})\rho(44')C_3(123, 1'2'3') \\
&+ Tr_{(4=4')}(P_{14} + P_{24} + P_{1'4'} \\
&+ P_{2'4'})C_2(34, 3'4')C_2(12, 1'2') \\
&+ Tr_{(4=4')}(P_{34} + P_{3'4'})C_2(24, 2'4')C_2(13, 1'3') \\
&- Tr_{(4=4')}(C_2(12, 3'4')C_2(34, 1'2') \\
&+ C_2(23, 1'4')C_2(14, 2'3') + C_2(24, 1'3')C_2(13, 2'4')).
\end{aligned} \tag{B.15}$$

This is readily rewritten in terms of matrix elements using the substitution

$$(1234, 1'2'3'4') \leftrightarrow (\alpha\beta\gamma\delta, \alpha'\beta'\gamma'\delta'). \tag{B.16}$$

Since the trace relations show that C^3 and C^4 must be nonvanishing in the general case one might introduce a specific Ansatz e.g. for C^3 in order to close the equations of motion and (partly) keep the trace relations fulfilled. When using the Ansatz

$$C^3_{\alpha\beta\gamma\alpha'\beta'\gamma'} = -\frac{1}{N-1}\sum_{\lambda=\lambda'}\left(\frac{2}{N} - P_{\alpha\lambda} - P_{\beta\lambda} - P_{\alpha'\lambda'} - P_{\beta'\lambda'}\right)\rho^2_{\gamma\lambda\gamma'\lambda'}C_{\alpha\beta\alpha'\beta'} \tag{B.17}$$

the sum over $\gamma = \gamma'$ gives (B.11) such that the trace relation (B.10) is fulfilled. This Ansatz for C^3 is antisymmetric in $(1, 2)$ and $(1', 2')$ but not in $(1, 3)$, $(2, 3)$, $(1', 3')$, and $(2', 3')$. Moreover, the trace relation (B.11) only holds for particle 3.

As an alternative one might consider the fully antisymmetric Ansatz

$$\begin{aligned}
C_3(123, 1'2'3') = &Tr_{(4=4')}((1 - P_{1'3'} - P_{2'3'})c_2(12, 1'2')c_2(34, 3'4') \tag{B.18} \\
&+ (1 - P_{1'2'} - P_{2'3'})c_2(13, 1'3')c_2(24, 2'4') \\
&+ (1 - P_{1'2'} - P_{1'3'})c_2(14, 1'4')c_2(23, 2'3'))
\end{aligned}$$

$$= Tr_{(4=4')}(c_2(12, 1'2')c_2(34, 3'4') - c_2(12, 3'2')c_2(34, 1'4')$$

$$- c_2(12, 1'3')c_2(34, 2'4')$$

$$+ c_2(13, 1'3')c_2(24, 2'4') - c_2(13, 2'3')c_2(24, 1'4') - c_2(13, 1'2')c_2(24, 3'4')$$

$$+ c_2(14, 1'4')c_2(23, 2'3') - c_2(14, 2'4')c_2(23, 1'3') - c_2(14, 3'4')c_2(23, 2'1')$$

$$= c_2(12, 1'2')(\rho^2(33') - \rho(33')) - c_2(12, 3'2')(\rho^2(31') - \rho(31'))$$

$$- c_2(13, 1'2')(\rho^2(23') - \rho(23'))$$

$$- c_2(12, 1'3')(\rho^2(32') - \rho(32')) + c_2(13, 1'3')(\rho^2(22') - \rho(22'))$$

$$- c_2(13, 2'3')(\rho^2(21') - \rho(21')) + (\rho^2(11') - \rho(11'))c_2(23, 2'3')$$

$$- (\rho^2(12') - \rho(12'))c_2(23, 1'3') - (\rho^2(13') - \rho(13'))c_2(23, 2'1').$$

Taking the trace over particles 1, 2, and 3 we obtain

$$Tr_{(1,2,3)}C^3(123, 123) \tag{B.19}$$

$$= Tr_{(1=1',3=3')}((\rho^2(11) - \rho(11))(\rho^2(33) - \rho(33))$$

$$- (\rho^2(13) - \rho(13))(\rho^2(31) - \rho(31))$$

$$- (\rho^2(13) - \rho(13))(\rho^2(31) - \rho(31)))$$

$$+ Tr_{(2=2',3=3')}(-(\rho^2(32) - \rho(32))(\rho^2(23) - \rho(23))$$

$$- (\rho^2(23) - \rho(23))(\rho^2(32) - \rho(32))$$

$$+ (\rho^2(33) - \rho(33))(\rho^2(22) - \rho(22)))$$

$$+ Tr_{(1=1',2=2')}(-(\rho^2(12) - \rho(12))(\rho^2(21) - \rho(21))$$

$$+ (\rho^2(11) - \rho(11))(\rho^2(22) - \rho(22))$$

$$- (\rho^2(12) - \rho(12))(\rho^2(21) - \rho(21))).$$

In case of a diagonal basis for ρ (B.7) we get

$$Tr_{(1,2,3)}C^3(123, 123) \tag{B.20}$$

$$= Tr_{(1,3)}((\rho^2(11) - \rho(11))(\rho^2(33) - \rho(33))$$

$$- \delta_{13}(\rho^2(13) - \rho(13))(\rho^2(31) - \rho(31))$$

$$- \delta_{13}(\rho^2(13) - \rho(13))(\rho^2(31) - \rho(31)))$$

$$+ Tr_{(2,3)}(-\delta_{23}(\rho^2(32) - \rho(32))(\rho^2(23) - \rho(23))$$

$$- \delta_{23}(\rho^2(23) - \rho(23))(\rho^2(32) - \rho(32))$$
$$+ (\rho^2(33) - \rho(33))(\rho^2(22) - \rho(22)))$$
$$+ Tr_{(1,2)}(-\delta_{12}(\rho^2(12) - \rho(12))(\rho^2(21) - \rho(21))$$
$$+ (\rho^2(11) - \rho(11))(\rho^2(22) - \rho(22))$$
$$- \delta_{12}(\rho^2(12) - \rho(12))(\rho^2(21) - \rho(21)))$$
$$= 3(N_2^2 - 2N_2 N + N^2) + 6Tr_{(1)}(n_1^4 - 2n_1^3 + n_1^2)$$

with $n_1 = \rho(11)$ and

$$N_2 = Tr_{(1)} n_1^2. \tag{B.21}$$

This differs, however, from (B.13) and thus does not fulfill the trace relation (B.11).[1] It is thus not possible to find an Ansatz for C^3 in leading order in ρ and C^2 which is fully antisymmetric, simultaneously closes the equations of motion for C^2 and fulfills the trace relation for C^3. It is thus of interest to obtain some explicit numbers for the traces of C^2 and C^3.

Some quantitative results for the traces can be obtained for homogenous Fermi systems in thermal and chemical equilibrium. To this end we consider infinite nuclear matter consisting of protons and neutrons with two spin projections each at finite temperature T and chemical potential μ which fixes the energy density $\mathcal{E}(T, \mu)$ as well as the Fermion density $n(T, \mu)$. The occupation numbers then read

$$n_\alpha(T, \mu) = (\exp((\epsilon_\alpha - \mu)/T) + 1)^{-1} \tag{B.22}$$

with the single-particle energy $\epsilon_\alpha = p_\alpha^2/(2M)$. In Fig. B.1 we display the results for

$$F_2(\epsilon) := n(\epsilon)(1 - n(\epsilon)), \tag{B.23}$$

corresponding to Eq. (B.6) (except for a $-$ sign) and

$$F_3(\epsilon) := 2\left((n(\epsilon)(1 - n(\epsilon))^2 - n(\epsilon)^2(1 - n(\epsilon))\right) \tag{B.24}$$

corresponding to Eq. (B.13) for a temperature of $5\,\mathrm{MeV}$ and $\mu = 38\,\mathrm{MeV}$. It is seen that F_2 has a maximum at the chemical potential ($\epsilon = \mu$) and F_3 is negative for $\epsilon < \mu$ and positive above, however, slightly smaller in magnitude. Due to the oscillating form of $F_3(\epsilon)$ the integral over ϵ—corresponding to the total trace (except for degeneracy numbers)—gives only a small value compared to the integral of $F_2(\epsilon)$ over ϵ.

---

[1] Alternative expressions are discussed in: E. Litvinova and P. Schuck, Phys. Rev. C 100 (2019) 064320; Phys. Rev. C 102 (2020) 034310.

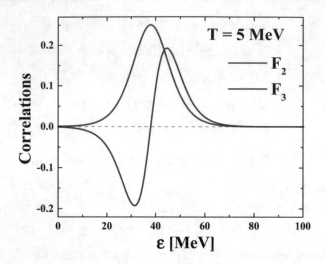

Fig. B.1 The two-body fluctuation (B.23) (blue line) and the 3-body fluctuation (B.24) (red line) as a function of the single-particle energy for nuclear matter at a temperature of 5 MeV and chemical potential $\mu = 38$ MeV

The Two-Body Problem in Vacuum

C

The stationary (two-body) Schrödinger equation for the energy $E = \hbar\omega = \omega$ ($\hbar = 1$) reads in the center-of-mass system (cms) after separating the constant center-of-mass motion:

$$\omega|\Psi\rangle = (H_0 + v)|\Psi\rangle, \tag{C.1}$$

where H_0 is the non-interacting Hamiltonian and the two-body interaction is denoted by v. Alternatively, one has to look for the "zeros" in ω of the equation

$$(\omega - H_0 - v)|\Psi\rangle =: G^{-1}(\omega)|\Psi\rangle = 0, \tag{C.2}$$

which in the nonrelativistic case of two-body interactions v and $H_0 = T = p^2/(2\mu)$ in the cms (with $\mu = m/2$ denoting the reduced mass) reads in four-momentum space

$$\left(\omega - \frac{\mathbf{p}^2}{2\mu} - v\right)|\Psi\rangle =: G^{-1}(\omega, \mathbf{p})|\Psi\rangle = 0. \tag{C.3}$$

In Hilbert space this defines the retarded propagator (in the limit $\epsilon \to 0^+$)

$$G^+(\omega + i\epsilon, \mathbf{p}) = \frac{1}{\omega - \frac{\mathbf{p}^2}{2\mu} - v + i\epsilon} = \frac{1}{(\omega - \frac{\mathbf{p}^2}{2\mu} + i\epsilon)[1 - v(\omega - \frac{\mathbf{p}^2}{2\mu} + i\epsilon)^{-1}]}, \tag{C.4}$$

© The Author(s), under exclusive license to Springer Nature Switzerland AG 2021
W. Cassing, *Transport Theories for Strongly-Interacting Systems*, Lecture Notes in Physics 989, https://doi.org/10.1007/978-3-030-80295-0

which can be rewritten in the geometric expansion (assuming convergence) as

$$G^+(\omega + i\epsilon, \mathbf{p}) = \frac{1}{\omega - \frac{\mathbf{p}^2}{2\mu} + i\epsilon} \sum_{n=0}^{\infty} \left(v \left(\frac{1}{\omega - \frac{\mathbf{p}^2}{2\mu} + i\epsilon} \right) \right)^n \tag{C.5}$$

$$=: G_0^+(\omega + i\epsilon, \mathbf{p}) \sum_{n=0}^{\infty} \left(v G_0^+(\omega + i\epsilon, \mathbf{p}) \right)^n,$$

with the free retarded Green's function (in the center-of-mass system)

$$G_0^+(\omega + i\epsilon, \mathbf{p}) = \frac{1}{\omega - \frac{\mathbf{p}^2}{2\mu} + i\epsilon}, \tag{C.6}$$

which is analytic in ω in the complex upper half plane since the poles in ω are in the lower half plane due to the limit $+i\epsilon$ in the denominator. Omitting the arguments we obtain the **Dyson equation**

$$G^+ = G_0^+ \sum_{n=0}^{\infty} \left(v G_0^+ \right)^n = G_0^+ + G_0^+ v G_0^+ + G_0^+ v G_0^+ v G_0^+ + \cdots = G_0^+ + G_0^+ v G^+$$

$$\tag{C.7}$$

$$= G_0^+ (1 + v G^+) = G_0^+ \frac{1}{(1 - v G_0^+)} = \frac{1}{(1 - G_0^+ v)} G_0^+ .$$

In momentum-space representation this Dyson equation reads as

$$G^+(\mathbf{p}' - \mathbf{p}) = \langle \mathbf{p}' | G^+(\omega) | \mathbf{p} \rangle = \langle \mathbf{p}' | G_0^+(\omega) | \mathbf{p} \rangle + \langle \mathbf{p}' | G_0^+(\omega) v G^+(\omega) | \mathbf{p} \rangle \tag{C.8}$$

$$= \frac{\delta^3(\mathbf{p}' - \mathbf{p})}{\omega - \mathbf{p}'^2/2\mu + i\epsilon} + \int d^3 p_1 \frac{1}{\omega - \mathbf{p}'^2/2\mu + i\epsilon} v(\mathbf{p}' - \mathbf{p}_1) G^+(\mathbf{p}_1 - \mathbf{p}) .$$

By inversion we obtain alternatively (omitting the $(+)$)

$$G^{-1} = G_0^{-1}(1 - v G_0) = (1 - G_0 v) G_0^{-1} \tag{C.9}$$

or

$$G_0 G^{-1} = 1 - v . \tag{C.10}$$

Note that (C.10) is an equation for the two-body Green's function which is of one-body type only after separation of the center-of-mass motion!

The (two-body) Green's function G^+ in scattering theory—where we have the boundary condition of an incoming undistorted wave $|\Phi_\omega\rangle$—generates the scattering state of energy ω:

$$|\Psi_\omega^+\rangle = |\Phi_\omega\rangle + G_0^+(\omega)v|\Psi_\omega^+\rangle = \sum_{n=0}^\infty (G_0^+(\omega)v)^n|\Phi_\omega\rangle =: \hat{\Omega}(\omega)|\Phi_\omega\rangle \quad (C.11)$$

with the unitary Moeller operator $\hat{\Omega}(\omega)$ which follows

$$\hat{\Omega}(\omega) = \sum_{n=0}^\infty (G_0^+(\omega)v)^n = \frac{1}{1 - G_0^+(\omega)v}. \quad (C.12)$$

Explicitly we obtain for the matrix elements in coordinate space

$$\langle\mathbf{r}|G_0^{(+)}|\mathbf{r}'\rangle = \int\int d^3q\, d^3q'\, \langle\mathbf{r}|\mathbf{q}\rangle\langle\mathbf{q}|G_0^{(+)}|\mathbf{q}'\rangle\langle\mathbf{q}'|\mathbf{r}'\rangle \quad (C.13)$$

$$= \frac{1}{(2\pi)^3}\int d^3q\, \exp(i\mathbf{q}\cdot\mathbf{r})\frac{1}{E - \frac{\mathbf{q}^2}{2\mu} + i\epsilon}\exp(-i\mathbf{q}\cdot\mathbf{r}')$$

$$= \frac{1}{(2\pi)^3}2\mu\int d^3q\, \frac{\exp(i\mathbf{q}\cdot(\mathbf{r}-\mathbf{r}'))}{\mathbf{k}^2 - \mathbf{q}^2 + i\epsilon} = 2\mu\, G_0^{(+)}(\mathbf{r},\mathbf{r}');$$

i.e. the familiar Green's function in coordinate-space representation. We have used, furthermore,

$$\langle\mathbf{r}|\mathbf{q}\rangle = \langle\mathbf{q}|\mathbf{r}\rangle^* = (2\pi)^{-3/2}\exp(i\mathbf{q}\cdot\mathbf{r}). \quad (C.14)$$

These operator equations allow to define a T-matrix via

$$T(\omega)|\Phi_\omega\rangle = v|\Psi_\omega^+\rangle, \quad (C.15)$$

which follows the T-matrix (or Born) series

$$T(\omega) = v + vG^+(\omega)T = v\sum_{n=0}^\infty (G_0^+(\omega)v)^n = v\hat{\Omega}(\omega). \quad (C.16)$$

Comment Here we only consider stationary systems without the presence of a third particle in the environment. The two-body density matrix in this case is a pure ensemble build up from the scattering or bound state $|\Psi_\omega^+\rangle$ by

$$\rho_2 = |\Psi_\omega^+\rangle\langle\Psi_\omega^+|, \qquad\qquad \rho_{20} = |\Phi_\omega\rangle\langle\Phi_\omega|, \quad (C.17)$$

i.e.

$$\rho_2 = \hat{\Omega}(\omega)|\Phi_\omega\rangle\langle\Phi_\omega|\hat{\Omega}(\omega)^\dagger = \hat{\Omega}(\omega)\rho_{20}\hat{\Omega}(\omega)^\dagger. \quad (C.18)$$

The matrix elements of the T-matrix in momentum space for the elastic scattering process $\mathbf{p}_1 + \mathbf{p}_2 \rightarrow \mathbf{p}_1' + \mathbf{p}_2'$ (omitting discrete quantum numbers) can be written as

$$\langle \mathbf{p}_1'\mathbf{p}_2'|T(\omega)|\mathbf{p}_1\mathbf{p}_2\rangle = \delta^3(\mathbf{p}_1 + \mathbf{p}_2 - \mathbf{p}_1' - \mathbf{p}_2')T(\omega, \mathbf{q}), \tag{C.19}$$

where $\mathbf{q} = \mathbf{p}_1' - \mathbf{p}_1$ is the momentum transfer in the collision and ω denotes the on-shell energy of the (elastic) collision. Note that in the center-of-mass system we have $\mathbf{p}_2 = -\mathbf{p}_1$ and $\mathbf{p}_2' = -\mathbf{p}_1'$.

In lowest order the T-matrix (C.15) is given by the bare interaction v and the differential cross section reads

$$\frac{d\sigma(\omega)}{d\Omega} = \frac{\mu^2}{(2\pi)^2}|v(\mathbf{q})v^*(\mathbf{q})_{\mathcal{A}}| \tag{C.20}$$

with the Fourier transform of the bare interaction

$$v(\mathbf{q}) = \int d^3r \, \exp(-i\mathbf{q} \cdot \mathbf{r}) \, v(\mathbf{r}). \tag{C.21}$$

In Eq. (C.20) the index $\mathcal{A}$ implies antisymmetrization for identical fermions. In case of rotational invariance of the interaction we have $v(\mathbf{q}) = v(|\mathbf{q}|) = v(q)$ and the momentum transfer q in elastic scattering is related to the scattering angle ϑ by

$$q = 2k\sin(\vartheta/2), \tag{C.22}$$

where k is the absolute momentum in the center-of-mass system, i.e. $k^2 = 2\mu E_{cm}$ or

$$E_{cm} = \frac{k^2}{2\mu}. \tag{C.23}$$

The question arises, however, if the lowest order Born approximation (C.20) is a suitable approximation in case on nucleon-nucleon scattering. In order to investigate this issue we study a simple model for the interaction $v(r)$ in case of proton-neutron scattering such that the antisymmetrization can be discarded:

$$v(r) = V_r \exp(-r^2/a^2) - V_a \exp(-r^2/b^2) \tag{C.24}$$

with a short-range repulsive and intermediate range attractive part. The Fourier transform is easily calculated to give

$$v(q) = V_r(\pi a^2)^{3/2}\exp(-q^2a^2/4) - V_a(\pi b^2)^{3/2}\exp(-q^2b^2/4) \tag{C.25}$$

and the differential cross section reads (after integration over the azimuthal angle ϕ)

$$\frac{d\sigma(E_{cm})}{d\vartheta} = \frac{\mu^2}{2\pi} |v(2k \sin(\vartheta/2))|^2. \tag{C.26}$$

Integration over the angle ϑ gives the total scattering cross section σ_B,

$$\sigma_B = \int_0^\pi d\vartheta \, \sin(\vartheta) \frac{d\sigma(E_{cm})}{d\vartheta} = \int_{-1}^1 d\cos\vartheta \, \frac{d\sigma(E_{cm})}{d\cos\vartheta}. \tag{C.27}$$

In case of low energy scattering only s-wave scattering contributes to the scattering amplitude $f_0(k)$ and the angle integrated cross section is given by

$$\sigma(E_{cm}) = \frac{4\pi}{k^2} \sin^2(\delta_0(k)) = \frac{4\pi}{2\mu E_{cm}} \sin^2(\delta_0(k)) \tag{C.28}$$

with the phase shift $\delta_0(k)$ which has to be extracted from the solution of the radial Schrödinger equation for angular momentum $l = 0$,

$$\left(\frac{d^2}{dr^2} + k^2 \right) \chi(r) = 2\mu v(r) \, \chi(r), \tag{C.29}$$

using the familiar decomposition $r\psi_0(r) = \chi(r)$. A regular solution for $\chi(r)$ then has to vanish for $r = 0$ since the radial momentum operator must be Hermitian.

For our model case we employ $V_r = 700 \, \text{MeV}$, $V_a = 350 \, \text{MeV}$, $a = 0.35 \, \text{fm}$ and $b = 0.7 \, \text{fm}$. The corresponding interaction is displayed in Fig. C.1 for orientation explicitly showing a strong short-range repulsion and intermediate range attraction.

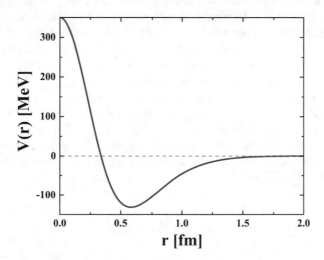

Fig. C.1 The model interaction $v(r)$ for the parameters described in the text

Fig. C.2 Comparison of the cross section in the Born approximation (C.26) (red line, integrated over the angle ϑ) with the result from the phase shift (C.28) (blue line) as a function of the center-of-mass energy E_{cm}

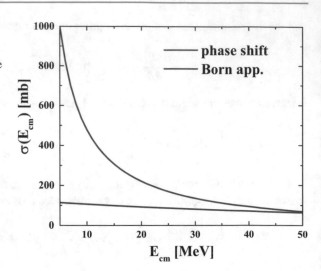

This choice of parameters does not lead to a bound state for negative energy and only scattering states for positive energy are physical states. We note in passing that a bound state of energy $\sim -2.25\,\text{MeV}$ is obtained when choosing the slightly more attractive $V_a \approx 410\,\text{MeV}$.

The homogenous (and regular) solution of (C.29) is given by $\sim \sin(kr)$ whereas $\chi(r) \sim \sin(kr + \delta_0(k))$ for $r \to \infty$. The actual numerical solution of (C.29) on a finite grid is straight forward and involves an iteration in the phase shift $\delta_0(k)$ for fixed momentum k. A comparison of the cross section (C.28) with the result in Born approximation is shown in Fig. C.2 which shows a huge difference between the results especially for low energies, thus illustrating the importance of a resummation of the interaction for low energy scattering. Note that with increasing energy also higher partial waves ($l > 0$) additionally contribute to the cross section σ.

The dramatic enhancement at $E_{CM} = 5\,\text{MeV}$ can be traced back to the sizeable enhancement of the exact solution $\psi_0(r)$ relative to the "free" solution in the range of the interaction ($<1.5\,\text{fm}$) as displayed in Fig. C.3 (l.h.s.). This enhancement decreases with increasing E_{cm} and almost vanishes for $E_{cm} = 50\,\text{MeV}$ as shown in Fig. C.3 (r.h.s.). Only a constant phase shift is seen here for $r > 2\,\text{fm}$. One thus can conclude that for low energy scattering the Born approximation gives a misleading result such that a resummation in Eq. (C.15) has to be performed or the Schrödinger equation to be solved explicitly. This conclusion does not change very much when considering two-body scattering in a nuclear environment as considered e.g. in Sects. 2.4 and 2.5.

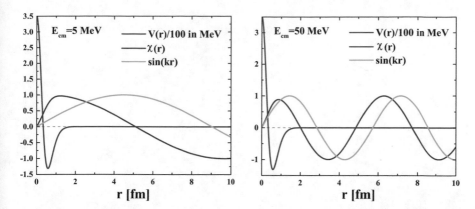

Fig. C.3 Comparison of the wavefunction $\chi(r)$ (red lines) with the free solution (green lines) for $E_{cm} = 5\,\text{MeV}$ (l.h.s.) and $E_{cm} = 50\,\text{MeV}$ (r.h.s.). The scaled potential $v(r)$ is depicted in terms of the blue lines for orientation

Periodic Boundary Conditions

D

In this Appendix we discuss a simple model system that allows to solve the off-shell collision term in Sect. 2.4 numerically and to compare solutions for off-shell and on-shell scattering. In case of periodic boundary conditions for a homogenous system in a finite volume of size $V = a^3$ the single-particle eigenstates in coordinate space (for nucleons) are

$$\langle \mathbf{r}|n_x n_y n_z, n_s, n_\tau \rangle = \langle \mathbf{r}|\alpha \rangle = \frac{1}{a^{3/2}} \exp\left(i \frac{2\pi}{a}(n_x x + n_y y + n_z z)\right) \chi(n_s) \tau(n_\tau),$$

$$(D.1)$$

where $\chi(n_s)$ denote the two orthogonal eigenstates for the spin and $\tau(n_\tau)$ for isospin, respectively. By construction these states are orthonormal in all quantum numbers. The momentum components of particles are given by ($i = 1, 2, 3$)

$$p_i(n_i) = \frac{2\pi \hbar}{a} n_i \qquad (D.2)$$

and the single-particle energies by

$$\epsilon(n_x, n_y, n_z) = \frac{2\hbar^2 \pi^2}{M_N a^2}(n_x^2 + n_y^2 + n_z^2) =: \bar{\omega}(n_x^2 + n_y^2 + n_z^2) \qquad (D.3)$$

when assuming no contributions from spin and isospin projections. Here M_N stands for the nucleon mass. Within this basis the conservation of energy and momentum in 2-body transitions can be strictly fulfilled and easily controlled via integer numbers n_i for momentum or n_i^2 for energy. Furthermore, the matrix elements of

© The Author(s), under exclusive license to Springer Nature Switzerland AG 2021
W. Cassing, *Transport Theories for Strongly-Interacting Systems*, Lecture Notes in Physics 989, https://doi.org/10.1007/978-3-030-80295-0

a $\delta^3(\mathbf{r}_1 - \mathbf{r}_2)$-type interaction are—assuming the interactions again to be diagonal in spin and isospin:

$$\langle n_x^3, n_y^3, n_z^3; n_x^4, n_y^4, n_z^4 | v | n_x^1, n_y^1, n_z^1; n_x^2, n_y^2, n_z^2 \rangle \tag{D.4}$$

$$= \frac{V_0}{a^6} \int_{-a/2}^{a/2} dx \int_{-a/2}^{a/2} dy \int_{-a/2}^{a/2} dz \, \exp\left(-i\frac{2\pi}{a}(n_x^1 + n_x^2 - n_x^3 - n_x^4)x\right)$$

$$\times \exp\left(-i\frac{2\pi}{a}(n_y^1 + n_y^2 - n_y^3 - n_y^4)y\right) \exp\left(-i\frac{2\pi}{a}(n_z^1 + n_z^2 - n_z^3 - n_z^4)z\right)$$

$$= \frac{V_0}{a^3} \, \delta(n_x^1 + n_x^2 - n_x^3 - n_x^4) \, \delta(n_y^1 + n_y^2 - n_y^3 - n_y^4) \, \delta(n_z^1 + n_z^2 - n_z^3 - n_z^4),$$

which implies momentum conservation in the transitions $1 + 2 \leftrightarrow 3 + 4$.

In case of a diagonal density matrix $\rho_{\alpha\alpha'} = \delta_{\alpha\alpha'} n_\alpha$ the change in the occupation numbers n_α—not to be mixed with the quantum numbers n_x, n_y, n_z—is given by the collision term $I_{\alpha\alpha}$,

$$\frac{d}{dt} n_\alpha(t) = I_{\alpha\alpha}(t) = -\frac{i}{\hbar} \sum_\beta \sum_{\lambda\gamma} [\langle \alpha\beta | v | \lambda\gamma \rangle C_{\lambda\gamma\alpha\beta}(t) - C_{\alpha\beta\lambda\gamma}(t) \langle \lambda\gamma | v | \alpha\beta \rangle]. \tag{D.5}$$

When inserting the off-shell solution for $C_2(t)$ from (2.103) we get

$$I_{\alpha\alpha}(t) = \frac{2}{\hbar^2} \sum_\beta \sum_{\lambda\gamma} \int_0^t dt' \, \cos\left(\frac{1}{\hbar}(\epsilon_\alpha + \epsilon_\beta - \epsilon_\lambda - \epsilon_\gamma)(t - t')\right) \tag{D.6}$$

$$\times \langle \alpha\beta | v | \lambda\gamma \rangle \langle \lambda\gamma | v | \alpha\beta \rangle_A [n_\lambda(t') n_\gamma(t') \bar{n}_\alpha(t') \bar{n}_\beta(t')$$

$$- n_\alpha(t') n_\beta(t') \bar{n}_\lambda(t') \bar{n}_\gamma(t')].$$

Alternatively, when inserting the on-shell solution for $C_2(t)$ we end up with

$$I_{\alpha\alpha}(t) = \frac{2\pi}{\hbar} \sum_\beta \sum_{\lambda\gamma} \delta(\epsilon_\alpha + \epsilon_\beta - \epsilon_\lambda - \epsilon_\gamma) \langle \alpha\beta | v | \lambda\gamma \rangle \langle \lambda\gamma | v | \alpha\beta \rangle_A \tag{D.7}$$

$$\times [n_\lambda(t) n_\gamma(t) \bar{n}_\alpha(t) \bar{n}_\beta(t) - n_\alpha(t) n_\beta(t) \bar{n}_\lambda(t) \bar{n}_\gamma(t)].$$

These expressions can be worked out further in case of periodic boundary conditions by using (D.3) and (D.4) giving (in the on-shell case)

$$
I_{\alpha\alpha}(t) = \frac{2\pi}{\hbar\bar{\omega}} \frac{3V_0^2}{4a^6} \sum_{\beta} \sum_{\lambda\gamma} \delta(m_\alpha^2 + m_\beta^2 - m_\lambda^2 - m_\gamma^2)
$$

$$
\times\ \delta(n_x^1 + n_x^2 - n_x^3 - n_x^4)\ \delta(n_y^1 + n_y^2 - n_y^3 - n_y^4) \tag{D.8}
$$

$$
\times\ \delta(n_z^1 + n_z^2 - n_z^3 - n_z^4)\ [n_\lambda(t)n_\gamma(t)\bar{n}_\alpha(t)\bar{n}_\beta(t) - n_\alpha(t)n_\beta(t)\bar{n}_\lambda(t)\bar{n}_\gamma(t)]
$$

using the notation

$$
m_\alpha^2 = (n_x^1)^2 + (n_y^1)^2 + (n_z^1)^2 \tag{D.9}
$$

etc. which (except for a factor $\bar{\omega}$) gives the discrete single-particle energies.

It is of general interest to investigate the validity of the on-shell approximation for the collision term (D.8) or the differences arising from the use of (D.6) or (D.7) for some physical observable. To this end we consider a cubic volume of side length $a = 10$ fm which defines the time-independent basis states and single-particle energies. As a two-body interaction we assume a local interaction of the form

$$
v(\mathbf{r}) = V_0\ \delta^3(\mathbf{r}) \tag{D.10}
$$

with some strength V_0. As initial condition we assume a deformed Fermi sphere with an (angular) averaged Fermi energy of 40 MeV, however, the ellipsoid in momentum space is twice elongated in z-direction as compared to the perpendicular directions. The average density amounts to 0.188 fm^{-3} which is slightly larger than normal nuclear matter density. Accordingly, the number of "nucleons" in the box is 188, i.e. 94 protons and neutrons each. As an observable we consider the quadrupole moment in momentum space

$$
Q_2(t) = \sum_{\alpha}(2p_{z,\alpha}^2 - p_{x,\alpha}^2 - p_{y,\alpha}^2)\,n_\alpha(t), \tag{D.11}
$$

which should approach $Q_2 = 0$ for $t \to \infty$ due to off-shell or on-shell collisions.

The numerical results for $Q_2(t)$ are displayed in Fig. D.1 on a logarithmic scale for the off-shell solution and the on-shell solution of the collision term employing $V_0 = 300$ MeV fm^3 (upper lines) and $V_0 = 500$ MeV fm^3 (lower lines). Here $V_0 = 300$ MeV fm^3 corresponds to an average interaction strength as used in model calculations for nuclear matter while $V_0 = 500$ MeV fm^3 is chosen as a "very strong" interaction case. As seen from Fig. D.1 in both cases the off-shell and on-shell results give very similar results for the quadrupole moment since the contributions from off-shell matrix elements in (D.6) give oscillating contributions

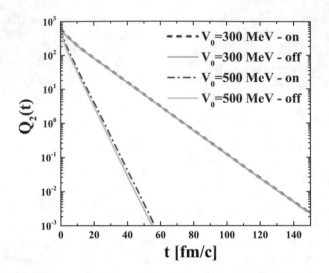

Fig. D.1 The quadrupole moment $Q_2(t)$ (D.11) in the on-shell and off-shell scattering limit for the couplings $V_0 = 300\,\text{MeV}\,\text{fm}^3$ (upper lines) and $V_0 = 500\,\text{MeV}\,\text{fm}^3$ (lower lines). The results for $V_0 = 300\,\text{MeV}\,\text{fm}^3$ are practically identical within the linewidth

in time and cancel out to a large extent[1] after summing over the states β, λ, γ. Accordingly, the on-shell collision limit holds well in this simple case with discrete energy differences, however, there might be physical examples—with a low number of basis states involved—that may show sizeable off-shell scattering effects.

It is worth noting that the results in Fig. D.1 have been obtained with a very small time step of $\Delta t = 0.0025\,\text{fm/c}$ in the off-shell case which is mandatory to achieve a good conservation of energy. On the other hand, in the on-shell calculation a time step of $0.5\,\text{fm/c}$ already leads to a very good energy conservation. For practical reasons the on-shell limit of the collision term is thus a very suitable approximation especially in view of the uncertainty in the strong interaction matrix elements and the huge gain in CPU performance.

Since the loss term in (D.8) is proportional to n_α we can determine the on-shell collision width for the state α by

$$\Gamma_\alpha = \frac{2\pi\hbar}{\hbar\omega}\frac{3V_0^2}{4a^6}\sum_\beta\sum_{\lambda\gamma}\delta(m_\alpha^2 + m_\beta^2 - m_\lambda^2 - m_\gamma^2)\delta(n_x^1 + n_x^2 - n_x^3 - n_x^4)$$

$$\times\,\delta(n_y^1 + n_y^2 - n_y^3 - n_y^4)\,\delta(n_z^1 + n_z^2 - n_z^3 - n_z^4)n_\beta\bar{n}_\lambda\bar{n}_\gamma. \qquad (D.12)$$

---

[1] Note that the states γ are fixed by momentum conservation but the summation over β, λ involves a large number of states in the present case.

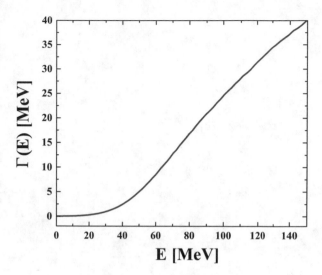

Fig. D.2 The collisional width (D.13) as a function of the energy E relative to nuclear matter of density $\rho \approx 0.167\,\text{fm}^{-3}$

In the continuum limit $a \to \infty$ the collision width of a particle with momentum $\mathbf{p}_\alpha$ reads for isotropic scattering with a constant cross section $\sigma = \frac{\mu^2}{2\hbar^4}\frac{3}{4}V_0^2$ ($\approx 32.4\,\text{mb}$ for $V_0 = 300\,\text{MeV}\,\text{fm}^3$)

$$\Gamma(\mathbf{p}_\alpha) = \frac{\hbar c}{(2\pi\hbar)^3}d\int d^3 p_\beta \int d\Omega\; v_{rel}(\mathbf{p}_\alpha - \mathbf{p}_\beta)\frac{\sigma}{4\pi}\,n(\mathbf{p}_\beta)(1 - n(\mathbf{p}_3))(1 - n(\mathbf{p}_4)),$$
(D.13)

where the final states $\mathbf{p}_3$ and $\mathbf{p}_4$ are fixed by energy and momentum conservation except for an angle $\Omega = (\cos(\vartheta), \phi)$ in the center-of-mass which has to be integrated over. The relative velocity—in the nonrelativistic limit—is given by

$$v_{rel}(\mathbf{p}_\alpha - \mathbf{p}_\beta) = \frac{|\mathbf{p}_\alpha - \mathbf{p}_\beta|}{M_N}$$
(D.14)

while the factor $d = 4$ in (D.13) stems from summation over spin and isospin. The width (D.13) is displayed in Fig. D.2 as a function of the kinetic energy $E = p^2/(2M_N)$ for the scattering of a nucleon on an occupied Fermi sphere with Fermi energy $E_F \approx 37\,\text{MeV}$ simulating symmetric nuclear matter of density $\rho \approx 0.167\,\text{fm}^{-3}$. For energies below E_F the width Γ practically vanishes due to Pauli blocking and smoothly increases up to about 40 MeV at $E = 150\,\text{MeV}$. This width, however, is small compared to the mass of the nucleon ($\approx 938\,\text{MeV}$) such that nucleons in this case can be considered as "good quasiparticles."

Phase-Space Integrals

The on-shell phase-space integrals incorporated in Sect. 3.2 cover the dynamics of a multi-particle system to a large extent. The n-body phase-space integral is generally defined by

$$R_n(P; m_1, \ldots, m_n) = \left(\frac{1}{(2\pi)^3}\right)^n \int \prod_{k=1}^{n} \frac{d^3 p_k}{2E_k} \, (2\pi)^4 \delta^4 \left(P - \sum_{j=1}^{n} p_j\right), \quad \text{(E.1)}$$

with P denoting the total four-momentum, m_j the masses of the particles and p_j their four-momenta. Since the phase-space integrals are Lorentz-invariant we will work in the center-of-mass system ($\mathbf{P} = 0$).

To show (as an example) the behavior of the different n-body phase-space integrals it is instructive to look e.g. at the consecutive decays in the reaction $p\bar{p} \to \pi\rho\rho \to 3\pi\rho \to 5\pi$ which are relevant in proton-antiproton annihilation. Also, this example connects the 3-, 4-, and 5-body phase-space integrals as a function of the invariant energy above threshold (see below).

For the sake of completeness, we start with the 1-body phase-space integral,

$$R_1(\sqrt{s}; m) = \frac{1}{(2\pi)^3} \int \frac{d^3 p}{2E} \, (2\pi)^4 \delta^4(\sqrt{s} - E) = \frac{\pi}{\sqrt{s}}, \quad \text{(E.2)}$$

where E is the on-shell energy $E = \sqrt{m^2 + \mathbf{p}^2}$ and the mass m of the particle is equal to the invariant energy $\sqrt{s}$. This result shows that the 1-body phase-space decreases with increasing $\sqrt{s}$. The 2-body phase-space integral can also be

© The Author(s), under exclusive license to Springer Nature Switzerland AG 2021
W. Cassing, *Transport Theories for Strongly-Interacting Systems*, Lecture Notes in Physics 989, https://doi.org/10.1007/978-3-030-80295-0

evaluated analytically,

$$R_2(\sqrt{s}; m_1, m_2) = \frac{1}{(2\pi)^2} \int \int \frac{d^3 p_1}{2E_1} \frac{d^3 p_2}{2E_2} \delta^3(\mathbf{p}_1 + \mathbf{p}_2)\delta(\sqrt{s} - E_1 - E_2)$$

(E.3)

$$= \frac{1}{4(2\pi)^2} \int \frac{d^3 p_1}{E_1 E_2} \delta(\sqrt{s} - E_1 - E_2)$$

$$= \frac{1}{4(2\pi)^2} \int_0^\infty dp_1 \int_0^\pi \int_0^{2\pi} \frac{d\phi d\theta \, p_1^2 \sin\theta}{E_1 E_2} \delta(\sqrt{s} - E_1 - E_2)$$

$$= \frac{1}{4\pi} \int_0^\infty \frac{dp_1 \, p_1^2}{\sqrt{m_1^2 + p_1^2}\sqrt{m_2^2 + p_1^2}} \delta\left(\sqrt{s} - E_1 - E_2\right).$$

The zeros of the δ-function are given by

$$p_0 = \pm \frac{\sqrt{\lambda(s, m_1^2, m_2^2)}}{2\sqrt{s}},$$

(E.4)

where only the positive value has to be taken into account. Rewriting the δ-function as

$$\delta(\sqrt{s} - E_1 - E_2) = \frac{\delta(p_1 - p_0)}{p_1/E_1 + p_1/E_2}$$

(E.5)

and inserting Eqs. (E.4) and (E.5) into Eq. (E.3) we obtain the two-body phase-space integral

$$R_2(\sqrt{s}; m_1, m_2) = \frac{1}{4\pi} \int_0^\infty \frac{dp_1 \, p_1}{E_1 E_2} \frac{E_1 E_2 \delta(p_1 - p_0)}{E_1 + E_2} = \frac{\sqrt{\lambda(s, m_1^2, m_2^2)}}{8\pi s},$$

(E.6)

with $E_1 + E_2 = \sqrt{s}$ from the original δ-function and $\lambda(x, y, z) = (x - y - z)^2 - 4yz$. The typical shape of $R_2(\sqrt{s}, m_1, m_2)$ is shown in Fig. E.1 for the masses $m_1 = 1\,\text{GeV}$ and $m_2 = 2\,\text{GeV}$ as a function of the invariant energy above threshold.[1] The upper limit is independent of the masses and is given by $1/(8\pi)$.

The on-shell three-body phase-space integral $R_3(\sqrt{s}, m_1, m_2, m_3)$ is the next in the hierarchy and a good example for the evaluation of phase-space integrals of higher order since the n-body decay can be considered as consecutive 2-body

---

[1] The figures in this Appendix are taken from: E. Seifert and W. Cassing, Phys. Rev. C97 (2018) 024913.

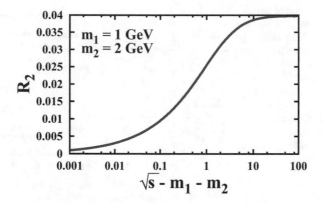

Fig. E.1 Two-body phase-space integral R_2 for particles with masses $m_1 = 1\,\mathrm{GeV}$ and $m_2 = 2\,\mathrm{GeV}$ as a function of the invariant energy above threshold

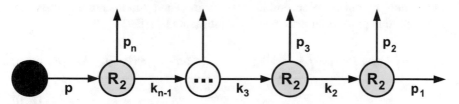

Fig. E.2 Illustration of the subsequent decay of an initial state (black dot) into n particles. The initial state may consist of m particles since only the invariant mass is relevant for the phase-space integral due to Lorentz invariance

decays (see Fig. E.2 for an illustration). Note that in Fig. E.2 $k_n = p$ and $k_1 = p_1$. A prerequisite in calculating the phase-space integral is that we do not have any incoming momenta in between the first and final 2-body decay. For the calculation of the process we employ the recursion relation for phase-space integrals,

$$R_n(P) = \int \frac{\mathrm{d}^4 p_n}{(2\pi)^3} \delta(p_n^2 - m_n^2)\, R_{n-1}(P - p_n),\qquad \text{(E.7)}$$

and also insert the identities

$$1 = \int \mathrm{d}M_{n-1}^2 \delta(M_{n-1}^2 - k_{n-1}^2),\qquad \text{(E.8)}$$

and

$$1 = \int \mathrm{d}^4 k_{n-1} \delta^4(P - p_n - k_{n-1}).\qquad \text{(E.9)}$$

Fig. E.3 Illustration of the 3-, 4-, and 5-body phase-space integrals as a function of the invariant energy above threshold. The red solid line shows the 3-body phase-space integral for $\pi\rho\rho$, the blue dashed line shows the 4-body phase-space integral for $3\pi\rho$ and the green dotted line shows the 5-body phase-space integral for 5 pions

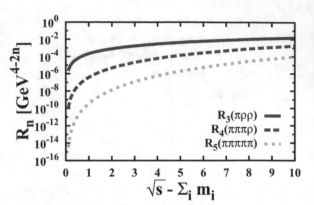

The first identity (E.8) gives the mass of the first cluster from which the 4-momentum p_n splits. The second identity ensures energy-momentum conservation in the splitting process. Inserting both identities into Eq. (E.7) we find

$$R_n(P) = \int dM_{n-1}^2 \underbrace{\int d^4k_{n-1} \int \frac{d^4p_n}{(2\pi)^3} \delta^4(k_{n-1}^2 - M_{n-1}^2)\delta^4(p_n^2 - m_n^2)\delta^4(P - p_n - k_{n-1})}_{R_2(P;m_n,M_{n-1})/(2\pi)} R_{n-1}(k_{n-1})$$

(E.10)

$$= \int_{(\sum_{i=1}^{n-1} m_i)^2}^{(M_n - m_n)^2} dM_{n-1}^2 \frac{R_2(P; m_n, M_{n-1})}{2\pi} R_{n-1}(k_{n-1}).$$

With this expression any n-particle phase-space integral can be calculated in a straight forward fashion as long as the masses m_i are known. Note that the last R_2, which one gets after applying Eq. (E.10) several times, has no additional factor $1/(2\pi)$. In Fig. E.3 the phase-space integrals for 3, 4 and 5 particles are shown as a function of the invariant energy above threshold for our example of initial $\pi\rho\rho$ with a subsequent decay into $3\pi\rho$ and a final decay to 5 pions. All phase-space integrals share a similar shape, only the magnitudes close to threshold vary substantially with the number of particles.

For practical purposes it is very helpful to have an analytical approximation for $R_3(\sqrt{s})$ with coefficients that can be fitted and tabulated for different mass combinations,

$$R_3(t, m_1, m_2, m_3) = a_1 t^{a_2} \left(1 - \frac{1}{a_3 t + 1 + a_4}\right),$$

(E.11)

where $t = \sqrt{s} - m_1 - m_2 - m_3$ denotes the invariant energy above threshold. The parameters a_1, a_2, a_3, and a_4 are fitted to the numerical results for R_3 from (E.10) which only have to be computed once.

Kramers–Kronig Relation

The Kramers–Kronig relations are often used to calculate the real part from the imaginary part (or vice versa) of response functions or retarded propagators in physical systems, because for stable systems causality implies the analyticity condition, and alternatively, analyticity implies causality of the corresponding stable physical system.

We consider a complex function $\chi(\omega) = \chi_1(\omega) + i\chi_2(\omega)$ of the complex variable ω, where $\chi_1(\omega)$ and $\chi_2(\omega)$ are real functions, respectively. Furthermore, we assume that $\chi(\omega)$ is analytic in the closed upper half plane of ω and vanishes like $1/|\omega|$ (or faster) for $|\omega| \to \infty$. The Kramers–Kronig relations then are given by

$$\chi_1(\omega) = \frac{1}{\pi} P \int_{-\infty}^{\infty} \frac{\chi_2(\omega')}{\omega' - \omega} \, d\omega' \tag{F.1}$$

and

$$\chi_2(\omega) = -\frac{1}{\pi} P \int_{-\infty}^{\infty} \frac{\chi_1(\omega')}{\omega' - \omega} \, d\omega', \tag{F.2}$$

where P denotes the Cauchy principal value (see below). The real and imaginary parts of $\chi(\omega)$ are thus closely connected and the full function can be reconstructed just knowing one of its parts.

The proof is based on Cauchy's residue theorem for complex integration. Let χ be any analytic function in the closed upper half plane, then the function $\omega' \to \chi(\omega')/(\omega' - \omega)$ (for real ω) is real and will also be analytic in the upper half of the plane. The residue theorem then states that

$$\oint \frac{\chi(\omega')}{\omega' - \omega} \, d\omega' = 0 \tag{F.3}$$

© The Author(s), under exclusive license to Springer Nature Switzerland AG 2021
W. Cassing, *Transport Theories for Strongly-Interacting Systems*, Lecture Notes in Physics 989, https://doi.org/10.1007/978-3-030-80295-0

for any contour within this region. Now choose the contour along the real axis (starting from $-\infty$), circumvent the pole at $\omega = \omega'$ in the upper half plane, and (for $\omega \to \infty$) close with a "large" semicircle in the upper half plane. Then we decompose the integral into its contributions along each of these three contour segments and pass them to proper limits. The length of the semicircular segment increases proportionally to $|\omega|$, but the integral over it vanishes because $\chi(\omega)$ vanishes faster than $1/|\omega|$. We are left with the segments along the real axis and the half-circle around the pole at ω'. Now we consider the limit of the half-circle radius r to go to zero which gives

$$0 = \oint \frac{\chi(\omega')}{\omega' - \omega} \, d\omega' = \mathcal{P} \int_{-\infty}^{\infty} \frac{\chi(\omega')}{\omega' - \omega} \, d\omega' - i\pi \chi(\omega), \qquad (F.4)$$

where the second term in the last expression is obtained using the residue theorem. By rearranging one arrives at the compact form of the Kramers–Kronig relations,

$$\chi(\omega) = \frac{1}{i\pi} \mathcal{P} \int_{-\infty}^{\infty} \frac{\chi(\omega')}{\omega' - \omega} \, d\omega'. \qquad (F.5)$$

When splitting $\chi(\omega)$ and equation (F.5) into real and imaginary parts we obtain the results in Eqs. (F.1) and (F.2).

The **Cauchy principal value** defines a way to circumvent singularities in the integrand $f(x)$ along the real x-axis. The Cauchy principal value is defined as follows: the finite number

$$\lim_{\varepsilon \to 0+} \left[\int_{a}^{b-\varepsilon} f(x) \, dx + \int_{b+\varepsilon}^{c} f(x) \, dx \right] \qquad (F.6)$$

where b is a point where

$$\int_{a}^{b} f(x) \, dx = \pm\infty \qquad (F.7)$$

for $a < b$ and

$$\int_{b}^{c} f(x) \, dx = \mp\infty \qquad (F.8)$$

for $c > b$.

Lorentz Transformations, γ-Matrices, and Dirac-Algebra

G

A covariant formulation of dynamics has to involve well-defined transformation properties of physical quantities (operators) under Lorentz transformations. The latter involve Lorentz boost with velocity $\vec{\beta}$ as well as spatial rotations with angle $\vec{\phi}$. A convenient formulation is achieved in a pseudo-Euclidian vector space of dimension 4 (Minkowski space) such that Lorentz transformations are represented by 4×4 matrices $\Lambda^{\mu}{}_{\nu}$. A suitable notation then is given by representing space-time and energy-momentum by four-vectors (in Minkowski space)

$$x = (ct, \mathbf{r})^T \equiv (ct, \mathbf{x})^T = \begin{pmatrix} ct \\ x \\ y \\ z \end{pmatrix}, \quad q = (\omega/c, \mathbf{q})^T = \begin{pmatrix} \omega/c \\ q_x \\ q_y \\ q_z \end{pmatrix} \tag{G.1}$$

with **contra-variant** components

$$x^0 = ct, \ x^1 = x, \ x^2 = y, \ x^3 = z; \qquad q^0 = \omega/c, \ q^1 = q_x, \ q^2 = q_y, \ q^3 = q_z \tag{G.2}$$

and **covariant** components

$$x_0 = x^0 = ct, \ x_1 = -x, \ x_2 = -y, \ x_3 = -z;$$
$$q_0 = \omega/c, \ q_1 = -q_x, \ q_2 = -q_y, \ q_3 = -q_z . \tag{G.3}$$

The transformation between the covariant and contra-variant vectors is given by

$$x_\mu = \sum_{\nu=0}^{3} g_{\mu\nu} x^\nu =: g_{\mu\nu} x^\nu \tag{G.4}$$

© The Author(s), under exclusive license to Springer Nature Switzerland AG 2021
W. Cassing, *Transport Theories for Strongly-Interacting Systems*, Lecture Notes
in Physics 989, https://doi.org/10.1007/978-3-030-80295-0

(for $\mu = 0, 1, 2, 3$) with the (Lorentz-invariant) **pseudo-metric tensor**

$$(g_{\mu\nu}) = (g^{\mu\nu}) = \begin{pmatrix} 1 & 0 & 0 & 0 \\ 0 & -1 & 0 & 0 \\ 0 & 0 & -1 & 0 \\ 0 & 0 & 0 & -1 \end{pmatrix}. \tag{G.5}$$

In order to further simplify the notation we have introduced in (G.4) the **Einstein convention** which implies to sum over all indices that appear twice. However, the indices have to be upper and lower indices separately! Nevertheless, we will keep the explicit summation in the following if not indicated else.

The squared length of a space-time vector then can be written as

$$\sum_{\mu,\nu=0}^{3} g_{\mu\nu} x^{\mu} x^{\nu} = \sum_{\mu=0}^{3} x^{\mu} x_{\mu} = \sum_{\mu=0}^{3} x'^{\mu} x'_{\mu}, \tag{G.6}$$

which has to be invariant under Lorentz transformations

$$x'^{\mu} = \sum_{\nu} \Lambda^{\mu}{}_{\nu} x^{\nu}; \quad \mu, \nu = 0, 1, 2, 3 \tag{G.7}$$

implying for the vector $(x^0, x^1, x^2, x^3)^T$:

$$\sum_{\mu=0}^{3} x^{\mu} x_{\mu} = c^2 t^2 - \mathbf{r}^2 = \text{const.} \tag{G.8}$$

This condition requires that

$$\sum_{\mu,\nu} g_{\mu\nu} \Lambda^{\mu}{}_{\rho} \Lambda^{\nu}{}_{\sigma} = \sum_{\mu,\nu} \Lambda^{\mu}{}_{\rho} g_{\mu\nu} \Lambda^{\nu}{}_{\sigma} = g_{\rho\sigma}. \tag{G.9}$$

When written as a matrix-multiplication this reads as

$$\Lambda^T g \Lambda = g. \tag{G.10}$$

By multiplication from the left with g we get (with $g^2 = 1_4$)

$$(g\Lambda^T g)\Lambda = 1_4 = \begin{pmatrix} 1 & 0 & 0 & 0 \\ 0 & 1 & 0 & 0 \\ 0 & 0 & 1 & 0 \\ 0 & 0 & 0 & 1 \end{pmatrix}, \tag{G.11}$$

which implies that the Lorentz transformation can be inverted and

$$\Lambda^{-1} = g\Lambda^T g. \tag{G.12}$$

Equation (G.9) follows from the requirement that the length squared (G.6) is invariant with respect to Lorentz transformations:

$$\sum_\mu x'^\mu x'_\mu = \sum_{\mu\nu} g_{\mu\nu} x'^\mu x'^\nu = \sum_{\mu,\nu,\rho,\sigma} g_{\mu\nu} \Lambda^\mu{}_\rho x^\rho \Lambda^\nu{}_\sigma x^\sigma = \sum_{\rho\sigma} g_{\rho\sigma} x^\rho x^\sigma = \sum_\rho x_\rho x^\rho. \tag{G.13}$$

For a Lorentz transformation in x^1-direction with velocity $\beta = v/c$ the matrix $\Lambda^\mu{}_\nu$ has the form

$$\Lambda^\mu{}_\nu = \begin{pmatrix} \gamma & -\gamma\beta & 0 & 0 \\ -\gamma\beta & \gamma & 0 & 0 \\ 0 & 0 & 1 & 0 \\ 0 & 0 & 0 & 1 \end{pmatrix} \tag{G.14}$$

with

$$\det \Lambda = \gamma^2 - \gamma^2\beta^2 = \gamma^2(1 - \beta^2) = 1 \tag{G.15}$$

for $\gamma^2 = 1/(1 - \beta^2)$. This "boost" in x_1-direction can also be parametrized with help of the "rapidity" ξ as:[1]

$$\Lambda^\mu{}_\nu(\xi) = \begin{pmatrix} \cosh\xi & -\sinh\xi & 0 & 0 \\ -\sinh\xi & \cosh\xi & 0 & 0 \\ 0 & 0 & 1 & 0 \\ 0 & 0 & 0 & 1 \end{pmatrix}. \tag{G.16}$$

A comparison with (G.14) gives the relation to the velocity:

$$\cosh\xi = \gamma, \quad \sinh\xi = \gamma\beta \Rightarrow \beta = \frac{\sinh\xi}{\cosh\xi} = \tanh\xi. \tag{G.17}$$

The advantage of this parametrization is that subsequent boosts are additive in ξ:

$$\Lambda(\xi_2)\Lambda(\xi_1) = \Lambda(\xi_1 + \xi_2). \tag{G.18}$$

---

[1] In the context of heavy-ion physics the rapidity conventionally is denoted by $y \equiv \xi$.

With (G.17) this directly gives the relation for the addition of the velocities:

$$\beta_{12} = \tanh(\xi_1 + \xi_2) = \frac{\sinh(\xi_1 + \xi_2)}{\cosh(\xi_1 + \xi_2)} = \frac{\sinh(\xi_1)\cosh(\xi_2) + \sinh(\xi_2)\cosh(\xi_1)}{\cosh(\xi_1)\cosh(\xi_2) + \sinh(\xi_1)\sinh(\xi_2)}$$

$$= \frac{\beta_1 + \beta_2}{1 + \beta_1\beta_2}. \tag{G.19}$$

The general form of the Lorentz transformation is obtained from (G.14) by a rotation of the spatial components in terms of a 3×3 matrix $R_3(\vec{\phi})$ since the Lorentz transformations form a 'group':

1. Two subsequent Lorentz transformations,

$$x'^\mu = \sum_\nu \Lambda^\mu{}_\nu x^\nu, \quad x''^\rho = \sum_\mu \Lambda'^\rho{}_\mu x'^\mu, \tag{G.20}$$

give

$$x''^\rho = \sum_{\mu,\nu} \Lambda'^\rho{}_\mu \Lambda^\mu{}_\nu x^\nu = \sum_\nu \Lambda''^\rho{}_\nu x^\nu, \tag{G.21}$$

which is again a Lorentz transformation since for the matrices Λ'', Λ' and Λ we get:

$$\Lambda''^T g \Lambda'' = (\Lambda'\Lambda)^T g (\Lambda'\Lambda) = \Lambda^T \underbrace{\Lambda'^T g \Lambda'}_{g} \Lambda = \Lambda^T g \Lambda = g, \tag{G.22}$$

which holds since

$$\Lambda'^T g \Lambda' = \Lambda^T g \Lambda = g. \tag{G.23}$$

2. The neutral element is the 4×4 unity matrix 1_4 for Lorentz-boosts with velocity $v = 0$.

3. As shown above there is an inverse transformation for each transformation Λ. This follows also from (G.9) with help of the determinant

$$\det(\Lambda^T g \Lambda) = \det \Lambda^T \ \det g \ \det \Lambda = (\det \Lambda)^2 \ \det g = \det g = -1, \tag{G.24}$$

i.e.

$$(\det \Lambda)^2 = 1 \Rightarrow \det \Lambda = \pm 1 \neq 0. \tag{G.25}$$

4. Since matrix multiplications are associative this also holds for Lorentz transformations.

Orthogonal transformations in 3 dimensions (rotations and reflections) constitute a subgroup,

$$d^{\mu}{}_{\nu} = \begin{pmatrix} 1 & 0 \\ 0 & d^i{}_k \end{pmatrix} \tag{G.26}$$

for $i, k = 1,2,3$ and

$$\sum_{m=1}^{3} d^i{}_m d^j{}_m = \delta_{ij}. \tag{G.27}$$

Further discrete transformations of the Lorentz-group are time-reflections

$$x'^i = x^i; \quad x'^0 = -x^0; \quad i = 1, 2, 3, \tag{G.28}$$

and spatial reflections

$$x'^0 = x^0, \quad x'^i = -x^i; \quad i = 1, 2, 3. \tag{G.29}$$

G.1 Lorentz-Goup: Scalars, Vectors, Tensors

A physical quantity Ψ is denoted as **Lorentz-scalar** if it is invariant under Lorentz transformations,

$$\Psi \rightarrow \Psi' = \Psi. \tag{G.30}$$

As an example we quote the electric charge e or the mass squared m^2.

A **Lorentz-vector or four-vector** A^{μ} transforms like the components of the space-time four-vector,

$$A^{\mu} \rightarrow A'^{\mu} = \sum_{\nu} \Lambda^{\mu}{}_{\nu} A^{\nu}. \tag{G.31}$$

The covariant components then transform as

$$
\begin{aligned}
A'_{\mu} &= \sum_{\nu} g_{\mu\nu} A'^{\nu} = \sum_{\nu,\rho} g_{\mu\nu} \Lambda^{\nu}{}_{\rho} A^{\rho} \\
&= \sum_{\nu,\rho,\sigma} g_{\mu\nu} \Lambda^{\nu}{}_{\rho} g^{\rho\sigma} A_{\sigma} = \sum_{\sigma} (g\Lambda g)_{\mu}{}^{\sigma} A_{\sigma} \\
&= \sum_{\sigma} (\Lambda^{-1})^{\sigma}{}_{\mu} A_{\sigma},
\end{aligned} \tag{G.32}
$$

where in the last step Eq. (G.12) has been used. Examples are the space-time four-vector x^μ, the energy-momentum four-vector q^μ, etc.

A **Lorentz-tensor** (of rank 2) is defined by the transformation properties of its contra-variant components as

$$T'^{\mu\nu} = \sum_{\rho,\sigma} \Lambda^\mu{}_\rho \Lambda^\nu{}_\sigma T^{\rho\sigma}. \tag{G.33}$$

An example is the energy-momentum tensor. Extensions to tensors of higher rank follow the same transformation law (with multiple Λ's).

By contraction with the metric tensor $g_{\mu\nu}$ single (or more) upper components can be converted to lower components, e.g.

$$T_\mu{}^\nu = \sum_\rho g_{\mu\rho} T^{\rho\nu}, \quad T^{\mu\nu} = \sum_{\rho,\sigma} g^{\mu\rho} g^{\nu\sigma} T_{\rho\sigma}. \tag{G.34}$$

A lower index has to be treated as in case of a covariant vector, e.g. for the tensor of rank 2,

$$T'^{\ \nu}_\mu = \sum_{\rho,\sigma} (\Lambda^{-1})^\rho{}_\mu \Lambda^\nu{}_\sigma T_\rho{}^\sigma. \tag{G.35}$$

Due to the linearity of the Lorentz transformation products of Lorentz-tensors (of any rank) are again Lorentz-tensors, e.g.

$$C_\mu{}^\nu{}_\rho = A_\mu T^\nu{}_\rho \tag{G.36}$$

is a tensor of rank 3 if A_μ is a four-vector and $T^\nu{}_\rho$ a Lorentz-tensor of rank 2.

The rank of a Lorentz-tensor n can be reduced to the rank $n-2$ by setting two indices equal and summing over this index. Accordingly the contraction of a tensor of rank 2 gives a Lorentz-scalar,

$$\Psi = \sum_\mu T_\mu{}^\mu, \tag{G.37}$$

since

$$\Psi' = \sum_\mu T'^\mu_\mu = \sum_{\mu,\nu,\rho} (\Lambda^{-1})^\nu{}_\mu \Lambda^\mu{}_\rho T_\nu{}^\rho = \sum_{\nu,\rho} \delta^\nu_\rho T_\nu{}^\rho = \sum_\nu T_\nu{}^\nu = \Psi . \tag{G.38}$$

Further Examples

1. The partial derivatives of a Lorentz-scalar Ψ with respect to x^μ gives the covariant components of a four-vector since

$$\partial'_\mu \Psi' =: \frac{\partial \Psi'}{\partial x'^\mu} = \sum_\nu \frac{\partial \Psi}{\partial x^\nu} \frac{\partial x^\nu}{\partial x'^\mu} = \sum_\nu (\Lambda^{-1})^\nu{}_\mu \frac{\partial \Psi}{\partial x^\nu} \tag{G.39}$$

employing the inverse of (G.7),

$$x^\mu = \sum_\nu (\Lambda^{-1})^\mu{}_\nu x'^\nu. \tag{G.40}$$

The index of this **four-gradient** thus has to be treated as a (lower) covariant index,

$$\partial_\mu \psi = \frac{\partial \psi}{\partial x^\mu}. \tag{G.41}$$

2. The **four-divergence** of a Lorentz-vector is a Lorentz-scalar since (due to (G.32))

$$\sum_\nu \partial'_\nu A'^\nu = \sum_{\mu,\nu,\rho} (\Lambda^{-1})^\mu{}_\nu \Lambda^\nu{}_\rho \partial_\mu A^\rho = \sum_{\mu,\rho} \delta^\mu_\rho \partial_\mu A^\rho = \sum_\mu \partial_\mu A^\mu . \tag{G.42}$$

3. Choosing the covariant components of a four-vector as

$$A_\mu = \partial_\mu \Psi = \frac{\partial \Psi}{\partial x^\mu}, \tag{G.43}$$

we get from (G.4) and (G.42)

$$\sum_{\mu,\nu} g^{\mu\nu} \partial_\mu A_\nu = \sum_{\mu,\nu} g^{\mu\nu} \partial_\mu \partial_\nu \Psi = \sum_{\mu,\nu} g^{\mu\nu} \partial'_\mu \partial'_\nu \Psi'. \tag{G.44}$$

The **D'Alembert-operator**,

$$\sum_\nu \partial_\nu \partial^\nu = \sum_{\mu,\nu} g^{\mu\nu} \partial_\mu \partial_\nu = \frac{1}{c^2} \frac{\partial^2}{\partial t^2} - \nabla^2, \quad \nabla^2 = \Delta = \frac{\partial^2}{\partial x^2} + \frac{\partial^2}{\partial y^2} + \frac{\partial^2}{\partial z^2}, \tag{G.45}$$

thus is invariant under Lorentz transformations.

For a four-vector with covariant components A_μ the covariant second derivatives

$$\sum_{\nu,\sigma} g^{\nu\sigma} \frac{\partial^2}{\partial x^\nu \partial x^\sigma} A_\mu = \sum_{\nu,\sigma} g^{\nu\sigma} \partial_\nu \partial_\sigma A_\mu = \sum_\nu \partial_\nu \partial^\nu A_\mu \tag{G.46}$$

transform like the covariant components of a four-vector. This is a central issue in electrodynamics.

4. The scalar product of two four-vectors is a Lorentz-scalar:

$$\sum_{\mu} A'^{\mu} B'_{\mu} = \sum_{\mu} \sum_{\sigma,\rho} \Lambda^{\mu}{}_{\rho} A^{\rho} (\Lambda^{-1})^{\sigma}{}_{\mu} B_{\sigma} = \sum_{\rho,\sigma} A^{\rho} \delta^{\sigma}_{\rho} B_{\sigma} = \sum_{\rho} A^{\rho} B_{\rho}.$$

(G.47)

G.2   Dirac Equation and Dirac-Clifford Algebra

The field equations for a free (spin $1/2\,\hbar$) Dirac particle are given by

$$\left(-i \sum_{\mu} \gamma^{\mu} \partial_{\mu} + M \right) \psi(x) = 0, \qquad i \sum_{\mu} \partial_{\mu} \bar{\psi}(x) \gamma^{\mu} + \bar{\psi}(x) M = 0,$$

(G.48)

where the spinor $\psi(x)$ is a four-component vector function of the space-time variable $x = (ct, \mathbf{r})$ and M denotes the bare mass of the fermion. The Pauli-adjoint spinor is given by

$$\bar{\psi}(x) = \psi^{\dagger}(x) \gamma^{0} \equiv (\psi_{1}^{*}(x), \ \psi_{2}^{*}(x), \ -\psi_{3}^{*}(x), \ -\psi_{4}^{*}(x)).$$

(G.49)

The 4×4 γ-matrices follow the anti-commutation relations

$$\{\gamma^{\mu}, \gamma^{\nu}\} = 2\, g^{\mu\nu} \cdot 1_{4} \qquad\qquad (\mu, \nu = 0, 1, 2, 3)$$

(G.50)

with the Lorentz-invariant pseudo-metric $g^{\mu\nu} = g_{\mu\nu}$. These matrices, furthermore, follow

$$(\gamma^{\mu})^{\dagger} = \gamma^{0} \gamma^{\mu} \gamma^{0},$$

(G.51)

and have the standard representation

$$\gamma^{k} = \begin{pmatrix} 0 & \sigma^{k} \\ -\sigma^{k} & 0 \end{pmatrix} \qquad\qquad k = 1, 2, 3$$

(G.52)

$$\gamma^{0} = \begin{pmatrix} 1_{2} & 0 \\ 0 & -1_{2} \end{pmatrix}$$

(G.53)

with the 2×2 Pauli matrices

$$\sigma^{1} = \begin{pmatrix} 0 & 1 \\ 1 & 0 \end{pmatrix}, \qquad \sigma^{2} = \begin{pmatrix} 0 & -i \\ i & 0 \end{pmatrix}, \qquad \sigma^{3} = \begin{pmatrix} 1 & 0 \\ 0 & -1 \end{pmatrix}$$

(G.54)

and the 2×2 unit matrix

$$1_2 = \begin{pmatrix} 1 & 0 \\ 0 & 1 \end{pmatrix}. \tag{G.55}$$

The four-current is given by

$$j^\mu(x) = \bar{\psi}(x)\gamma^\mu\psi(x) \tag{G.56}$$

(with $\mu = 0, 1, 2, 3$) and follows the continuity equation

$$\sum_{\mu=0}^{3} \partial_\mu j^\mu(x) = \left(\frac{\partial}{x^0} j^0(x) + \sum_{k=1}^{3} \frac{\partial}{x^k} j^k(x) \right) = 0 \tag{G.57}$$

thus expressing a vanishing four-divergence. The latter implies that the quantity

$$N_F = \int d^3r \; j^0(x) \tag{G.58}$$

is constant in time and describes the total net-fermion number.

Now starting with the 4 γ-matrices one can construct (by multiplication and linear combination) a complete system of 4×4 matrices which provide a "basis" in the space of complex 4×4 matrices. Since a 4×4 matrix has 16 elements one needs also 16 "basis" elements. The standard choice for the basis matrices Γ_i is:

$$1_4 : \; 4 \times 4 \text{ unit matrix} \tag{G.59}$$

$$\gamma^\mu \; (\mu = 0, 1, 2, 3) : \; 4 \text{ matrices}$$

$$\sigma^{\mu\nu} = \frac{i}{2}[\gamma^\mu, \gamma^\nu] : \; 6 \text{ matrices (antisymmetric)}$$

$$\gamma^\mu\gamma^5 \; (\mu = 0, \ldots, 3) : \; 4 \text{ matrices}$$

$$\gamma^5 : \; 1 \text{ matrix}$$

with the "chirality" matrix γ^5:

$$\gamma^5 = \gamma_5 = i\gamma^0\gamma^1\gamma^2\gamma^3 = \begin{pmatrix} 0 & 0 & 1 & 0 \\ 0 & 0 & 0 & 1 \\ 1 & 0 & 0 & 0 \\ 0 & 1 & 0 & 0 \end{pmatrix}. \tag{G.60}$$

This matrix can also be written as

$$\gamma^5 = \frac{i}{4!}\epsilon_{\mu\nu\rho\sigma}\gamma^\mu\gamma^\nu\gamma^\rho\gamma^\sigma \tag{G.61}$$

with the antisymmetric unit tensor of rank 4

$$\epsilon_{\mu\nu\rho\sigma} = \begin{cases} 1 \text{ if } (\mu, \nu, \rho, \sigma) \text{ is an even permutation of } (0, 1, 2, 3) \\ -1 \text{ if } (\mu, \nu, \rho, \sigma) \text{ is an odd permutation of } (0, 1, 2, 3) \\ 0 \text{ else.} \end{cases}$$

It is easy to show that γ^5 has the properties

$$(\gamma^5)^2 = 1_4, \qquad (\gamma^5)^\dagger = \gamma^5, \qquad \{\gamma^\mu, \gamma^5\} = 0 \qquad \mu = 0, \ldots, 3. \tag{G.62}$$

The elements of $\sigma^{\mu\nu} \neq 0$ are the spatial nondiagonal components $(\mu, \nu) = (k, l)$ with $k \neq l$ and $(\mu, \nu) = (0, k)$. They can be contracted to three-vectors

$$\Sigma^m = \frac{1}{2}\sum_{k,l=1}^{3}\epsilon^{mkl}\sigma_{kl} = \frac{i}{2}\sum_{k,l=1}^{3}\epsilon^{mkl}\gamma_k\gamma_l \qquad (m = 1, \ldots, 3) \tag{G.63}$$

and

$$\sigma^{0k} = i\gamma^0\gamma^k = i\alpha^k \qquad (k = 1, \ldots, 3), \tag{G.64}$$

with

$$\epsilon^{mkl} = \epsilon^{mkl0} \tag{G.65}$$

denoting the ϵ-tensor in three dimensions. Σ^m has the properties of spin-matrices, i.e.

$$\left[\frac{1}{2}\Sigma^m, \frac{1}{2}\Sigma^n\right] = i\,\epsilon^{mnp}\left(\frac{1}{2}\Sigma^p\right) \text{ and } \sum_{m=1}^{3}\left(\frac{1}{2}\Sigma^m\right)^2 = \frac{1}{2}\left(\frac{1}{2}+1\right)\cdot 1_4. \tag{G.66}$$

This is easy to show within the standard representation

$$\Sigma^m = \begin{pmatrix} \sigma^m & 0 \\ 0 & \sigma^m \end{pmatrix} \qquad (m = 1, 2, 3) \tag{G.67}$$

with the Pauli matrices σ^m. Furthermore,

$$\gamma^k \gamma^5 = \begin{pmatrix} \sigma^k & 0 \\ 0 & -\sigma^k \end{pmatrix} \quad \text{and} \quad \gamma^0 \gamma^5 = \begin{pmatrix} 0_2 \ 1_2 \\ -1_2 \ 0_2 \end{pmatrix}. \tag{G.68}$$

The hermicity properties are given by (G.51), (G.62) and read

$$(\sigma^{\mu\nu})^\dagger = \gamma^0 \sigma^{\mu\nu} \gamma^0, \qquad\qquad (\gamma^\mu \gamma^5)^\dagger = \gamma^0 (\gamma^\mu \gamma^5) \gamma^0. \tag{G.69}$$

The 16 matrices

$$\Gamma_1 = 1_4, \ \Gamma_2 = \gamma^0, \dots, \Gamma_{16} = \gamma^5 \tag{G.70}$$

are traceless (except for Γ_1) and define an orthonormal basis for 4×4 matrices with the scalar product

$$(\Gamma_A, \Gamma_B) = \frac{1}{4} Tr(\Gamma_A^\dagger \Gamma_B), \tag{G.71}$$

$$\frac{1}{4} Tr \, \Gamma_A = \delta_{A,1}, \qquad \frac{1}{4} Tr \, (\Gamma_A^\dagger \Gamma_B) = \delta_{A,B} \qquad (A, B = 1, \dots, 16). \tag{G.72}$$

These matrices thus are linear independent and an arbitrary 4×4 matrix M can be written as a linear combination of the system (G.70):

$$M = \sum_{A=1}^{16} C_A^M \Gamma_A \quad \text{with} \quad C_A^M = \frac{1}{4} Tr(\Gamma_A^\dagger M). \tag{G.73}$$

Since products (or multiple products) of Γ-matrices again are a 4×4 matrix the product can also be written as a linear combination of Γ-matrices. This leads to the following 'multiplication table':

$$\gamma^\mu \gamma^\nu = g^{\mu\nu} \cdot 1_4 - i \, \sigma^{\mu\nu} \tag{G.74}$$

$$\gamma^\lambda \sigma^{\mu\nu} = i \, (g^{\lambda\mu} \gamma^\nu - g^{\lambda\nu} \gamma^\mu - \epsilon^{\lambda\mu\nu}{}_\eta \, (\gamma^\eta \gamma^5))$$

$$\gamma^\mu (\gamma^\nu \gamma^5) = g^{\mu\nu} \gamma^5 - \frac{1}{2} \epsilon^{\mu\nu}{}_{\kappa\lambda} \, \sigma^{\kappa\lambda}$$

$$(\gamma^\lambda \gamma^5) \sigma^{\mu\nu} = i \, [g^{\lambda\mu} (\gamma^\nu \gamma^5) - g^{\lambda\nu} (\gamma^\mu \gamma^5)] - \epsilon^{\lambda\mu\nu}{}_\eta \, \gamma^\nu$$

$$\sigma^{\kappa\lambda} \sigma^{\mu\nu} = (g^{\kappa\mu} g^{\lambda\nu} - g^{\kappa\nu} g^{\lambda\mu}) \cdot 1_4 - i \, \epsilon^{\kappa\lambda\mu\nu} \gamma^5$$

$$-i \, [g^{\kappa\mu} \sigma^{\lambda\nu} + g^{\lambda\nu} \sigma^{\kappa\mu} - g^{\kappa\nu} \sigma^{\lambda\mu} - g^{\lambda\mu} \sigma^{\kappa\nu}]$$

$$\sigma^{\mu\nu}\gamma^5 = -\frac{i}{2}\,\epsilon^{\mu\nu}{}_{\kappa\lambda}\,\sigma^{\kappa\lambda}(=\gamma^5\,\sigma^{\mu\nu})$$

$$(\gamma^\mu\,\gamma^5)\,(\gamma^\nu\,\gamma^5) = -g^{\mu\nu}\cdot 1_4 + i\sigma^{\mu\nu}$$

$$\gamma^5\,(\gamma^\mu\,\gamma^5) = -\gamma^\mu.$$

The system of matrices (G.70) with the multiplication laws (G.74) is denoted as **Dirac-Clifford algebra**.

An application of (G.74) are transition matrix elements (squared) $|M|^2$ where traces over products of γ-matrices have to be performed. Further examples are:

$$\frac{1}{4}\,Tr(\gamma^\mu\gamma^\nu) = g^{\mu\nu} \tag{G.75}$$

$$\frac{1}{4}\,Tr(\gamma^{\mu_1}\gamma^{\mu_2}\dots\gamma^{\mu_n}) = 0 \qquad\qquad n \text{ odd}$$

$$\frac{1}{4}\,Tr(\gamma^\mu\gamma^\nu\gamma^\rho\gamma^\sigma) = g^{\mu\nu}g^{\rho\sigma} + g^{\mu\sigma}g^{\nu\rho} - g^{\mu\rho}g^{\nu\sigma}$$

$$\frac{1}{4}\,Tr(\gamma^\mu\gamma^\nu\gamma^5) = 0$$

$$\frac{1}{4}\,Tr(\gamma^\mu\gamma^\nu\gamma^\rho\gamma^\sigma\gamma^5) = i\,\epsilon^{\mu\nu\rho\sigma}.$$

Conventionally such products—by contraction with arbitrary four-vectors $a_\mu, b_\nu, \dots$—are written in shorthand notation as:

$$\sum_\mu a_\mu\gamma^\mu = \not{a}, \tag{G.76}$$

i.e.

$$Tr(\not{a}\,\not{b}) = 4\,a\cdot b, \tag{G.77}$$

etc. In this context we also recall the identities:

$$\sum_\mu \gamma_\mu\,\not{a}\,\gamma^\mu = -2\not{a} \tag{G.78}$$

$$\sum_\mu \gamma_\mu\,\not{a}\,\not{b}\,\gamma^\mu = 4\,a\cdot b\cdot 1_4$$

$$\sum_\mu \gamma_\mu\,\not{a}\,\not{b}\,\not{c}\,\gamma^\mu = -2\not{c}\,\not{b}\,\not{a}.$$

G.3   Solutions of the Free Dirac Equation

We note in passing that the solution of the Dirac equation (G.48) implies that each component of the Dirac spinor ψ_j ($j = (1, 2, 3, 4)$) fulfills the Klein-Gordon equation:

$$(-\nabla^2 + M^2)\psi_j(x) = -\frac{\partial^2}{\partial t^2}\psi_j(x) \tag{G.79}$$

for a fermion with mass M and thus has the dispersion relation

$$p_0^2 = \mathbf{p}^2 + M^2 \tag{G.80}$$

as appropriate for a stable free particle in vacuum. Accordingly, the free Dirac equation has the plane-wave solution

$$\psi(x) = \psi(0)\exp(\pm ip \cdot x) . \tag{G.81}$$

In order to find an appropriate basis for the spinor $\psi(0)$ we consider separately solutions for positive and negative energy:

$$\psi_+(x) = u(\mathbf{p})\exp\{-i(Et - \mathbf{p} \cdot \mathbf{x})\}, \qquad \psi_-(x) = v(\mathbf{p})\exp\{+i(Et - \mathbf{p} \cdot \mathbf{x})\} \tag{G.82}$$

The Dirac equation (G.48) then gives:

$$\left(-\sum_\mu p_\mu\gamma^\mu + M \cdot 1_4\right)u(\mathbf{p}) = (p_\mu\gamma^\mu + M \cdot 1_4)v(\mathbf{p}) = 0. \tag{G.83}$$

To specify the spinors $u(\mathbf{p})$ and $v(\mathbf{p})$ we first consider a particle at rest, i.e. for $\mathbf{p} = 0$. Eq. (G.83) then reads

$$\gamma^0 u(0) = u(0), \qquad \gamma^0 v(0) = -v(0). \tag{G.84}$$

Now we may choose for $u(0)$ and $v(0)$ two orthonormal eigenvectors to γ^0 with eigenvalues $+1$ and -1, which we will denote by $u_r(0)$, $v_r(0)$, $r = \pm$, respectively. Since the matrices (G.63) commute with γ^0 the spinors $u_r(0)$ and $v_r(0)$ may simultaneously be eigenspinors of $\Sigma \cdot \mathbf{a}$ with eigenvalues ± 1, where $\mathbf{a}$ is an arbitrary unit vector, since

$$(\Sigma \cdot \mathbf{a})^2 = (\mathbf{a} \cdot \mathbf{a}) \cdot 1_4 = 1_4. \tag{G.85}$$

The standard choice is $\mathbf{a} = (0, 0, 1)$ which gives

$$\Sigma^3 u_s(0) = s\, u_s(0), \qquad\qquad \Sigma^3 v_s(0) = s\, v_s(0). \qquad\qquad (G.86)$$

These 4 basis spinors are normalized according to

$$\bar{u}_r(0)u_s(0) = \delta_{rs}, \qquad\qquad -\bar{v}_r(0)v_s(0) = \delta_{rs}. \qquad\qquad (G.87)$$

It follows that

$$\bar{u}_r(0)v_s(0) = \bar{v}_r(0)u_s(0) = 0, \qquad\qquad (G.88)$$

since u and v are eigen spinors for different eigenvalues of γ^0. The completeness relation then can be written as

$$\sum_{s=\pm}[u_s(0)\bar{u}_s(0) - v_s(0)\bar{v}_s(0)] = 1_4 \qquad\qquad (G.89)$$

and the expansion of γ^0 in terms of eigenspinors reads

$$\sum_{s=\pm}[u_s(0)\bar{u}_s(0) + v_s(0)\bar{v}_s(0)] = \gamma^0. \qquad\qquad (G.90)$$

As a consequence the matrices

$$\sum_s u_s(0)\bar{u}_s(0) = \frac{1}{2}(1_4 + \gamma^0) \equiv \Lambda_+(0), \quad -\sum_s v_s(0)\bar{v}_s(0) = \frac{1}{2}(1_4 - \gamma^0) \equiv \Lambda_-(0),$$

$$(G.91)$$

with

$$\Lambda_r(0)\Lambda_s(0) = \delta_{rs}\Lambda_r(0) \qquad\qquad r, s = \pm \qquad\qquad (G.92)$$

have the properties of projection operators on the solutions of positive and negative energy $\pm M$ for $\mathbf{p} = 0$.

In standard representation we get:

$$u_r(0) = \begin{pmatrix} \chi_r \\ 0 \end{pmatrix}, \qquad v_r = \begin{pmatrix} 0 \\ \chi_r \end{pmatrix} \qquad \text{with } \chi_+ = \begin{pmatrix} 1 \\ 0 \end{pmatrix}, \ \chi_- = \begin{pmatrix} 0 \\ 1 \end{pmatrix} \qquad (G.93)$$

and (in terms of 4×4 matrices)

$$\Lambda_+ = \begin{pmatrix} 1 & 0 & 0 & 0 \\ 0 & 1 & 0 & 0 \\ 0 & 0 & 0 & 0 \\ 0 & 0 & 0 & 0 \end{pmatrix}, \quad \Lambda_- = \begin{pmatrix} 0 & 0 & 0 & 0 \\ 0 & 0 & 0 & 0 \\ 0 & 0 & 1 & 0 \\ 0 & 0 & 0 & 1 \end{pmatrix}. \tag{G.94}$$

In order to obtain the result for $\mathbf{p} \neq 0$ we employ a Lorentz transformation with the velocity

$$\mathbf{v} = \frac{\mathbf{p}}{E_p} \tag{G.95}$$

to achieve

$$\Lambda : (M, 0, 0, 0) \rightarrow \left(\sqrt{p^2 + M^2}, \mathbf{p} \right). \tag{G.96}$$

With $\mathbf{p}_e = \mathbf{p}/|\mathbf{p}|$ and $\cosh(b) = E_p/M$, we have alternatively

$$\cosh \left(\frac{b}{2} \right) = \sqrt{\frac{1}{2}(\cosh(b) + 1)} = N_p (E_p + M), \tag{G.97}$$

$$\sinh \left(\frac{b}{2} \right) = \sqrt{\frac{1}{2}(\cosh(b) - 1)} = N_p |\mathbf{p}|,$$

with

$$N_p = \frac{1}{\sqrt{2M(E_p + M)}}. \tag{G.98}$$

Furthermore, we use

$$S_p = N_p \left(M \cdot 1_4 + \sum_\mu p_\mu \gamma^\mu \gamma^0 \right)_{p^0 = E_p}. \tag{G.99}$$

This leads to the following spinors for Dirac particles with momentum $\mathbf{p}$

$$u_r(\mathbf{p}) = N_p \left(M \cdot 1_4 + \sum_\mu p_\mu \gamma^\mu \right) u_r(0), \qquad v_r(\mathbf{p}) = N_p \left(M \cdot 1_4 - \sum_\mu p_\mu \gamma^\mu \right) v_r(0),$$

$$\tag{G.100}$$

which are a solution of Eq. (G.83) since

$$(M + \sum_\mu p_\mu \gamma^\mu)(M - \sum_\mu p_\mu \gamma^\mu) = M^2 - \sum_{\mu,\nu} p_\mu p_\nu \frac{1}{2}\{\gamma^\mu, \gamma^\nu\} = 0. \quad \text{(G.101)}$$

The normalization is

$$\bar{u}_r(\mathbf{p})u_s(\mathbf{p}) = -\bar{v}_r(\mathbf{p})v_s(\mathbf{p}) = \delta_{rs}, \qquad \bar{u}_r(\mathbf{p})v_s(\mathbf{p}) = \bar{v}_r(\mathbf{p})u_s(\mathbf{p}) = 0.$$
$$\text{(G.102)}$$

In order to evaluate the projectors on spinors with positive and negative energy at finite $\mathbf{p}$,

$$\Lambda_+(\mathbf{p}) = S_p \Lambda_+(0) S_p^{-1} = \sum_s u_s(\mathbf{p})\bar{u}_s(\mathbf{p}),$$

$$\Lambda_-(\mathbf{p}) = S_p \Lambda_-(0) S_p^{-1} = -\sum_s v_s(\mathbf{p})\bar{v}_s(\mathbf{p}) \quad \text{(G.103)}$$

we use

$$S_p \gamma^0 S_p^{-1} = N_p^2 (M \cdot 1_4 + \sum_\mu p_\mu \gamma^\mu \gamma^0)\gamma^0 (M \cdot 1_4 + \sum_\mu \gamma^0 p_\mu \gamma^\mu) \quad \text{(G.104)}$$

$$= N_p^2 \left[M^2 \gamma^0 + 2M \sum_\mu p_\mu \gamma^\mu + \sum_{\mu,\nu} p_\mu \gamma^\mu p_\nu (2g^{0\nu} - \gamma^\nu \gamma^0) \right]$$

$$= N_p^2 \left[2M \sum_\mu p_\mu \gamma^\mu + 2 \sum_\mu p_\mu \gamma^\mu p^0 + (M^2 - \sum_{\mu,\nu} p_\mu \gamma^\mu p_\nu \gamma^\nu)\gamma^0 \right]$$

$$= \frac{1}{M} \sum_\mu p_\mu \gamma^\mu$$

and get

$$\Lambda_\pm(\mathbf{p}) = \frac{1}{2M}(M \cdot 1_4 \pm \sum_\mu p_\mu \gamma^\mu)_{p^0 = E_p}. \quad \text{(G.105)}$$

The completeness relation (G.89) then reads

$$\sum_s [u_s(\mathbf{p})\bar{u}_s(\mathbf{p}) - v_s(\mathbf{p})\bar{v}_s(\mathbf{p})] = 1_4. \quad \text{(G.106)}$$

In standard representation the 4 basis spinors are given by

$$u_s(\mathbf{p}) = \frac{1}{\sqrt{2M(E_p + M)}} \begin{pmatrix} (E_p + M)\chi_s \\ (\vec{\sigma} \cdot \mathbf{p})\chi_s \end{pmatrix}, \quad v_s(\mathbf{p}) = \frac{1}{\sqrt{2M(E_p + M)}} \begin{pmatrix} (\vec{\sigma} \cdot \mathbf{p})\chi_s \\ (E_p + M)\chi_s \end{pmatrix}$$
$$\text{(G.107)}$$

or explicitly ($p^1 = p_x$, $p^2 = p_y$, $p^3 = p_z$)

$$u_+(\mathbf{p}) = \frac{1}{\sqrt{2M(E_p + M)}} \begin{pmatrix} E_p + M \\ 0 \\ p^3 \\ p^1 + ip^2 \end{pmatrix}; \quad u_-(\mathbf{p}) = \frac{1}{\sqrt{2M(E_p + M)}} \begin{pmatrix} 0 \\ E_p + M \\ p^1 - ip^2 \\ -p^3 \end{pmatrix};$$
$$\text{(G.108)}$$

$$v_+(\mathbf{p}) = \frac{1}{\sqrt{2M(E_p + M)}} \begin{pmatrix} p^3 \\ p^1 + ip^2 \\ E_p + M \\ 0 \end{pmatrix}; \quad v_-(\mathbf{p}) = \frac{1}{\sqrt{2M(E_p + M)}} \begin{pmatrix} p^1 - ip^2 \\ -p^3 \\ 0 \\ E_p + M \end{pmatrix}.$$
$$\text{(G.109)}$$

It is straight forward to evaluate the matrix elements

$$\bar{u}_s(\mathbf{p})\gamma^\mu u_s(\mathbf{p}) = \frac{p^\mu}{M} = \bar{v}_s(\mathbf{p})\gamma^\mu v_s(\mathbf{p}), \quad\quad \text{(G.110)}$$

such that the Dirac equation (multiplied by M) reads

$$M\bar{u}_s(\mathbf{p})\left(-\sum_\mu p_\mu \gamma^\mu + M \cdot 1_4\right)u_s(\mathbf{p}) = -\sum_\mu p_\mu p^\mu + M^2 = 0, \quad \text{(G.111)}$$

which is equivalent to the free dispersion relation $p_0^2 = \mathbf{p}^2 + M^2$.

In the standard representation of u_s the lower components become smaller by a factor $|\mathbf{p}|/(2M)$ in the nonrelativistic limit $|\mathbf{p}| \ll M$ while the relations are opposite for v_s.

Furthermore, the relation between u_s and v_s is given by

$$\gamma^5 u_s(\mathbf{p}) = v_s(\mathbf{p}), \quad\quad\quad \gamma^5 v_s(\mathbf{p}) = u_s(\mathbf{p}); \quad\quad \text{(G.112)}$$

the chirality matrix γ^5 thus exchanges u- and v-spinors.

G.4   Quantization of the Free Dirac Field

In this subsection we briefly recall the quantization of the free Dirac field. With the abbreviations

$$\varepsilon^+(\mathbf{p}) = \sqrt{\mathbf{p}^2 + M^2}, \tag{G.113}$$

$$\varepsilon^-(\mathbf{p}) = -\sqrt{\mathbf{p}^2 + M^2} \tag{G.114}$$

an arbitrary solution of the free Dirac equation can be expanded as:

$$\Psi(x) = \int \frac{d^3p}{(2\pi)^3} \frac{M}{\omega(\mathbf{p})} \times \sum_{\lambda=1}^{2} [c_\lambda(\mathbf{p})u_\lambda(\mathbf{p}) \exp(-i(\varepsilon^+(\mathbf{p})t - \mathbf{p} \cdot \mathbf{x})) \tag{G.115}$$

$$+ d_\lambda^\dagger(\mathbf{p})v_\lambda(\mathbf{p}) \exp(i(\varepsilon^-(\mathbf{p})t - \mathbf{p} \cdot \mathbf{x}))],$$

$$\bar{\Psi}(x) = \int \frac{d^3p}{(2\pi)^3} \frac{M}{\omega(\mathbf{p})} \times \sum_{\lambda=1}^{2} [c_\lambda^\dagger(\mathbf{p})\bar{u}_\lambda(\mathbf{p}) \exp(+i(\varepsilon^+(\mathbf{p})t - \mathbf{p} \cdot \mathbf{x})) \tag{G.116}$$

$$+ d_\lambda(\mathbf{p})\bar{v}_\lambda(\mathbf{p}) \exp(-i(\varepsilon^-(\mathbf{p})t - \mathbf{p} \cdot \mathbf{x}))],$$

with $\omega(\mathbf{p}) = \sqrt{\mathbf{p}^2 + M^2} = E_p$ as the positive defined on-shell single-particle energy while the sum over λ denotes the summation over the two spin projections. Here the expansion coefficients $c_\lambda, c_\lambda^\dagger, d_\lambda, d_\lambda^\dagger$—after quantization—are creation and annihilation operators for particles and antiparticles, respectively, (with spin projections λ) following the anti-commutator relations ($\lambda \equiv r, s$)

$$\{c_r(\mathbf{p}), c_s^\dagger(\mathbf{p}')\} = \{d_r(\mathbf{p}), d_s^\dagger(\mathbf{p}')\} = \frac{E_p}{M} \delta_{rs} \, \delta^3(\mathbf{p} - \mathbf{p}'), \tag{G.117}$$

while all other anti-commutators vanish:

$$\{c_r(\mathbf{p}), c_s(\mathbf{p}')\} = \{d_r(\mathbf{p}), d_s(\mathbf{p}')\} = \{c_r^\dagger(\mathbf{p}), c_s^\dagger(\mathbf{p}')\} = \{d_r^\dagger(\mathbf{p}), d_s^\dagger(\mathbf{p}')\} = 0 \tag{G.118}$$

as well as the mixed anti-commutators between $c, c^\dagger$- and $d, d^\dagger$-operators. We recall that the particle number operator for fermions with momentum $\mathbf{p}$ and spin projection r is given by

$$N_r(\mathbf{p}) = \frac{M}{E_p} c_r^\dagger(\mathbf{p})c_r(\mathbf{p}) \tag{G.119}$$

in the normalization of Bjorken and Drell.

G.5   Transformation Properties of Dirac Spinors

We briefly recall the transformation properties of spinors with respect to Lorentz transformations.[2] In case of space-time transformations the components of a spinor $\psi(x)$ are not mixed since each component is transformed separately. For the four-momentum we have

$$[P^\mu, \psi(x)] = -i \, \partial^\mu \psi(x) \qquad (G.120)$$

and the unitary transformation $U(a)$ in Hilbert space must follow

$$U^{-1}(a)\psi(x+a)U(a) = \psi(x) \qquad (G.121)$$

for a transformation $x^\mu \to x'^\mu = x^\mu + a^\mu$ in Minkowski space with

$$U(a) = \exp(ia_\mu P^\mu) = (U^{-1}(a))^\dagger. \qquad (G.122)$$

We recall that (G.120) are the equations of motion in the Heisenberg picture.

In case of **homogenous Lorentz transformations** $x \to x' = \Lambda x$ the transformation in Hilbert space is more complicated. Here the requirement for the transformation $U(\Lambda)$—according to the Ehrenfest theorem —must fulfill the condition

$$\langle \Xi'|\psi(\Lambda x)|\Xi'\rangle = \langle \Xi|U^\dagger(\Lambda)\psi(\Lambda x)U(\Lambda)|\Xi\rangle$$
$$= S(\Lambda)\langle \Xi|\psi(x)|\Xi\rangle \text{ with } |\Xi'\rangle = U(\Lambda)|\Xi\rangle, \qquad (G.123)$$

where $U(\Lambda)$ denotes the operator in Hilbert space induced by the transformation Λ in Minkowski space. Since $|\Xi\rangle$ is arbitrary, we get:

$$U^\dagger(\Lambda)\psi(\Lambda x)U(\Lambda) = S(\Lambda)\psi(x) \text{ or } U(\Lambda)\psi(x)U^\dagger(\Lambda) = S^{-1}(\Lambda)\psi(\Lambda x).$$
$$(G.124)$$

Since $U(\Lambda)$ must be a unitary operator, we may write this operator as

$$U(\Lambda) = \exp\left(\frac{i}{2}\alpha_{\mu\nu}M^{\mu\nu}\right), \qquad \Lambda = \Lambda(\alpha) \qquad (G.125)$$

---

[2] In this subsection we will use the Einstein convention throughout.

with an antisymmetric (real) parameter matrix $\alpha_{\mu\nu}$[3] and an antisymmetric (Hermitian) operator matrix $M^{\mu\nu}$. In case of infinitesimal transformations we have

$$S^{-1}(\Lambda) \approx 1_4 + \frac{i}{4}\alpha_{\mu\nu}\sigma^{\mu\nu} \qquad (G.126)$$

and with (G.124) we obtain

$$\psi(x) + \frac{i}{2}\alpha_{\mu\nu}[M^{\mu\nu}, \psi(x)] = \left(1_4 + \frac{i}{4}\alpha_{\mu\nu}\sigma^{\mu\nu}\right)\psi(x^\rho + \alpha^\rho_{\ \delta}\,x^\delta) \qquad (G.127)$$

$$= \left(1_4 + \frac{i}{4}\alpha_{\mu\nu}\sigma^{\mu\nu}\right)(1_4 + \alpha^\rho_{\ \delta}\,x^\delta\partial_\rho)\psi(x)$$

$$= \left(1_4 + \frac{i}{4}\alpha_{\mu\nu}\sigma^{\mu\nu}\right)$$

$$\times \left[1_4 + \frac{i}{2}\alpha_{\rho\delta}\left(\frac{1}{i}x^\delta\partial^\rho - \frac{1}{i}x^\rho\partial^\delta\right)\right]\psi(x),$$

which implies

$$[M^{\mu\nu}, \psi(x)] = \left[\frac{1}{2}\sigma^{\mu\nu} + \frac{1}{i}(-x^\mu\partial^\nu + x^\nu\partial^\mu)\right]\psi(x). \qquad (G.128)$$

This is different from the case of Klein-Gordon (Bose-) fields due to the extra term $1/2\,\sigma^{\mu\nu}$, which mixes the spinor components. In the special case of rotations we obtain:

$$\frac{1}{2}\sigma^{kl} = \epsilon^{klm}\left(\frac{1}{2}\Sigma_m\right), \qquad (G.129)$$

where the three-vector Σ_m has the commutation relations of angular momentum $1/2$ $\hbar$. When considering a single-particle wavefunction in the rest frame with $\mathbf{p} = 0$, independent of x, we get in case of a rotation $\Lambda = R = R(\vec{\varphi})$

$$\langle 0|\psi(x)U(R(\vec{\varphi}))|\mathbf{p} = 0, s; c\rangle = \exp\left(-\frac{i}{2}\vec{\varphi}\cdot\Sigma\right)\langle 0|\psi(x)|\mathbf{p} = 0, s; c\rangle. \qquad (G.130)$$

Accordingly the spin of a Dirac particle is $1/2\hbar$ since a rotation by the angle 2π gives a $(-)$-sign.

---

[3] Due to 6 independent parameters of the Lorentz transformation one needs an antisymmetric 4×4 parameter matrix $\alpha_{\mu\nu}$ and operator matrix $M^{\mu\nu}$.

Discrete transformations leave the field equations

$$(-i\gamma^\mu \partial_\mu + M \cdot 1_4)\psi(x) = 0, \qquad i\partial_\mu \bar{\psi}(x)\gamma^\mu + \bar{\psi}(x)M \cdot 1_4 = 0 \qquad \text{(G.131)}$$

and the anticommutator-relations for the field operators invariant, however, cannot be expanded infinitesimally around 1_4.

1. **Parity-transformation** U_p

In this case we have in analogy to Eq. (G.124)

$$U_p \psi(x) U_p^{-1} = \eta_p \gamma^0 \psi(t, -\mathbf{x}), \qquad \qquad \text{(G.132)}$$

$$U_p \bar{\psi}(x) U_p^{-1} = \eta_p^* \bar{\psi}(t, -\mathbf{x})\gamma^0$$

with $|\eta_p| = 1$. Using

$$\gamma^0 u_s(\mathbf{p}) = u_s(-\mathbf{p}), \qquad \gamma^0 v_s(\mathbf{p}) = -v_s(-\mathbf{p}) \qquad \text{(G.133)}$$

this gives for the particle operators

$$U_p c_s^\dagger(\mathbf{p}) U_p^{-1} = \eta_p^* c_s^\dagger(-\mathbf{p}), \qquad \qquad \text{(G.134)}$$

$$U_p d_s^\dagger(\mathbf{p}) U_p^{-1} = -\eta_p d_s^\dagger(-\mathbf{p}).$$

Thus Fermi particles and antiparticles have opposite internal parity! By convention we set $\eta_p = +1$ for protons. Then (from electromagnetic processes) one finds $\eta_p = +1$ for the electron e^-, muon μ^-, neutron and all other baryons $\eta_p = +1$, while e^+, μ^+ and all antibaryons have $\eta_p = -1$.

Independent of convention bound states of a fermion and antifermion—in a state with even angular momentum ($l = 0, 2, 4, \ldots$)—have $\eta_p = -1$.

2. **Time-reflection** τ

This transformation is anti-unitary and can be written as

$$\tau = U_T K \qquad \qquad \text{(G.135)}$$

with a unitary U_T while K denotes the operator of complex conjugation. The action is

$$\tau \, \psi_\alpha(x) \tau^{-1} = \sum_{\beta=1}^{4} T_{\alpha\beta} \, \psi_\beta(-t, \mathbf{x}) \qquad \qquad \text{(G.136)}$$

for $\alpha = 1, \dots, 4$, with a 4×4 matrix T, which must fulfill (being form invariant and unitary)

$$T^{-1} = T^\dagger, \; T\gamma^\mu T^{-1} = (\gamma^\mu)^* . \tag{G.137}$$

In standard representation T reads as:

$$T = i\gamma^1\gamma^3 = -\Sigma^2 = -\begin{pmatrix} \sigma^2 & 0 \\ 0 & \sigma^2 \end{pmatrix} (= -T^*), \tag{G.138}$$

and its action on the particle operators is:

$$T\,u_s(\mathbf{p}) = u^*_{-s}(-\mathbf{p}), \qquad T\,v_s(\mathbf{p}) = v^*_{-s}(-\mathbf{p}). \tag{G.139}$$

From Eq. (G.136) we obtain the transformation properties for $c^\dagger_s, d^\dagger_s$

$$\tau\,c^\dagger_s(\mathbf{p})\tau^{-1} = -c^\dagger_{-s}(-\mathbf{p}), \tag{G.140}$$

$$\tau\,d^\dagger_s(\mathbf{p})\tau^{-1} = -d^\dagger_{-s}(-\mathbf{p}). \tag{G.141}$$

3. Charge conjugation U_c

This unitary operator changes the sign of the charge current and acts as follows:

$$U_c\psi(x)U_c^{-1} = \eta_c C\bar{\psi}^T(x), \qquad U_c\bar{\psi}(x)U_c^{-1} = -\eta^*_c \psi^T(x)C^{-1}, \tag{G.142}$$

with $|\eta_c| = 1$ and a 4×4 Matrix C with the properties

$$C^{-1} = C^\dagger, \qquad C\gamma^\mu C^{-1} = -(\gamma^\mu)^T, \tag{G.143}$$

where T here denotes "transposition." In standard representation C is given by

$$C = i\gamma^2\gamma^0 = -i\vec{\alpha}^2 = -i\begin{pmatrix} 0 & \sigma^2 \\ \sigma^2 & 0 \end{pmatrix} (= -C^{-1} = -C^\dagger = -C^T = C^*). \tag{G.144}$$

The action on the basis spinors of $C\gamma^0 = i\,\gamma^2$ is:

$$(C\gamma^0)u^*_s(\mathbf{p}) = v_s(\mathbf{p}), \qquad (C\gamma^0)v^*_s(\mathbf{p}) = u_s(\mathbf{p}) \tag{G.145}$$

and the action on the particle operators follows as:

$$U_c c^\dagger_s(\mathbf{p})U_c^{-1} = \eta^*_c d^\dagger_s(\mathbf{p}), \qquad U_c d^\dagger_s(\mathbf{p})U_c^{-1} = \eta_c c^\dagger_s(\mathbf{p}). \tag{G.146}$$

Consequently a particle-antiparticle system (with even orbital angular momentum $l = 0, 2, 4, \ldots$) has the charge conjugation parity -1.

G.6   Green's Function of the Free Dirac Field

The **two-point function** of the free Dirac field is defined by the vacuum expectation value

$$G^0(x, y) = -i \langle 0 | \hat{T}[\psi(x)\bar{\psi}(y)] | 0 \rangle, \tag{G.147}$$

which is a 4×4 matrix. The prescription for time-ordering $\hat{T}$ places all annihilation operators to the right, however, with a relative $(-)$-sign for each permutation. Thus one gets an additional prefactor $(-)^p$, where p denotes the number of permutations needed to achieve a time-ordered sequence of Fermion-operators, e.g.

$$\hat{T}[\psi(x)\bar{\psi}(y)] = \Theta(x^0 - y^0)\psi(x)\bar{\psi}(y) - \Theta(y^0 - x^0)\bar{\psi}(y)\psi(x). \tag{G.148}$$

We then obtain

$$G^0(x, y) = -i \left[\Theta(x^0 - y^0)(-i\, S^{(+)}(x - y; M)) - \Theta(y^0 - x^0)(-i\, S^{(-)}(x - y, M) \right] \tag{G.149}$$

$$= (M \cdot 1_4 + i\gamma^\mu \partial_\mu) \left[\Theta(x^0 - y^0)\Delta^{(+)}(x - y) - \Theta(y^0 - x^0)\Delta^{(-)}(y - x) \right]$$

$$= (M \cdot 1_4 + i\gamma^\mu \partial_\mu)\Delta_F(x - y)$$

$$= S_F(x - y; M)$$

with the Schwinger Δ_F-function

$$\Delta_F(x, M) = -\frac{i}{(2\pi)^3} \int d^4q \; \epsilon(q)\delta(q^2 - M^2) \exp(-iq \cdot x). \tag{G.150}$$

In (G.150) the prefactor $\epsilon(q)$ is given by

$$\epsilon(q) = \Theta(q^0) - \Theta(-q^0) = \frac{q^0}{|q^0|} \tag{G.151}$$

and gives a positive sign for positive frequencies and a negative sign for negative frequencies. Note that the combination $\epsilon(q)\delta(q^2 - M^2)$ is invariant under homogenous Lorentz transformations that do not mix four-vectors in the forward light cone with those from the backward light cone. Thus Δ_F is Lorentz-invariant with respect to transformations with $\det(\Lambda) = 1$:

$$\Delta_F(x') = \Delta_F(\Lambda x) = -\frac{i}{(2\pi)^3} \int d^4q \, \epsilon(q)\delta(q^2 - M^2) \exp(-i(\Lambda^{-1}q) \cdot x)$$

$$\text{(G.152)}$$

$$= -\frac{i}{(2\pi)^3} \int d^4q' \, \epsilon(\Lambda q') \, \delta((\Lambda q')^2 - M^2) \exp(-iq' \cdot x) \left| \frac{\partial(q^0..q^3)}{\partial(q'^0..q'^3)} \right|$$

with $q' = \Lambda^{-1}q$. The functional-determinant is $\equiv 1$ and due to the invariance of the $\epsilon\delta$-factor we get

$$\Delta_F(\Lambda x) = \Delta_F(x) \text{ if } \det(\Lambda) = 1. \qquad \text{(G.153)}$$

On the other hand $\Delta_F(x)$ changes sign for discrete transformations $x \to -x$ since $\epsilon(-q) = -\epsilon(q)$, i.e.

$$\Delta_F(-x) = -\Delta_F(x). \qquad \text{(G.154)}$$

For spacelike x $((x^0)^2 < \mathbf{x}^2$ or $x^2 < 0)$ Eqs. ((G.153) and (G.154)) imply that Δ_F vanishes,

$$\Delta_F(x) = 0 \text{ if } x^2 < 0. \qquad \text{(G.155)}$$

This property of the Δ-function implies "micro-causality" which also holds for $S_F(x - y)$ according to (G.149).

In (G.149) the operators $S^{(+)}$, $S^{(-)}$ denote the fraction with only positive (or negative) frequencies, i.e.

$$S^{(\pm)}(x; M) = -(M \cdot 1_4 + i\gamma^\mu \partial_\mu) \, \Delta^{(\pm)}(x, M) \qquad \text{(G.156)}$$

$$= i(2\pi)^{-3} \int d^3p \, \frac{M}{E_p} \, [\pm\Lambda_\pm(p) \exp(-\pm ip \cdot x)]_{p^0=E_p}$$

$$= i(2\pi)^{-3} \int d^4p \, \delta(p^2 - M^2) \, [\pm\Theta(\pm p^0) \, (M \cdot 1_4 + \gamma^\mu p_\mu)]$$

$$\times \exp(-ip \cdot x)$$

with

$$S^{(+)}(x; M) + S^{(-)}(x; M) = S(x; M) \qquad \text{(G.157)}$$

with the help of the projectors $\Lambda_\pm$. The causal (Feynman) propagator (G.149) for Dirac fields then can be written as,

$$S_F(x; M) = (M \cdot 1_4 + i\,\gamma^\mu \partial_\mu)\,\Delta_F(x; M),$$ (G.158)

and has the Fourier representation

$$S_F(x; M) = (2\pi)^{-4} \int d^4p\, S_F(p; M) \exp(-ip \cdot x)$$ (G.159)

with

$$S_F(p; M) = \frac{M \cdot 1_4 + p_\mu \gamma^\mu}{p^2 - M^2 + i\eta}.$$ (G.160)

Alternatively one can write (for $\eta \to 0^+$)

$$S_F(p) = \frac{1}{p_\mu \gamma^\mu - M \cdot 1_4}$$ (G.161)

using

$$(M \cdot 1_4 + p_\mu \gamma^\mu)\,(M \cdot 1_4 - p_\mu \gamma^\mu) = M^2 - p^2.$$ (G.162)

Since

$$(-i\gamma^\mu \partial_\mu + M \cdot 1_4)\,(i\gamma^\mu \partial_\mu + M \cdot 1_4) = (\partial_\mu \partial^\mu + M^2) \cdot 1_4,$$ (G.163)

$S_F(x)$ is the Green's function of the Dirac-problem in the sense of

$$(-i\gamma^\mu \partial_\mu + M \cdot 1_4)S_F(x - x'; M) = -\delta^4(x - x') \cdot 1_4.$$ (G.164)

Density-Dependent Relativistic Mean-Field Theory

The density-dependent relativistic mean-field theory[1] is an extension of the relativistic mean-field theory proposed first by Serot and Walecka[2] and has been widely employed in the literature on different levels of sophistication to describe finite nuclei and nuclear matter on the basis of a covariant Lagrangian density $\mathcal{L}$ and is briefly repeated here for completeness. A reminder of relativistic field theory for Dirac particles is given in Appendix G along with the notation used here. If the reader is familiar with these aspects Appendix G can be skipped. Furthermore, in this section we will use the Einstein convention.

The Lagrangian of the isospin symmetric Quantum-Hadro-Dynamics (QHD) consists of the free Dirac Lagrangian for the nucleons, the Lagrangian for the scalar σ-field and the vector ω-field with self-interactions and an interaction part of Yukawa type (in the stationary limit) for the nucleon-meson interactions. The couplings are taken as a function of a Lorentz-scalar $\hat{\rho}_0$, that depends on the nucleon fields $\bar{\Psi}$ and Ψ:

$$\mathcal{L} = \mathcal{L}_B + \mathcal{L}_M + \mathcal{L}_I, \tag{H.1}$$

$$\mathcal{L}_B = \bar{\Psi} \left(i\gamma_\mu \partial^\mu - M \right) \Psi, \tag{H.2}$$

$$\mathcal{L}_M = \frac{1}{2}\partial_\mu\sigma\partial^\mu\sigma - U(\sigma) - \frac{1}{4}F_{\mu\nu}F^{\mu\nu} + O(\omega^\mu\omega_\mu), \tag{H.3}$$

$$\mathcal{L}_I = \Gamma_\sigma(\hat{\rho}_0)\bar{\Psi}\sigma\Psi - \Gamma_\omega(\hat{\rho}_0)\bar{\Psi}\gamma^\mu\omega_\mu\Psi. \tag{H.4}$$

---

[1] C. Fuchs, H. Lenske, and H. H. Wolter, Phys. Rev. C 52 (1995) 3043.

[2] J. D. Walecka, Ann. Phys. 83 (1974) 491; B. D. Serot and J. D. Walecka, Adv. Nucl. Phys. 16 (1986) 1.

© The Author(s), under exclusive license to Springer Nature Switzerland AG 2021   235
W. Cassing, *Transport Theories for Strongly-Interacting Systems*, Lecture Notes in Physics 989, https://doi.org/10.1007/978-3-030-80295-0

The mesonic energies $U(\sigma)$ and $O(\omega^\mu \omega_\mu)$ are not specified here (see below). When discarding self-interactions the latter only consist of the mass terms

$$U(\sigma) = \frac{m_\sigma^2}{2}\sigma^2, \qquad O(\omega^\mu \omega_\mu) = \frac{m_\omega^2}{2}\omega^\mu \omega_\mu \qquad (H.5)$$

with m_σ and m_ω denoting the "mass" of the σ and ω field, respectively. Furthermore, the antisymmetric tensor $F^{\mu\nu}$ is given by

$$F^{\mu\nu} = \partial^\mu \omega^\nu - \partial^\nu \omega^\mu \qquad (H.6)$$

in close analogy to electrodynamics.

If the Lorentz-scalar is taken as $\hat{\rho}_0 = \bar{\Psi}\Psi$ in (H.1) the couplings will depend on the scalar density (SDD) and if it is taken as $\hat{\rho}_0 = \bar{\Psi}u_\mu \gamma^\mu \Psi$ they will depend on the baryon density (VDD).

H.1   Equations of Motion

The equations of motion—derived from the Euler-Lagrange equations—read,[3]

$$\partial_\mu \partial^\mu \sigma + \frac{\partial U}{\partial \sigma} = \Gamma_\sigma(\hat{\rho}_0)\bar{\Psi}\Psi, \qquad (H.7)$$

$$\partial_\nu F^{\mu\nu} + \frac{\partial O}{\partial \omega_\mu} = \Gamma_\omega(\hat{\rho}_0)\bar{\Psi}\gamma^\mu \Psi, \qquad (H.8)$$

for the meson fields and

$$(i\gamma_\mu \partial^\mu - M)\Psi + \Gamma_\sigma(\hat{\rho}_0)\sigma\Psi - \Gamma_\omega(\hat{\rho}_0)\gamma_\mu \omega^\mu \Psi$$

$$+\frac{\partial \Gamma_\sigma(\hat{\rho}_0)}{\partial \hat{\rho}_0}\frac{\partial \hat{\rho}_0}{\partial \bar{\Psi}}\bar{\Psi}\sigma\Psi - \frac{\partial \Gamma_\omega(\hat{\rho}_0)}{\partial \hat{\rho}_0}\frac{\partial \hat{\rho}_0}{\partial \bar{\Psi}}\bar{\Psi}\gamma^\nu \omega_\nu \Psi = 0 \qquad (H.9)$$

for the nucleon field. One can rearrange the equation of motion for the nucleons and write it in the form of the free Dirac equation[4] as

$$0 = \left(\gamma_\mu \left(i\partial^\mu - \hat{\Sigma}^\mu\right) - \left(M - \hat{\Sigma}^s\right)\right)\Psi \qquad (H.10)$$

$$= \left(\gamma_\mu \left(i\partial^\mu - \hat{\Sigma}^{\mu(0)} - \hat{\Sigma}^{\mu(r)}\right) - \left(M - \hat{\Sigma}^{s(0)} - \hat{\Sigma}^{s(r)}\right)\right)\Psi.$$

---

[3] T. Steinert and W. Cassing, Phys. Rev. C 98 (2018) 014908.
[4] A reminder of Dirac spinors and the Dirac-algebra is given in Appendix G.

In (H.10) $\hat{\Sigma}^s$ is the scalar selfenergy and modifies the mass while $\hat{\Sigma}^\mu$ is the vector selfenergy that modifies the four-momentum of the nucleons. The selfenergies are divided into two parts $\hat{\Sigma} = \hat{\Sigma}^{(0)} + \hat{\Sigma}^{(r)}$. The first one is the regular selfenergy $\hat{\Sigma}^{(0)}$, the second is denoted as rearrangement selfenergy $\hat{\Sigma}^{(r)}$. The rearrangement selfenergies are the result of the density dependence of the couplings and contain terms that are included in the equations of motion but not in the Lagrangian! They arise from the differentiation of the couplings with respect to the $\bar{\Psi}$-field.

The nature of the rearrangement selfenergies depends on the choice of the density $\hat{\rho}_0$. In case of SDD couplings we get

$$\frac{\partial \hat{\rho}_0}{\partial \bar{\Psi}} = \Psi, \qquad (H.11)$$

and for VDD couplings

$$\frac{\partial \hat{\rho}_0}{\partial \bar{\Psi}} = u_\mu \gamma^\mu \Psi. \qquad (H.12)$$

The first choice leads to a vanishing vector rearrangement selfenergy and the second to a vanishing scalar rearrangement selfenergy. We will specify $\hat{\rho}_0$ later and keep both the vector and the scalar rearrangement selfenergies in the following.

The equations of motion (H.7), (H.8) and (H.10) are too complicated to be solved on the many-body level.[5] We therefore introduce the mean-field approximation to simplify the equations and to allow for actual calculations. In this approximation the quantum fluctuations in the mesonic equations of motion are neglected which is justified if the source terms become large. The right side of Eq. (H.7) and (H.8) are then replaced by their normal ordered expectation values. This leads to

$$\Gamma_\sigma(\hat{\rho}_0)\bar{\Psi}\Psi \;\; \to \Gamma_\sigma(\rho_0)\langle: \bar{\Psi}\Psi :\rangle, \qquad (H.13)$$

$$\Gamma_\omega(\hat{\rho}_0)\bar{\Psi}\gamma^\mu\Psi \to \Gamma_\omega(\rho_0)\langle: \bar{\Psi}\gamma^\mu\Psi :\rangle, \qquad (H.14)$$

where $\langle: \bar{\Psi}\Psi :\rangle = \rho_s$ is the scalar density and $\langle: \bar{\Psi}\gamma^\mu\Psi :\rangle = j^\mu$ is the baryon current with $\langle: \bar{\Psi}\gamma^0\Psi :\rangle = j^0 = \rho_B$ as the baryon density. The couplings depend now on the normal ordered expectation value of $\hat{\rho}_0$.

To further simplify the equations we introduce the local-density approximation (LDA). In the case that the density of the system is locally approximately constant one can neglect the spatial derivatives of the meson fields. If one is only interested in the thermodynamics of homogenous systems, this approximation is definitively justified. Furthermore, we can additionally neglect the time derivatives of the meson fields and the spatial components of the baryon current $\langle: \bar{\Psi}\gamma^i\Psi :\rangle$ if one investigates a stationary and homogenous system in equilibrium. The equations of

---

[5] Since the Lagrangian has to be considered as an effective one it is not clear if higher order terms are physically meaningful.

motion (H.7), (H.8), and (H.10) simplify drastically within these approximations. The meson fields drop out as independent degrees of freedom since they are completely defined by the scalar density and the baryon density,

$$\frac{\partial U}{\partial \sigma} = \Gamma_\sigma(\rho_0)\rho_s, \tag{H.15}$$

$$\frac{\partial O}{\partial w^0} = \Gamma_\omega(\rho_0)\rho_B. \tag{H.16}$$

The spatial part of the ω-field vanishes ($\vec{\omega} = 0$) and the field is defined by its zeroth component denoted by ω further on. The spatial part of the normal vector selfenergy $\Sigma^{\mu(0)}$ also vanishes in this approximation because it is proportional to the $\vec{\omega}$-field and if we choose the nuclear rest frame with $u^\mu = (1,0,0,0)$ also the spatial vector rearrangement selfenergies vanish due to Eq. (H.12). The complete vector selfenergy then is given by its zeroth component.

Within these approximations the selfenergies of the nucleons no longer depend on any field operators $\hat{\Sigma} \to \Sigma$ and are now simple complex numbers. This allows us to use the known free Dirac spinors in the further evaluation. We write the equation of motion for the nucleons (H.10) as,

$$\left(\gamma^\mu \Pi_\mu - M^*\right) \Psi(x) = 0, \tag{H.17}$$

with $\Pi^0 = p^{*0} = p^0 - \Sigma^0$, $\vec{\Pi} = \vec{p}$ and $M^* = M - \Sigma^s$. In momentum space the equation reads

$$\left(\gamma^\mu p_\mu^* - M^*\right) u^*(p) = 0, \qquad \left(\gamma^\mu p_\mu^* + M^*\right) v^*(p) = 0 \tag{H.18}$$

with $u^*(p)$ as the effective spinor for particles and $v^*(p)$ as the effective spinor for antiparticles. This leads to the mass-shell condition

$$p^{*\mu} p_\mu^* - M^{*2} = 0 \quad \Rightarrow \quad p^{*0} = \pm\sqrt{\mathbf{p}^2 + M^{*2}}. \tag{H.19}$$

The effective spinors are obtained by replacing the mass and the energy with their effective values in the free Dirac spinors $u(p)$ and $v(p)$ and fulfill the relations,

$$\bar{u}_r^* u_s^* = \delta_{rs} = -\bar{v}_r^* v_s^*, \qquad \bar{u}_s^* \gamma^\mu u_s^* = \bar{v}_s^* \gamma^\mu v_s^* = \frac{\Pi^\mu}{M^*}. \tag{H.20}$$

With the abbreviations

$$\epsilon^+(\mathbf{p}) = \Pi^{0+} + \Sigma^0 = \sqrt{\mathbf{p}^2 + M^{*2}} + \Sigma^0, \tag{H.21}$$

$$\epsilon^-(\mathbf{p}) = \Pi^{0-} + \Sigma^0 = -\sqrt{\mathbf{p}^2 + M^{*2}} + \Sigma^0 \tag{H.22}$$

the spinor Ψ and the Pauli-adjoint spinor $\bar{\Psi}$ can be expanded as

$$\Psi(x) = \int \frac{d^3 p}{(2\pi)^3} \frac{M^*}{E^*(\mathbf{p})} \times \sum_{\lambda=1}^{2} [c_\lambda(\mathbf{p}) u_\lambda(\mathbf{p}) \exp(-i(\varepsilon^+(\mathbf{p})t - \mathbf{p} \cdot \mathbf{x})) \tag{H.23}$$

$$+ d_\lambda^\dagger(\mathbf{p}) v_\lambda(\mathbf{p}) \exp(i(\varepsilon^-(\mathbf{p})t - \mathbf{p} \cdot \mathbf{x}))],$$

$$\bar{\Psi}(x) = \int \frac{d^3 p}{(2\pi)^3} \frac{M^*}{E^*(\mathbf{p})} \times \sum_{\lambda=1}^{2} [c_\lambda^\dagger(\mathbf{p}) \bar{u}_\lambda(\mathbf{p}) \exp(+i(\varepsilon^+(\mathbf{p})t - \mathbf{p} \cdot \mathbf{x})) \tag{H.24}$$

$$+ d_\lambda(\mathbf{p}) \bar{v}_\lambda(\mathbf{p}) \exp(-i(\varepsilon^-(\mathbf{p})t - \mathbf{p} \cdot \mathbf{x}))],$$

with $E^*(\mathbf{p}) = \sqrt{\mathbf{p}^2 + M^{*2}}$ as the positive defined single-particle energy. This allows to calculate the baryon density and the scalar density in equilibrium as

$$\rho_B = \langle : \bar{\Psi} \gamma^0 \Psi : \rangle = d \int \frac{d^3 p}{(2\pi)^3} \left(\tilde{f}^*(\mathbf{p}) - \bar{\tilde{f}}^*(\mathbf{p}) \right), \tag{H.25}$$

$$\rho_s = \langle : \bar{\Psi} \Psi : \rangle = d \int \frac{d^3 p}{(2\pi)^3} \frac{M^*}{E^*(\mathbf{p})} \left(\tilde{f}^*(\mathbf{p}) + \bar{\tilde{f}}^*(\mathbf{p}) \right), \tag{H.26}$$

with the degeneracy factor $d = 4$ for spin and isospin symmetric nuclear matter. The functions $\tilde{f}$ and $\bar{\tilde{f}}$ are the equilibrium distribution functions for fermions, respectively, antifermions

$$\tilde{f}(\mathbf{p}) = \left(\exp\left((\epsilon^+ - \mu)/T \right) + 1 \right)^{-1} = \left(\exp\left(\left(E^*(\mathbf{p}) + \Sigma^0 - \mu \right)/T \right) + 1 \right)^{-1}$$

$$= \left(\exp\left((E^*(\mathbf{p}) - \mu^*)/T \right) + 1 \right)^{-1} = n_F(T, \mu^*, M^*), \tag{H.27}$$

$$\bar{\tilde{f}}(\mathbf{p}) = \left(\exp\left((-\epsilon^- + \mu)/T \right) + 1 \right)^{-1} = \left(\exp\left(\left(E^*(\mathbf{p}) - \Sigma^0 + \mu \right)/T \right) + 1 \right)^{-1}$$

$$= \left(\exp\left((E^*(\mathbf{p}) + \mu^*)/T \right) + 1 \right)^{-1} = n_{\bar{F}}(T, \mu^*, M^*), \tag{H.28}$$

and are related to the regular Fermi-distribution functions but with the effective chemical potential $\mu^* = \mu - \Sigma^0$ and the energy $E^*(\mathbf{p}) = E_p^* = \sqrt{\mathbf{p}^2 + M^{*2}}$.

This allows to evaluate the equations for the meson fields, Eqs. (H.15) and (H.16), giving two coupled selfconsistent equations that have to be solved simultaneously:

$$\frac{\partial U}{\partial \sigma} = \Gamma_\sigma(\rho_0) \, \rho_s(T, \mu^*, M^*), \tag{H.29}$$

$$\frac{\partial O}{\partial \omega} = \Gamma_\omega(\rho_0) \, \rho_B(T, \mu^*, M^*). \tag{H.30}$$

H.2   Thermodynamics and Thermodynamic Consistency

With these selfconsistent equations fixed we can evaluate the energy-momentum tensor,

$$T^{\mu\nu} = \frac{\partial \mathcal{L}}{\partial \left(\partial_\mu \Psi\right)} \frac{\partial \Psi}{\partial x_\nu} - g^{\mu\nu} \mathcal{L}, \tag{H.31}$$

where the energy density $\mathcal{E}$ and the pressure P of a system are given as normal ordered expectation values from the diagonal elements of the tensor,

$$\mathcal{E} = \langle : T^{00} : \rangle, \tag{H.32}$$

$$P = \langle : T^{ii} : \rangle = \frac{1}{3} \sum_{i=1}^{3} \langle : T^{ii} : \rangle. \tag{H.33}$$

For the following it is of advantage to write the selfenergies split into the normal and the rearrangement part. This is due to the fact that the rearrangement selfenergies appear in the equation of motion but not in the Lagrangian. The energy density in mean-field approximation is then given by (including a factor of 2 for isospin symmetric matter)

$$\begin{aligned}
\mathcal{E} &= U(\sigma) - O(\omega) + 2\langle : i\bar{\Psi}\gamma^0\partial_0\Psi : \rangle \\
&\quad - 2\langle : \bar{\Psi}\left(\gamma_\mu\left(i\partial^\mu - \Sigma^{\mu(0)}\right) - \left(M - \Sigma^{s(0)}\right)\right)\Psi : \rangle \\
&= U(\sigma) - O(\omega) + 2\langle : i\bar{\Psi}\gamma^0\partial_0\Psi : \rangle - 2\underbrace{\langle : \bar{\Psi}\left(\gamma_\mu\Pi^\mu - M^*\right)\Psi : \rangle}_{=0} \\
&\quad - 2\langle : \bar{\Psi}\left(\gamma_0\Sigma^{0(r)} - \Sigma^{s(r)}\right)\Psi : \rangle \\
&= U(\sigma) - O(\omega) + 2\langle : i\bar{\Psi}\gamma^0\partial_0\Psi : \rangle - \Sigma^{0(r)}\rho_B + \Sigma^{s(r)}\rho_s, \tag{H.34}
\end{aligned}$$

where the equation of motion (H.10) has been employed to simplify the expression. With the solutions for Ψ (H.23) and $\bar{\Psi}$ (H.24) the first term gives:

$$\mathcal{E} = U(\sigma) - O(\omega) - \Sigma^{0(r)}\rho_B + \Sigma^{s(r)}\rho_s \tag{H.35}$$

$$+ 2\sum_{\lambda=1}^{2}\int \frac{d^3p}{(2\pi)^3}\frac{M^*}{E_p^*}\left(\epsilon^+(\mathbf{p})\langle : c_\lambda^\dagger(\mathbf{p})c_\lambda(\mathbf{p}) : \rangle - \epsilon^-(\mathbf{p})\langle : d_\lambda(\mathbf{p})d_\lambda^\dagger(\mathbf{p}) : \rangle\right)$$

$$= U(\sigma) - O(\omega) - \Sigma^{0(r)}\rho_B + \Sigma^{s(r)}\rho_s$$

$$+ d\int \frac{d^3p}{(2\pi)^3}\left(\left(E_p^* + \Sigma^0\right)n_F(T, \mu^*, M^*) - \left(-E_p^* + \Sigma^0\right)n_{\bar{F}}(T, \mu^*, M^*)\right)$$

$$= U(\sigma) - O(\omega) - \Sigma^{0(r)}\rho_B + \Sigma^{s(r)}\rho_s + \Sigma^{0(0)}\rho_B + \Sigma^{0(r)}\rho_B$$

$$+ d \int \frac{d^3p}{(2\pi)^3} E_p^* \left(n_F(T, \mu^*, M^*) + n_{\bar{F}}(T, \mu^*, M^*)\right)$$

$$= U(\sigma) - O(\omega) + \Sigma^{s(r)}\rho_s + \Sigma^{0(0)}\rho_B$$

$$+ d \int \frac{d^3p}{(2\pi)^3} E_p^* \left(n_F(T, \mu^*, M^*) + n_{\bar{F}}(T, \mu^*, M^*)\right)$$

$$= U(\sigma) - O(\omega) + \Sigma^{s(r)}\rho_s + \Sigma^{0(0)}\rho_B + E_0(T, \mu^*, M^*)$$

with $d = 4$ for the sum over spin and isospin. Note that the vector rearrangement term has been cancelled and gives no direct contribution to the energy density. In (H.35) E_0 is the energy density for a non-interacting particle evaluated at the effective chemical potential μ^* with the effective mass M^*.

Using the equation of motion for the nucleon field (H.10), the pressure in mean-field approximation reads

$$P = -U(\sigma) + O(\omega) + \Sigma^{0(r)}\rho_B - \Sigma^{s(r)}\rho_s + \frac{2}{3} \sum_{i=1}^{3} \langle : i\bar{\Psi}\gamma^i\partial_i\Psi : \rangle. \qquad (H.36)$$

The further evaluation is analogue to the energy density and gives

$$P = -U(\sigma) + O(\omega) + \Sigma^{0(r)}\rho_B - \Sigma^{s(r)}\rho_s \qquad (H.37)$$

$$+ \frac{2}{3} \sum_{\lambda=1}^{2} \sum_{i=1}^{3} \int \frac{d^3p}{(2\pi)^3} \left(\frac{M^*}{E_p^*}\right)^2 \left(\frac{p^i}{M^*}\langle : c_\lambda^\dagger(\mathbf{p})c_\lambda(\mathbf{p}) : \rangle p^i \right.$$

$$\left. + \frac{p^i}{M^*}\langle : d_\lambda(\mathbf{p})d_\lambda^\dagger(\mathbf{p}) : \rangle p^i \right)$$

$$= -U(\sigma) + O(\omega) + \Sigma^{0(r)}\rho_B - \Sigma^{s(r)}\rho_s$$

$$+ \frac{d}{3} \int \frac{d^3p}{(2\pi)^3} \frac{\mathbf{p}^2}{E_p^*} \left(n_F(T, \mu^*, M^*) + n_{\bar{F}}(T, \mu^*, M^*)\right)$$

$$= -U(\sigma) + O(\omega) + \Sigma^{0(r)}\rho_B - \Sigma^{s(r)}\rho_s + P_0(T, \mu^*, M^*).$$

In (H.37) P_0 is the pressure for a non-interacting particle with the effective quantities μ^* and M^*. Contrary to the energy density (H.35) we get a direct contribution from the vector rearrangement term in the pressure!

One, furthermore, has to check if the model is thermodynamic consistent. In nuclear matter at temperature $T = 0$ this is tested by comparing the thermodynamic definition of the pressure to the mechanical definition via the energy-momentum tensor,

$$\rho_B^2 \frac{\partial}{\partial \rho_B} \left(\frac{E}{\rho_B} \right) = P = \frac{1}{3} \sum_{i=1}^{3} \langle : T^{ii} : \rangle. \tag{H.38}$$

This method is only sufficient in the canonical ensemble at $T = 0$ where the energy density is proportional to the thermodynamic potential. In the grand-canonical ensemble, where the thermodynamic potential is proportional to the pressure ($\Omega = -P$), it is better to show that the thermodynamic definition of the energy density is identical to its mechanical definition,

$$E = Ts - P + \mu_B \rho_B \stackrel{!}{=} \langle : T^{00} : \rangle = \mathcal{E}. \tag{H.39}$$

Another important check concerns the differential form of the grand-canonical potential. The potential/pressure derived above depends via the selfenergies on the additional parameters σ, ω, ρ_s, and ρ_B, that also enter the differential form. It is therefore necessary that the derivatives of the pressure with respect to these parameters vanish to regain the known differential form of the thermodynamic potential, i.e.

$$\frac{\partial P}{\partial \sigma} = \frac{\partial P}{\partial \omega} = \frac{\partial P}{\partial \rho_s} = \frac{\partial P}{\partial \rho_B} = 0. \tag{H.40}$$

We assume further on that the conditions (H.40) are fulfilled to prove Eq. (H.39). The entropy density is defined as the differential of the pressure with respect to the temperature

$$s = \frac{\partial P}{\partial T}\bigg|_{\sigma,\omega} + \underbrace{\frac{\partial P}{\partial \sigma} \frac{\partial \sigma}{\partial T}}_{=0} + \underbrace{\frac{\partial P}{\partial \omega} \frac{\partial \omega}{\partial T}}_{=0} + \underbrace{\frac{\partial P}{\partial \rho_s} \frac{\partial \rho_s}{\partial T}}_{=0} + \underbrace{\frac{\partial P}{\partial \rho_B} \frac{\partial \rho_B}{\partial T}}_{=0} = \frac{\partial P_0}{\partial T} = s_0(T, \mu^*, M^*)$$

$$\tag{H.41}$$

and takes the form of the non-interacting entropy density with the effective quantities μ^* and M^*. The same holds for the particle density,

$$n = \frac{\partial P}{\partial \mu}\bigg|_{\sigma,\omega} + \underbrace{\frac{\partial P}{\partial \sigma} \frac{\partial \sigma}{\partial \mu}}_{=0} + \underbrace{\frac{\partial P}{\partial \omega} \frac{\partial \omega}{\partial \mu}}_{=0} + \underbrace{\frac{\partial P}{\partial \rho_s} \frac{\partial \rho_s}{\partial \mu}}_{=0} + \underbrace{\frac{\partial P}{\partial \rho_B} \frac{\partial \rho_B}{\partial \mu}}_{=0} = \frac{\partial P_0}{\partial \mu}$$

$$= n_0(T, \mu^*, M^*) = \rho_B, \tag{H.42}$$

which is identical to its definition in Eq. (H.25). The thermodynamical definition of the energy density is then

$$
\begin{aligned}
E &= T s_0(T, \mu^*, M^*) + U(\sigma) - O(\omega) - \Sigma^{0(r)}\rho_B + \Sigma^{s(r)}\rho_s \\
&\quad - P_0(T, \mu^*, M^*) + \mu^* n_0(T, \mu^*, M^*) + (\mu - \mu^*)\rho_B \qquad (H.43) \\
&= U(\sigma) - O(\omega) - \Sigma^{0(r)}\rho_B + \Sigma^{s(r)}\rho_s + \Sigma^0 \rho_B + E_0(T, \mu^*, M^*) \\
&= U(\sigma) - O(\omega) + \Sigma^{s(r)}\rho_s + \Sigma^{0(0)}\rho_B + E_0(T, \mu^*, M^*)
\end{aligned}
$$

and thus equal to the mechanical definition (H.32). This proves the thermodynamic consistency of the theory. Naturally, this holds also in the canonical ensemble for $T = 0$.

Furthermore, one has to specify the explicit form of the selfenergies. To cover also the most general cases we assume that the scalar and the vector coupling depend on different densities $\hat{\rho}_\sigma$ and $\hat{\rho}_\omega$, that we take as

$$
\hat{\rho}_\sigma = \alpha \bar{\Psi}\Psi + \beta \bar{\Psi} u^\mu \gamma_\mu \Psi, \qquad \hat{\rho}_\omega = \gamma \bar{\Psi}\Psi + \delta \bar{\Psi} u^\mu \gamma_\mu \Psi, \qquad (H.44)
$$

and translate in mean-field approximation to

$$
\rho_\sigma = \alpha \rho_s + \beta \rho_B, \qquad \rho_\omega = \gamma \rho_s + \delta \rho_B. \qquad (H.45)
$$

The choice $\alpha = \gamma = 1$ and $\beta = \delta = 0$ leads to SDD couplings and the choice $\alpha = \gamma = 0$ and $\beta = \delta = 1$ to VDD couplings. The special case of $\alpha = \beta = \gamma = \delta = 0$ is the standard relativistic mean-field model with constant couplings that has been widely used in the literature and is employed in Sect. 3.1.

The normal selfenergies are fixed by the Lagrangian and read in mean-field approximation:

$$
\Sigma^{s(0)} = \Gamma_\sigma(\rho_\sigma)\sigma, \qquad \Sigma^{\mu(0)} = \Gamma_\omega(\rho_\omega)\omega\delta^{\mu 0}. \qquad (H.46)
$$

The rearrangement selfenergies follow from the equation of motion of the nucleons (H.9) and, using the densities from Eq. (H.44), are given by

$$
\hat{\Sigma}^{s(r)} = \alpha \, \hat{\Gamma}'_\sigma \bar{\Psi}\sigma\Psi - \gamma \, \hat{\Gamma}'_\omega \bar{\Psi}\gamma^\mu \omega_\mu \Psi, \qquad (H.47)
$$

$$
\hat{\Sigma}^{\mu(r)} = \left(-\beta \, \hat{\Gamma}'_\sigma \bar{\Psi}\sigma\Psi + \delta \, \hat{\Gamma}'_\omega \bar{\Psi}\gamma^\mu \omega_\mu \Psi\right) u^\mu. \qquad (H.48)
$$

In mean-field approximation they translate to

$$
\Sigma^{s(r)} = \alpha \, \Gamma'_\sigma \sigma \rho_s - \gamma \, \Gamma'_\omega \omega \rho_B, \qquad (H.49)
$$

$$
\Sigma^{0(r)} = -\beta \, \Gamma'_\sigma \sigma \rho_s + \delta \, \Gamma'_\omega \omega \rho_B, \qquad (H.50)
$$

where Γ' stands for $\partial\Gamma(\rho_0)/\partial\rho_0$. Some comment is useful in the special case of SDD couplings. If α or γ is different from zero the effective mass M^* or the effective chemical potential μ^* depend on the scalar density. The scalar density ρ_s therefore depends on itself implicitly and Eq. (H.26) becomes a selfconsistent equation, but unlike the other two selfconsistent equations this one does not follow from the mesonic equations of motion! Consequently, if ρ_s is defined once the selfconsistent equations (H.29) and (H.30) are solved.

With the selfenergies and the densities defined we can finally prove Eq. (H.40) and the thermodynamic consistency of the model. On the level of the selfenergies the derivative of the pressure with respect to $x\epsilon\{\sigma,\omega,\rho_s,\rho_B\}$ is

$$
\frac{\partial P}{\partial x} = -\frac{\partial U}{\partial x} + \frac{\partial O}{\partial x} + \frac{\partial \Sigma^{0(r)}}{\partial x}\rho_B + \Sigma^{0(r)}\frac{\partial \rho_B}{\partial x} - \frac{\partial \Sigma^{s(r)}}{\partial x}\rho_s - \Sigma^{s(r)}\frac{\partial \rho_s}{\partial x} + \frac{\partial P_0}{\partial x}
$$

(H.51)

$$
= -\frac{\partial U}{\partial x} + \frac{\partial O}{\partial x} + \frac{\partial \Sigma^{0(r)}}{\partial x}\rho_B + \Sigma^{0(r)}\frac{\partial \rho_B}{\partial x} - \frac{\partial \Sigma^{s(r)}}{\partial x}\rho_s - \Sigma^{s(r)}\frac{\partial \rho_s}{\partial x}
$$

$$
+ \underbrace{\frac{\partial P_0}{\partial \mu^*}\frac{\partial \mu^*}{\partial x}}_{-\rho_B\partial_x(\Sigma^0)} + \underbrace{\frac{\partial P_0}{\partial M^*}\frac{\partial M^*}{\partial x}}_{\rho_s\partial_x(\Sigma^s)}
$$

$$
= -\frac{\partial U}{\partial x} + \frac{\partial O}{\partial x} + \Sigma^{0(r)}\frac{\partial \rho_B}{\partial x} - \Sigma^{s(r)}\frac{\partial \rho_s}{\partial x} - \rho_B\frac{\partial \Sigma^{0(0)}}{\partial x} + \rho_s\frac{\partial \Sigma^{s(0)}}{\partial x}.
$$

Here we have used the relations $\frac{\partial M^*}{\partial M} = 1 = \frac{\partial \mu^*}{\partial \mu}$, which holds true since we treat ρ_s and ρ_B as variables, as well as $\frac{\partial P_0}{\partial \mu} = \rho_B$ and $\frac{\partial P_0}{\partial M} = -\rho_s$ that correspond to Eq. (H.25) and (H.26).

For the further evaluation we have to use the exact forms of the selfenergies and to include the density dependence of the couplings while taking the derivatives:

$$
\frac{\partial P}{\partial x} = -\frac{\partial U}{\partial \sigma}\frac{\partial \sigma}{\partial x} + \frac{\partial O}{\partial \omega}\frac{\partial \omega}{\partial x} + \left(-\beta\Gamma'_\sigma\sigma\rho_s + \delta\Gamma'_\omega\omega\rho_B\right)\frac{\partial \rho_B}{\partial x}
$$

$$
- \left(\alpha\Gamma'_\sigma\sigma\rho_s - \gamma\Gamma'_\omega\omega\rho_B\right)\frac{\partial \rho_s}{\partial x}
$$

$$
- \rho_B\left(\Gamma_\omega\frac{\partial \omega}{\partial x} + \Gamma'_\omega\omega\left(\gamma\frac{\partial \rho_s}{\partial x} + \delta\frac{\partial \rho_B}{\partial x}\right)\right)
$$

$$
+ \rho_s\left(\Gamma_\sigma\frac{\partial \sigma}{\partial x} + \Gamma'_\sigma\sigma\left(\alpha\frac{\partial \rho_s}{\partial x} + \beta\frac{\partial \rho_B}{\partial x}\right)\right)
$$

$$
= \frac{\partial \sigma}{\partial x}\left(\Gamma_\sigma\rho_s - \frac{\partial U}{\partial \sigma}\right) + \frac{\partial \omega}{\partial x}\left(\frac{\partial O}{\partial \omega} - \Gamma_\omega\rho_B\right).
$$

(H.52)

The derivatives vanish if the selfconsistent equations (H.29) and (H.30)—following from the mesonic equations of motion (H.7), (H.8)—are fulfilled. This finally proves Eq. (H.40), thus demonstrating the thermodynamic consistency of the model.

Although the density-dependent relativistic mean-field theory is fully Lorentz-invariant , thermodynamically consistent and allows for the description of a wide class of equations of state for nuclear matter, it is essentially an on-shell theory due to the mass-shell condition (H.19).

Index

© The Author(s), under exclusive license to Springer Nature Switzerland AG 2021
W. Cassing, *Transport Theories for Strongly-Interacting Systems*, Lecture Notes
in Physics 989, https://doi.org/10.1007/978-3-030-80295-0

Printed in the United States
by Baker & Taylor Publisher Services